Darwin in the Twenty-First Century

Studies in Science and the Humanities
from the Reilly Center
for Science, Technology, and Values

DARWIN IN THE TWENTY-FIRST CENTURY

NATURE, HUMANITY, AND GOD

EDITED BY

Phillip R. Sloan | **Gerald McKenny** | **Kathleen Eggleson**

University of Notre Dame Press

Notre Dame, Indiana

The Press gratefully acknowledges the support of the Institute for Scholarship
in the Liberal Arts, University of Notre Dame, in the publication of this book.

Manufactured in the United States of America

Library of Congress Cataloging-in-Publication Data

Darwin in the twenty-first century : nature, humanity, and God /
edited by Phillip R. Sloan, Gerald McKenny, Kathleen Eggleson.
     pages   cm
  Includes bibliographical references and index.
  ISBN 978-0-268-04147-2 (pbk. : alk. paper) —
  ISBN 0-268-04147-4 (pbk. : alk. paper)
  1. Evolution (Biology)—Religious aspects.   2. Evolution (Biology)
I. Sloan, Phillip R.   II. McKenny, Gerald P.   III. Eggleson, Kathleen.
  BL263.D25   2015
  201'.65768—dc23
                                                                    2015007743

This volume is dedicated to the memory of
Archbishop Józef Życiński
(September 1, 1948–February 10, 2011),
a model of a scholar and churchman
engaged at the deepest levels
in the dialogue of
science and theology.

CONTENTS

FIGURES AND TABLES

PREFACE

Phillip R. Sloan, Gerald McKenny,
and Kathleen Eggleson

In November of 2009, the University of Notre Dame hosted the conference "Darwin in the Twenty-First Century: Nature, Humanity, and God." Sponsored primarily by the John J. Reilly Center for Science, Technology, and Values at Notre Dame, and the Science, Theology, and the Ontological Quest (STOQ) project within the Vatican Pontifical Council for Culture, it was one of the last major academic conferences conducted during the international commemoration year of the 150th anniversary of the publication of Charles Darwin's *On the Origin of Species* on November 24, 1859, and the bicentennial of Darwin's birth on February 12, 1809.[1] The conference at Notre Dame was also coordinated with the larger international symposium "Biological Evolution: Facts and Theories," which took place at the Gregorian University in Rome, 9–11 March 2009, under STOQ auspices (Auletta, LeClerc, and Martinez 2011). These two complementary events provided a sustained interdisciplinary examination of the heritage of evolutionary theory and its implications for human, social, and religious concerns both in the present and in the future. This volume is intended both as a product of the Notre Dame conference and also as an advancement toward further maturity of the field in the twenty-first century. To accomplish this goal, it has made a selection of the papers most relevant to this aim presented at the Notre Dame conference, with the addition of three papers revised from their version presented at the Rome conference (Gilbert, Depew, Newman). One additional paper has been written explicitly for this volume (chap. 10 by Sloan).

The focus of this collection is on present and future developments in evolutionary science and its impact on the humanities, rather than on the commemoration of a historical achievement, although papers of a historical nature were delivered at the Notre Dame conference. This volume is not, for this reason, intended to capture the full range of issues and topics discussed at the conference. It is intended to build on these discussions and to some degree anticipate the future of interdisciplinary scientific and philosophical inquiry into evolutionary biology. These projections include the directions of new scientific developments, the implications of evolutionary theory for the human sciences, and new directions in theological reflection, with particular emphasis on views within the Roman Catholic tradition.

The monumental ramifications of evolutionary theory for a wide range of fields are obvious to all who follow these topics. The scope and diversity of evolution-focused research is extensive, including all of the life sciences (especially molecular biology, microbiology, genetics, and ecology), physical and cultural anthropology, philosophy and history of science, the social sciences, and theology. Comprehensive coverage of this vast and dynamic landscape cannot be accomplished in a single event or volume, and we have not attempted to do so. In the area of science, for example, we have chosen to focus particularly on areas of science that have emerged subsequent to the 1959 centennial of the publication of the *Origin of Species*, particularly in dimensions of developmental biology and developmental genetics.

In the area of theology, the Notre Dame and Rome conferences both display the considerable developments that have taken place since 1959 within Roman Catholic circles concerning the dialogue between evolutionary theory and theology. This recent response to Darwin in Catholic theology has taken two principal forms, both of which are partially represented in this volume. One form focuses on the distinction, formulated most rigorously by St. Thomas Aquinas, on the difference between primary and secondary causation. According to this distinction, God upholds natural causes, including those described by evolutionary theory, enabling them to operate with their own integrity. This first position is represented in this volume by chapters by William Carroll and John O'Callaghan.

Some Catholic theologians have thought that this account, while sound as far as it goes, does not capture the theological significance of what occurs through natural causes, including those of biological evolution. They seek to find theological meaning in evolutionary processes. Examples of this type of response to Darwin include Teilhard de Chardin and those theologians who have made use of process thought. This approach, albeit in a version that owes more to process thought than to Teilhard, is represented in this volume by Archbishop Józef Życiński. In accordance with its forward-looking theme, this volume also includes what may turn out to be a third principal form of response to Darwin in Catholic theology, one that is indebted to the theological vision of Hans Urs von Balthasar and more specifically to his category of theo-drama. This emerging approach is represented in this volume by Celia Deane-Drummond.

The division of this collection into the three main topic areas of "Nature," "Humanity," and "God" reflects both the structure of the conference itself and also the major areas which evolutionary theory inevitably impacts and connects: science and broader "natural philosophy"; the place of the human being in the natural world; the impact of evolutionary theory on ethical reflection; and the relation of an evolutionary historical account of origins to the claims of traditional monotheistic religions.

In the development of this broadly interdisciplinary approach, the editors first must thank the committee that assisted in the initial organization of the Notre Dame conference. This included, beyond the editors, Professors Hope Hollocher and Jeffrey Feder of the Notre Dame biology department, Agustin Fuentes of the Notre Dame anthropology department, and John O'Callaghan and Grant Ramsey of the Notre Dame philosophy department. The conference would not have been possible without the substantial financial support of the John J. Reilly Center for Science, Technology, and Values at Notre Dame, developed by the generosity of John D. Reilly. Additional support was given by the Henkels Family Lecture Fund of the Notre Dame College of Arts and Letters, administered by the Notre Dame Institute for Scholarship in the Liberal Arts, with additional support from the Integrative Graduate Educational Research Traineeship

(IGERT) fund and Global Linkages of Biology, Environment, and Society (GLOBES) Project Fund, administered by Professor Jeffrey Feder of the Notre Dame biology department. Dr. Kenneth Filchak of the University of Notre Dame biology department made it possible for Professor Simon Conway Morris to participate in the conference under this fund. We were also encouraged to undertake this conference by the Vatican STOQ project, which made it possible for two members of our organizing committee to attend the preplanning conference for the March 2009 conference in Rome in September of 2008. Four members then attended the Rome conference itself. Particular appreciation is due to His Excellency, Giovanni Cardinal Ravasi, president of the Pontifical Council for Culture, to Professor Gennaro Auletta, then scientific director of the STOQ program, and to Monsignors Tomascz Trafny and Melchor Sánchez de Toca, of the Pontifical Council for Culture, for their financial support and encouragement. Through the collaboration of Monsignor Trafny and Gennaro Auletta, we have been given permission to reprint the three papers originally delivered at the Rome conference, "Biological Evolution: Facts and Theories," and to reprint the chapter by Auletta, Colagé, and D'Ambrosio originally delivered at the Notre Dame conference but subsequently published in an Italian journal before its appearance here. Monsignor Trafny also kindly served as editor of the manuscript of Archbishop Życiński, who passed away unexpectedly in February of 2011.

The editors would like to thank all the participants in this conference, including those not represented in this volume. Presenters ranged from graduate students to senior scholars of international distinction. No collection of resultant papers can capture the dynamic interactions, floor discussions, and full range of conversations generated by a conference that brought together such a widely interdisciplinary audience. This event included nine plenary sessions with sixteen major papers, seven parallel contributed paper sessions with a total of twenty-one presentations, and an interdisciplinary summary panel discussion composed of nine responses to the question "Are We beyond the Conflict between Science and Religion?"

Special appreciation also goes to several local individuals who assisted the conference in many ways. The administrative assistant to the Reilly Center, Kimberly Milewski, provided valuable assistance in organizing the conference and keeping up with the correspondence as we planned the event. Cheryl Reed of the Notre Dame secretarial support staff assisted considerably in the final editing of the manuscript. Additional assistance in this final editing was also given by Reilly Center Notre Dame student assistant Christina Mondi. Both Professor Gerald McKenny, the Reilly Center Director during the conference, and his successor, Professor Don Howard, assisted in the completion of the project, including supplying student assistance. For this we are most appreciative.

All Notre Dame faculty who plan such a conference are indebted to Harriet Baldwin, emerita director of academic conferences for the College of Arts and Letters. Her wise advice and deep experience assisted us in making this a successful conference. We are deeply in her debt. We also acknowledge the considerable support given by the office of Marie Blakey, at the time director of the Office for Communications for the College of Arts and Letters, and by graphic designer Chantelle Snyder for preparation of excellent publicity materials and the design of the cover image. We also thank for help in the reviews and preparation of the final manuscript the assistance of Charles Van Hof, Matthew Dowd, and Wendy McMillen of the University of Notre Dame Press.

In offering this volume as a generally prospective, rather than retrospective, view on evolution to the scholarly community, we hope to encourage a greater degree of interdisciplinary discussion in the future between all parties engaged with the questions raised by Darwin's great work. It will be interesting to look back in 2059 at the inevitable commemoration of the second century of Darwinian studies to assess progress in existing fields and the emergence of novel ones.

December 3, 2014

NOTE

1. According to the University of Cambridge Darwin Project website that kept track of the 2009 Darwin year events, there were forty-three multiday conferences, seventeen one-day workshops and symposia, and numerous

other dinners, art exhibits, musical performances, poetry readings, museum and library displays, and innumerable individual lectures that in some way internationally honored the work of Charles Darwin or his heritage (Whye 2009).

REFERENCES

Auletta, G., M. LeClerc, and R. A. Martinez, eds. (2011). *Biological Evolution: Facts and Theories; A Critical Appraisal 150 Years after "The Origin of Species."* Rome: Gregorian and Biblical Press.

Whye, J., ed. (2009). *University of Cambridge Darwin Website.* http://www.darwin-online.org.uk/2009.html.

ONE

INTRODUCTION
Restructuring an Interdisciplinary Dialogue

Phillip R. Sloan

1959: THE APOTHEOSIS OF THE MODERN SYNTHESIS

Almost exactly fifty years before the Notre Dame conference, the world's largest centenary commemoration of Darwin's legacy was held at nearby University of Chicago. This event, organized by a committee spearheaded by University of Chicago anthropologist Sol Tax, drew nearly 2,500 registrants. In attendance were the primary leaders of evolutionary biology, paleontology, anthropology, and genetics, with a representation of sociologists, philosophers, and theologians (Smocovitis 1999). Two hundred and fifty delegates from universities and professional societies representing fourteen countries added an unusually high level of general academic prestige to the event as well.

Although other conferences were held in 1959, coordinated by an international super-committee formed in the mid-fifties—the Darwin Anniversary Committee—none equaled the University of Chicago symposium either in scope or in the assemblage of intellectual accomplishment. Accompanied by civic events, a grand procession to Rockefeller Chapel, and even an operetta written for the occasion, the Chicago celebration is an archetypal example of how scientific commemorations create, as well as celebrate, a major scientific consensus, in this case the theoretical framework of neo-Darwinism, often termed since a classic statement of Julian Huxley (Huxley 1942), as the "Modern Synthesis" or simply "Synthesis."[1]

The character of this neo-Darwinian theoretical framework, as it was highlighted at Chicago, was the outcome of a four-decade process of theoretical unification of two conflicting strands emerging from Darwin's original work, both of which could receive justification from the texts of Darwin and the arguments of his early interpreters. One was a theory of descent from common ancestors by means of natural selection, in which, as Darwin himself eventually came to explain the process (Darwin 1876), modifiable traits are transmitted from generation to generation by physical atomistic units—"gemmules" in Darwin's original formulation. These were subsequently named "pangenes" by Hugo DeVries in 1889, with the designator later shortened to "genes" by Wilhelm Johannsen in 1909. As it entered biological discourse, this "gene" concept was combined after 1900 with Mendelism, particularly as interpreted after 1900 by Hugo DeVries and William Bateson. From this point it was combined with the chromosome theory by Thomas Hunt Morgan and his students through their landmark work on the fruit fly *Drosophila* (Schwartz 2008; Beurton, Falk, and Rheinberger 2000; Keller 2000).

The initial implication drawn by many from this theory of genetic inheritance for evolutionary change was that such change must be discontinuous, that normal genetic assortment processes in accord with Mendelian principles could only lead to recombinatorial novelty, and that only by means of discontinuous "mutations" of atomistic genes could genuinely new properties relevant to the evolution of species be introduced into a population. This strand deemphasized the role of natural selection in the genesis of new species. Into the 1930s, this theory, in modified form, had prominent supporters, even receiving endorsement in such influential works as Erwin Schrödinger's 1944 *What is Life?* lectures (Bowler 1983, chap. 8; Schrödinger 1974, chap. 3).

The other strand emphasized evolutionary change as a continuous and gradualistic process, in which species are conceived as ephemeral entities that evolve slowly under the action of natural selection, often seen as a force operating directly on organisms. In conjunction with a sophisticated mathematical interpretation by the creators of modern statistics, most notably Karl Pearson, this interpretation placed itself in open opposition to the new Mendelism in the 1910s and '20s, resulting in one of the great theoretical controversies within modern biology (Pence 2011; Gayon 1998; Olby 1987; Depew and Weber 1995, chaps. 8–9; Provine 1971). Genetical determinism or statistical population analysis; saltational stepwise change or evolutionary gradualism; discontinuous mutations or the gradual accumulation of slight morphological changes in response to external selective pressure exerted by the environment—these were the oppositions that the Synthesis sought to reconcile.

Through the complex mathematical analysis of population genetics developed in papers of the 1910s and '20s, and in detail in his major work of 1930, *The Genetical Theory of Natural Selection*, Ronald A. Fisher was able to deal with both sides of this controversy and successfully combine the statistical mathematics of large number populations with a theory of atomistic genes subject to micromutational modifications (Okasha 2012; Depew and Weber 1995, Provine 1971). Such small mutations, preserved within the population by Hardy-Weinberg equilibrium, could now form the raw material upon which natural selection operated. Natural selection was conceptualized as a deforming pressure analogous to a Newtonian force acting upon genes in populations rather than as an external selection process operating primarily on whole organisms in nature. The result of this mathematical theory of natural selection was a reworking of the traditional Darwinian theory of natural selection. It enabled one to maintain both a gradualistic theory of species change over time and the particulate theory of inheritance operating according to Mendelian principles. Random micromutations of discrete and atomistic genes formed the foundation of evolutionary novelty. Operating over sufficiently expansive time scales, this was deemed competent both to explain the microevolution of species accessible to field and laboratory observation, and also the longer-term macroevolution of major groups in geological time. The theoretical resolution of several of these issues that emerged in the 1930s and '40s, associated with the names of Ronald A. Fisher, Julian Huxley, J. B. S. Haldane, Ernst Mayr, G. Ledyard Stebbins, Theodosius Dobzhansky, George Gaylord Simpson, and Sewall Wright, constituted "the Modern Synthesis." The assumption that such a theoretical synthesis had been achieved, in which had been resolved the main historic issues surrounding Darwin's theory of evolution by natural selection, formed the implicit framework of the proceedings of the 1959 Chicago symposium.

The 1959 Chicago commemoration provided a forum for all the main strands of development within the tacit consensus as it had coalesced to that point, with major papers by most of those typically identified as architects of the Synthesis: Ernst Mayr, George Gaylord Simpson, E. B. Ford, Alfred Emerson, G. Ledyard Stebbins, Sewall Wright, and Theodosius Dobzhansky. At the same time, there were indications of some of the tensions within the theoretical structure of neo-Darwinian theory that have emerged to greater prominence in the past half-century. These bear on the shape evolutionary theory may take in the next decades.

TENSIONS WITHIN THE SYNTHESIS

Two groups of scientists—paleontologists and developmental biologists—had often not been comfortable with the solutions to issues claimed by the architects of the Synthesis, with the notable exception of paleontologist George Gaylord Simpson, who continued to argue in his Chicago presentation for the smooth theoretical coherence of paleontology and neo-selectionist Darwinism that he had advocated in the 1940s (Simpson 1944, 1960; Amundson 2007, chaps. 8–10). In spite of Simpson's defense, however, concerns from the side of paleontology were raised by the presentation at the conference by University of Chicago paleontologist Everett C. Olson, who saw the need for new and radical thinking to deal with the nagging issue of major steps in evolutionary history: "Just what directions these [new reflections] might take is uncertain, but it will require persons able to think in radical terms, outside the current framework, to undertake the early steps" in a theoretical reform (Olson 1960, 543).

Edinburgh developmental biologist Conrad Hal Waddington, who had long been concerned with issues beyond simple gene inheritance, for which he had coined the term "epigenetics" (Waddington

1942), also posed some objections. Waddington's contribution, devoted to adaptation within the evolutionary process, is notable for its surprising willingness to consider presumably discarded "Lamarckian" issues surrounding the interplay of environment, genome, and heritable evolutionary change. In his view, a new role needed to be given to external causes of changes in the genetic structure through the effects of ionizing radiation, chemical factors, and even variations in normal environmental conditions. Waddington was also one of the few speakers to make reference to the possible importance of external effects on the underlying DNA as a source of evolutionary novelty, a discovery that was only slowly being assimilated into biological communities outside of biochemistry and microbiology by that date (Olby 2003; Strasser 2003; Gingras 2011, Sloan 2014). In his contribution, Waddington displayed considerable interest in finding theoretical explanations for the phenomenon of *coordinated* morphological change in evolutionary history (Waddington 1960). Organisms were not, in Waddington's view, simply loose assemblages of parts and functions governed by atomistic genes, and the coordination of morphology and development displayed by real organisms seemed little explained by the reigning population genetical interpretation of natural selection (Amundson 2007, chap. 9).

A review of the scientific developments in evolutionary theory since the Chicago commemoration highlights the importance of these complaints from paleontologists and developmental biologists, complicating the understanding of the history of evolutionary theory since 1959 (Depew and Weber 1995; Gilbert, Opitz, and Raff 1996). In retrospect, the reservations of paleontologists and embryologists a half century ago now seem prescient of the subsequent theoretical developments considered in this volume. What has occurred most prominently in the interim has been a new awareness of complications introduced into the understanding of evolution by paleontology, molecular genetics, and developmental biology. Whether these complications constitute a fundamental alteration in evolutionary theory, warranting the language of a "paradigm shift" as claimed by some, or if they are only a cause for minor adjustments within the theoretical structure of the Synthesis, is an issue that is currently under debate and can only be assessed by future history (Gilbert, Newman, and Gayon chaps. in this volume; Leland et al. 2014; Bateson and Gluckman 2011; Pigliucci and Müller 2010, esp. chaps. 1, 16, 17),

The dramatic introduction in 1972 of the so-called "punctuated equilibrium" theory by the paleontologists Niles Eldredge and Stephen Jay Gould was the first in a series of programmatic statements to announce itself as offering a new framework for understanding evolutionary change (Eldredge and Gould 1972). As Eldredge and Gould presented their original arguments, the discontinuities displayed by the fossil record were not to be covered over by appeals to the traditional "imperfection of the geological record" argument employed by the Synthesis tradition. These gaps were instead taken as real, requiring additional explanations of evolutionary change beyond the reigning genetical theory of natural selection. Although some parties latched on to these arguments to defend supernaturalistic accounts or saltationist models of evolution, the concern of Eldredge and Gould was rather to suggest an alternative naturalistic explanation that better fit the empirical geological evidence than the gradualist assumptions of the genetical theory of natural selection. The explanatory structure of this new perspective drew heavily on developmental biology, as Stephen Gould attempted in his landmark *Ontogeny and Phylogeny* (Gould 1977).[2]

The development of complex regulatory genetics in the late 1950s, led by the French school associated with the names of François Jacob, André Lwoff, Élie Wollman, and Jacques Monod, raised further conceptual problems for the atomistic "gene" concept assumed as foundational in theoretical population genetics and the genetical natural selection theory built upon this (Gilbert 2000; Morange 2000; Moss 2003, chap. 5). The gene concept assumed in the mathematical models of population genetics proved difficult to reconcile with the description of the enormously complex biochemical pathways and feedback and repressor mechanisms of regulatory genetics. These processes linked the nucleotide base sequences in DNA to the production of proteins and structures and functions of real organisms in the process of development. The complex relationship between the concept of the "gene" employed in statistical-population genetics and that utilized in developmental biology and molecular genetics continues to define an area of philosophical inquiry in life science (Falk 2000, Neumann-Held and Rehmann-Sutter 2006; Rheinberger 2010, chap. 8; Waters 2013).

In summary, the strong emergence of developmentalist and epigenetic perspectives in modern evolutionary biology in recent decades has reemphasized concerns that for various reasons, including practical ones, were not considered in the classic Synthesis. New attention is now given to factors in evolution typically ignored in the evolutionary theory of 1959, including such issues as the influence on evolutionary history created by the restrictions imposed by morphology and the coordination of form and function in organisms. It also has led to reconsideration of the active interactions of organisms with their environment, the importance of learned behavioral traits and their effect on the genome, and even the possibility that environmental factors can play a more significant causal role in evolutionary change than previously allowed, viewpoints long excluded as "Lamarckian" but now developed in more subtle ways by several biological theorists (Gissis and Jablonka 2011; Bateson and Gluckman 2011). The complexity of the "gene" concept introduced by developmental genetics and molecular biology has also challenged the hard determinism assumed between elements of a sequestered germ plasm and the phenotype (West-Eberhard 2003; Moss 2003, 2006). We now see experimental work that suggests the importance of a looser connection in the phenotype-genotype relationship, and how this may explain the rapidity of evolutionary change in some biological groups in paleontological history under new environmental conditions, such as the famous "Cambrian explosion." With these developments, a new way of understanding "macro" evolution—the evolution of major new groups and body plans in history—has come into play (Müller and Newman 2003).

IMPLICATIONS FOR AN INTERDISCIPLINARY DIALOGUE

There are also additional reasons for drawing out the consequences of the new developmentalist, organismic, and systems perspectives that have emerged in recent biology for broader issues. These bear on their consequences for the anthropological and theological issues raised by evolutionary theory that are explored in limited ways in this volume. One of the greatest sources of difficulty in developing a productive cross-disciplinary discussion, particularly between working scientists and theologians interested in evolution, has been a direct consequence of certain ground assumptions of the theory of natural selection imbedded in neo-Darwinian theory that have been elevated to metaphysical claims in the philosophy of nature. This particularly concerns two interrelated issues: One is the denial of some kind of natural teleology operative in nature and in organic life that flows from the high priority given within the classic Synthesis to purely chance-like events in evolutionary history. The second, and related, issue concerns the link of these interpretations to the well-worn "evolution and creation" debates. The need for greater clarity in the academic, as well as the public, discussion of these issues is urgent, and it is envisioned that the papers in this volume provide a contribution to a more illuminating cross-disciplinary conversation in the future.

The theoretical framework that lies behind this strong emphasis on stochastic and indeterminist principles in neo-Darwinian theory is a topic of considerable discussion in the philosophy of biology, with competing positions drawn that reach a high level of sophistication and involve the interpretations of the interactions of concepts of fitness and probability theory (Ramsey and Pence forthcoming; Pence and Ramsey 2013). As historical studies have shown, the elevation of the concept of "chance" in evolutionary theory to the importance it attained in the classic reconciliation of genetics and natural selection theory in the 1930s is the product of a conceptual development of the early twentieth century. It is not itself something we find emphasized in the texts of Darwin, who, writing before the "probabilistic revolution" of the nineteenth century, mainly referred to "unknown laws" that governed the causes of variation.[3]

In formulating the genetical theory of natural selection, R. A. Fisher was concerned to develop an interpretation of evolutionary theory modeled on statistical mechanics of Ludwig Boltzmann, Willard Gibbs, and James Clerk Maxwell (Depew and Weber 1995, chap. 10; Hodge 1992). This physical theory accounted for such phenomena as the macrobehavior of gases in accord with standard deterministic laws of pressure and volume as mass-action effects underlain by the randomized behavior of component atoms that were themselves governed only by probabilistic laws in their individual behavior. By the explicit transposition of an analogy between the behavior of molecules in gases and the behavior of genes in populations acted upon by selective forces, it became possible to analyze evolutionary change with statistical models, utilizing the idealizations of parametric statistics, one of which is mathematical randomness. Interbreeding within populations was modeled as a random process; micromutation, as the fundamental source of evolutionary novelty, was envisioned as undirected, unpredictable, and likely caused by such events as randomized cosmic ray bombardment. As a consequence, evolutionary change was presumed to be governed in its theoretical foundations by undirected causal chains eventually reaching back to randomized mutations acted upon by natural selection.

The axiomatic status of this claim in common interpretations of neo-Darwinian theory has subsequently come to underlie the recurrent claim, encountered in the textbooks as well as the writings of popularizers, that evolution is governed by blind chance, and that it thereby destroys all traditional notions of natural teleology and, by implication, theology (Dawkins 2008, 1987; Monod 1971; Stamos 2001).[4] In the words of geneticist Peter Medawar, "it is upon the notion of *randomness* that geneticists have based their case against a benevolent or malevolent deity and against their being any overall purpose or design in nature" (Medawar and Medawar 1977, 167). The predictable pushback against these claims from various constituencies has been the main generating source of the ongoing controversy over "creation science" and "intelligent design" that has spilled over into decisions of school boards, court trials, legislative decrees, and eventually electoral politics in the United States and even internationally.

Gaining greater clarity on these matters is urgently important, but the heated character of the public debate makes this difficult. It also requires clarity over the issues at stake in the debate. Are these issues of probability theory, of scientific realism, or of the fundamental metaphysics of nature? The deficiencies of the traditional conceptualization of this debate on philosophical grounds are addressed in limited ways in this volume in the chapters by David Depew; Gennaro Auletta, Ivan Colagè, and Paolo D'Ambrosio; William Carroll; John O'Callaghan; and Józef Życiński; and deeper clarification hinges on developing a more accurate understanding of the meaning of "creation" in relation to "purpose" as it has functioned in the tradition of Abrahamic religions.[5] But in addition to arguments that can be raised on a philosophical level, there are also conceptual developments emerging from within modern evolutionary biology itself that require some rethinking of the role of undirected natural events as a fundamental explanatory category in evolutionary theory. These developments open up new territory for conversation between evolutionary scientists with philosophers, ethicists, and theologians, which this volume will explore in a limited way.

The new concepts and perspectives that have accompanied the evolutionary-developmental and systems-theoretical perspectives of "evo-devo," and the post–Human Genome Project concern with "epigenetics," have forced back onto the table consideration of issues of form, structure, immanent teleological directedness, and the dynamic interaction of organisms and their environments (West-Eberhard 2003). These developments are taking place often through the introduction of a different conceptual vocabulary—concepts such as "robustness," "system," "plasticity," "niche construction," and "teleonomy" (Bateson and Gluckman 2011; Pigliucci and Müller 2010; Nijhout 2002)—and do much of the work that commentators in touch with a longer tradition of discourse might understand as being appeals to "formal" and even "immanent final" causes (Grene 1967).[6] The fact that such conceptual translations can now be made, even if imperfectly, facilitated by developments emerging from within evolutionary science itself, opens up space for a new level of conversation between ethicists, theologians, philosophers, and scientists concerning such crucial issues as the place of the human being in an evolutionary cosmos, the possibility of a normative ethics with some connection to biology, and new ways of conceptualizing human transcendence-in-continuity. This can be conducted without entry into the controversies over "intelligent design," which inevitably return us to the heritage of the Platonic-Stoic and British natural theology traditions that Darwin's arguments discredited.

HUMAN BEINGS IN AN EVOLUTIONARY WORLD

The Notre Dame conference devoted the least attention to issues in human evolution in view of the fact that human evolution was a major topic of the March 2009 Rome conference with which our conference was in collaboration. That conference presented a wide array of perspectives on physical, cultural, philosophical, and theological anthropology to which the reader is referred (Auletta, LeClerc, and Martinez 2011). Additional insights are developed in this volume that were not explored in the Rome conference.

The issues raised by the extension of evolutionary biology to human origins, and to the analysis of other human properties, are immense, touching culture, ethics, sociology, human relations, and, of course, theology and religious belief. These extensions are also the source of the continued interest in, and debate over, Darwin's theory and its historical progeny. The *Origin of Species* devoted only one cryptic sentence to human questions, but any attentive reader in 1859 could see the inevitable extension to human origins of the theory of common descent and divergence under the action of natural selection. Although Charles Lyell, Thomas Huxley, Alfred Russel Wallace, and Ernst Haeckel beat Darwin into press with reflections on evolution and human origins, Darwin's public encounter with this question was of much greater range and ambition. His willingness in the *Descent of Man* of 1871, followed quickly by his *Expression of the Emotions in Man and the Animals* in 1872, to take on not only human physical evolution, but also to engage the wide range of issues of social life, ethics, religion, race, mental properties, aesthetics, and all other traditional features used to define human uniqueness "exclusively from the side of natural history" (Darwin [1871] 1981, 1:71), raised the stakes to a new level. The debate over the adequacy of Darwin's theory as an explanation of human existence in all these dimensions has waxed and waned over the subsequent 155 years (Hodge and Radick 2009, parts 3–4).

In view of this long history of debate, one might ask what new can be said about these topics? By looking back again at the Chicago commemoration for perspective, we gain some valuable insights. One of the issues most important to the individual who spearheaded the organization of that celebration, University of Chicago anthropologist Sol Tax, was the *reintegration* of anthropology and evolutionary theory (Smocovitis 1999; Depew, personal communication).[7] In his view, in part as a result of the pathological developments of the twentieth century around the concept of biological race and "social" Darwinism, "a generation of students grew up convinced that biological and cultural anthropology needed each other mainly to demonstrate the limitations of the biological in man" (Tax 1960a, 271–72). This reconnection of evolutionary biology and anthropology being sought at the Chicago centenary was one that had been incubating for at least a decade in encounters between theoreticians of the Synthesis, such as Theodosius Dobzhansky, and anthropologists, such as Sol Tax's University of Chicago colleague and co-organizer Sherwood Washburn. Tax and Washburn aimed to overcome this split by bringing together a group of scientists, sociologists, behavioral ecologists, and anthropologists generally sympathetic to a reintegration of these two wings of the scientific study of humanity. Panels were organized around the topics of "Man as an Organism," "The Evolution of Mind," and "Social and Cultural Evolution." But in Tax's cautious view,

> for me, the Centennial brought Darwin and evolution back into anthropology, not by resurrecting analogies, but by distinguishing man as a still-evolving species, characterized by the possession of cultures which change and grow non-genetically. Human evolution includes the addition of culture to man's biology; "cultural evolution" at the human level is quite a different matter. Anthropologists accept the first without question; they are divided about the second. (Tax 1960a, 282).

Since 1959, it can be claimed that the effort to reintegrate evolutionary biology and anthropology, in large measure by means of a resurrection of Darwin's zoological perspective on all aspects of human existence, represents one of the most dramatic developments in evolutionary science in the last half-century, and it promises to define much of the research project in many dimensions of the human sciences into the near future. Emerging from a subunit of ecology, animal behavior studies, and the analysis of social insects under the banner of "sociobiology," and spearheaded by the 1975 publication of *Sociobiology: The New Synthesis* by Harvard entomologist E. O. Wilson, this broad effort has aimed to supply an adaptationist analysis of all major human phenomena, from ethics and religion, to culture, sexuality, art, and sociality. These efforts at reintegration, far exceeding the cautious efforts of Tax and Washburn, have generated a vast literature and a substantial debate over the adequacy of Darwinian explanations. For some of the main proponents of sociobiology, this was a logical expansion of the integration of population dynamics and the genetical theory of natural selection of the classic Synthesis into the domain of the human sciences. In the words of E. O. Wilson,

> Without [Darwinian evolution] the humanities and social sciences are the limited descriptors of surface phenomena. . . . With it, human nature can be laid open as an object of fully empirical research, biology can be put to the service of liberal education, and our self-conception can be enormously and truthfully enriched. (Wilson 1978, 2; see also Wilson 1998)

Although it would be incorrect to consider Wilson's voice definitive of the ambitions of the large body of work currently taking place under the less-contentious labels of "human behavioral ecology," "evolutionary psychology," and "ethnoprimatology" (Barkow 2006; Buss 2004; Waal 1996, 2006), or the restricted philosophical inquiries into topics like evolutionary ethics (Rosenberg 2009; Katz 2000; Sober and Wilson 1998), these enterprises are united by the assumed adequacy of evolutionary accounts to deal with human questions long considered the domain of philosophy, theology, cultural anthropology, and sociology.

The Notre Dame conference and this volume have limited focus to a few key issues in the human sciences. Physical anthropologist Bernard Wood offers an update on the current views on the human lineage as a scientific baseline for some of these reflections. Following this is an examination of some key issues surrounding the relation of evolutionary theory to ethics, led off by a defender of the possibility, historian and philosopher Robert Richards. Philosophers of biology Paul Griffiths and John Wilkins in a joint paper offer a less optimistic view of evolutionary ethics in analyzing the degree to which evolutionary accounts can bridge the gap between beliefs created as evolutionarily useful adaptations and objectively true beliefs of the kind presumed by versions of ethical realism.

Richards, Griffiths, and Wilkins all address a topic that has been under contention since the writings of Darwin and Herbert Spencer in the nineteenth century. These early efforts to derive an ethics from evolutionary process had been subjected to the criticisms of Thomas Henry Huxley in 1893 in a famous lecture, and then fashioned into a more influential philosophical critique by philosopher G. E. Moore in 1905 in his *Principia Ethica*. Moore in particular developed the criticism that such attempts to blend genetic process and normative judgments involve a "naturalistic fallacy" that undermines the enterprise (Farber 1994). The claim that such arguments actually do involve a fallacy has itself generated a considerable literature, and the argument hinges in part on the degree to which "nature" itself is conceived to have or not have inherent teleological purposes (Clayton and Schloss 2004, part 2).

In a fourth contribution to this section, co-editor Phillip Sloan offers a combined historical and philosophical analysis of the epistemological issues involved in applying Darwin's zoological vision to human beings. Drawing on some developments in Continental philosophical anthropology, he offers some suggestions for a transformation of perspective in the scientific analysis of human beings that repositions a concept of human transcendence and human dignity within the phenomenological givenness of the human experience of the world as an upright creature.

EVOLUTION AND THEOLOGY

Theological and religious questions surrounding Darwin's theory constitute the most contentious issues in the public mind, and these disputes have, if anything, been more prominent in the last fifty-five years than they were in the half-century previous to the 1959 centennial. On the surface this is a somewhat surprising claim. Memories of the earlier public controversies raised by the strident materialism of Ernst Haeckel, by the debates over "social" Darwinism in the early decades of the twentieth century, and, in the United States, by the public conflict of science and biblical fundamentalism reaching its most public manifestation in the famous trial of John Scopes in Tennessee in 1925, were fresher in 1959 (Numbers 2006; Larson 1998). But again, looking back to the Chicago celebration of 1959 gives us some perspective on what has happened to define the current situation.

According to some analyses, the Chicago celebration actually reinvigorated the popular religious conflicts in the United States over evolution and defined the form it has taken in the last fifty-five years. The vast media campaign mounted by Sol Tax to highlight the Chicago event drew considerable public attention to this gathering from outside the confines of the Chicago area and the boundaries of specific scientific disciplines. Through these efforts by Tax to draw such public interest in the Chicago gathering, Julian Huxley's closing convocation address in Rockefeller Chapel, delivered on November 26—Thanksgiving Day—under the title "The Evolutionary Vision," received wide coverage. In his address, Huxley sketched out a grand cosmological vision dominated by evolution, in which he claimed that all reality "from atoms and stars to fish and flowers to human societies and values" was governed by a single evolutionary process. Then as one of the implications of this evolutionary vision, he proclaimed that

> in the evolutionary pattern of thought there is no longer either need or room for the supernatural. The earth was not created, it evolved. So did all the animals and plants that inhabit it, including our human selves, mind and soul as well as brain and body. So did religion. . . . Evolutionary man can no longer take refuge from his loneliness in the arms of a divinized father-figure whom he has himself created. (Huxley 1960, 252–53)

The predictable controversy created by Huxley's sermon in the cause of secular evolutionary humanism from the pulpit of what has been called the only Baptist Gothic cathedral in the world, has been seen by at least one historian of these events as a watershed that triggered the wave of fundamentalist attacks on evolutionary biology that began in many respects with the publication shortly afterward (1961) of John Whitcomb and Henry Morris's *The Genesis Flood* (Smocovitis 1999, 303–5, 315–16). Negative responses from mainline Catholic and Protestant sources also appeared in newspapers and magazines. However we date the origins of the current conflicts over evolution and creation,[8] they represent an increasingly vociferous controversy that has reached to school curricula, legislative actions, court decisions, and political campaigns.

It would be incorrect to conclude that an anti-religious polemic was an intention of the organizers of the Chicago symposium. But contributions from professional theologians and philosophers of religion, and representatives of major religious traditions formed only a minor component of the Chicago celebration. Two contributions— one an important article on creation and causality by the rising Lutheran church historian and theologian Jaroslav Pelikan of the University of Chicago Divinity School, the other by J. Franklin Ewing, S.J., of the Anthropology Department of Fordham University— constitute the only input from Christian voices. There was no effort beyond an essay by Chicago historian of medicine Ilza Veith on Eastern concepts of creation and evolution to deal with the response to evolution of other religious traditions.

Pelikan's article, similar to those in this volume by William Carroll and John O'Callaghan, sought to clarify some of the developments in the concept of creation within the Christian tradition. He spoke in a conciliatory way about how "from quiet corners all over Christendom . . . , theologians are listening to scientists with seriousness and humility" (Pelikan 1960, 28). Father Ewing's essay provides a useful benchmark for assessing Roman Catholic responses to evolutionary theory and the changes in these since 1959. Ewing was writing in the wake of the modest opening up of official Catholic thought to evolutionary biology after a long period of silence and tacit repression of discussion of the topic in the wake of the Modernist controversies of the early century (Artigas, Glick, and Martinez 2006;

Appleby 1987). The opening of Catholic thought to modern biblical scholarship with Pius XII's encyclical *Divino afflante spiritu* of 1943 allowed much greater freedom for Catholic scholars to depart from literal readings of such texts as Genesis 1. With his subsequent encyclical *Humani generis* of 1950, Pius XII also permitted Catholic scholars to consider evolutionary theory cautiously, and as a "hypothesis." This was followed by the gradual introduction of the teaching of evolutionary biology into Catholic universities, with the first such course at the University of Notre Dame, for example, dating from 1956 (Sloan 2009).

Father Ewing's contribution summarized the main points in the recent attitudes within Catholicism toward evolutionary science and commented on the cogency of some of the main points of *Humani generis*. In closing, he lamented the fact that "too few Catholic thinkers have been really coming to grips with evolution," and argued that contribution by them to these discussions "could supply the philosophy of evolution with the mentally satisfying components of God as Creator and final end of all things in the universe outside himself and God as the Conserver" (Ewing 1960, 27–28).

At the time Ewing wrote these words, the posthumously published writings of his fellow Jesuit, the French human paleontologist Pierre Teilhard de Chardin (d. 1955), were entering discussions of evolution and theology, with the first English translation of the *Phenomenon of Man* appearing in the 1959 Darwin commemorative year. Teilhard's views are also briefly discussed at the end of the Ewing article. But with a favorable introductory essay to the English translation of Teilhard's *Phenomenon* by none other than Julian Huxley, who then incorporated Teilhard's evolutionary vision into his provocative Thanksgiving Day Chicago address (Huxley 1960, 253), Teilhard's works entered the discussions of theology and science with novel insights that did not always sit well with traditional theology. Since that time, his writings have generated polarized discussions within and outside Catholic theological and ecclesiological circles both over their scientific merit and their theological orthodoxy.

Whatever the deficiencies of Teilhard's ambitious efforts—and subsequent assessments from both scientists and theologians have often been critical—they did represent an effort to create a sweeping cosmological vision of the history of life, combining an orthogenetic interpretation of evolutionary biology with a theistic cosmology.[9] At least for a period of time, Teilhard's writings had a profound effect on many Catholic and Protestant thinkers interested in moving the encounter of theology and evolutionary biology beyond the level of a tentative exploration of "hypotheses." Dimensions of Teilhard's vision have affected the thought of important Catholic theologians and scientists in diverse ways since 1959, and it is unquestionable that Teilhard's indirect influence has stimulated a much deeper engagement of Catholic intellectuals with the questions raised by modern science generally, whether or not they are sympathetic to his approach (Rahner 1966; Haught 2000).[10] Coupled with the official encouragement of critical scriptural scholarship under Pius XII, and with the return of many theologians to the fathers of the church for new theological insights that became some of the ingredients in the theological developments of the Second Vatican Council (1962–65), a new openness to evolutionary theory within Catholic thought is evident, illustrated by the Rome 2009 conference sponsored by the Pontifical Council for Culture.[11]

Although it does not have the status of an official encyclical letter, the address of the late Pope John Paul II to the Pontifical Academy of Science in October of 1996 on the topic of evolution also spurred a new level of conversation about these issues within Roman Catholic circles. Reflecting back on Pius XII's permission in 1950 for Catholic thinkers to consider evolutionary theory as a worthy scientific "hypothesis," John Paul remarked that,

> Today, almost half a century after the publication of the Encyclical [*Humani generis*], new knowledge has led to the recognition of the theory of evolution as more than a hypothesis. It is indeed remarkable that this theory has been progressively accepted by researchers, following a series of discoveries in various fields of knowledge. The convergence, neither sought nor fabricated, of the results of work that was conducted independently is in itself a significant argument in favor of this theory. (John Paul II 1996)[12]

The Notre Dame and Rome conferences display some of the considerable developments that have taken place since 1959 within Roman Catholic circles concerning the dialogue between evolutionary theory and theology. Two of the four papers represented in this volume (O'Callaghan, Carroll) develop mainly in dialogue with the Thomistic tradition. The other two (Życiński, Deane-Drummond) are more oriented to interpretations affected either by process theology or by some of the work of Hans Urs von Balthasar. Given the richness of contemporary theological reflection on questions surrounding such large topics as theodicy, the role of chance in relation to creation, and the relation of divine action to the world, one expects further developments in the coming decades on these issues. The reader is directed to both the Rome volume and the recent volume edited by Louis Caruana for additional discussions (Auletta, LeClerc, and Martinez 2011; Caruana 2009). The contributions in this volume display the way in which the theological questions have moved to a carefully delimited ground centered on two main issues, teleology and creation. On both issues, considerable clarification is required to move us beyond the stick-figure characterizations that have defined so much of the discussion since 1859.

As a philosophical issue, the concept of teleology in relation to evolutionary theory is addressed in two contributions that appear in this volume. The first from philosopher of biology David Depew, originally delivered at the Rome conference, examines the issue of teleology in relation to evolution through a generally historical approach, reaching back to the debates that emerged around this issue in Darwin's day. These are then interfaced by Depew with contemporary philosophical discussions in the philosophy of biology. This analysis is followed by a collaborative chapter delivered at the Notre Dame conference by Rome philosophers Gennaro Auletta, Paolo D'Ambrosio, and Ivan Colagè, who deal with the topic from a contemporary perspective. They draw on a considerable body of experimental work, and examine the issue from the standpoint of epigenetics and developmental biology. In their analysis of "teleonomical" issues, it is also clear that they have carved out space that moves us beyond the characterizations of these concepts in the classic papers of Ernst Mayr (1974, 1992).

It should be evident that neither of these contributions defends the view, commonly held in popular controversies, that teleology in relation to evolutionary theory necessarily means the imposition of purposes on things by external intelligent design. In this common interpretation of teleological explanation in discussions of evolution, the concept has been an easy target of evolutionary critiques.

This heritage of interpretation, owing more to Platonic, Stoic, and Galenic sources than to the Judeo-Christian tradition (Sedley 2007; Sloan 1985), has provided an easy foil against which naturalistic alternatives could then be developed, running from David Hume in the eighteenth century to Richard Dawkins in the twenty-first. The heritage of this interpretation of teleology is ancient, and arguments against it formed the rhetorical structure Darwin employed in the *Origin of Species*, resting upon a contrast made between his naturalistic evolutionary accounts and those of a "special creationism," in which an intelligent agent has designed each species and function for specific ends (Hanby 2013).

This heritage of interpretation of the meaning of creation has given us both the intelligent design movement and its counter-point, the "blind watchmaker" alternatives, now joined in combat in contemporary public debates. As discussions in this volume highlight, this familiar reading has little to do with another tradition of teleological reasoning that follows a line from Aristotle to the great scholastics, and subsequently into Leibniz, Kant, and a primarily German tradition. It is this tradition that requires deeper discussion if we are to get some greater clarity on "teleological realism" in the future (Barham 2012a; Sloan 2012).

It is to be acknowledged that even this "immanent" teleology has been a bone of contention in some of the debates of the last fifty years, with many philosophers of biology claiming that internal, as well as external teleology, can be eliminated by a proper understanding of evolution in combination with the concept of "teleonomy" in the senses defined by Ernst Mayr in his influential discussions (Mayr 1974, 1992). But as discussed above, the introduction of concepts of "robustness" and "plasticity," the renewed attention to the concept of "organism" and "organization," some of the insights of systems-dynamics theory in relation to evolutionary biology with the rise of evo-devo, and the greater attention now paid to epigenetics push the issue of teleological realism back onto the table (Walsh 2009). And with these developments, new lines of inquiry across disciplines and traditional boundaries have been opened up.

On the theological plane, the need to clarify the concept of creation is of primary concern if there is to be a more productive dialogue between Abrahamic traditions and contemporary evolutionary science. Within Roman Catholic thought, need for such clarification was a particular concern of Pope-Emeritus Benedict XVI, who addressed these issues through his sponsorship of a series of discourses with his former students in late summer of 2006 (Horn and Wiedenhofer 2008). Scholarly explorations of these issues by several theologians interested in a productive dialogue with contemporary science have also emerged in recent decades (Hanby 2013).

Contributions to this volume by philosopher John O'Callaghan and historian and philosopher William Carroll highlight, albeit in different ways, aspects of the traditional Christian conception of *creatio ex nihilo* and its implications for evolutionary science generally within the framework of Thomistic philosophy. These analyses make it evident that the intelligent design controversies do not significantly engage the issue of relevance, and that a deeper understanding of the rich theological tradition deriving from the scholastic tradition in fact has many resources that help clarify this issue. These analyses show how the common debates over design, chance, and the existence of pain and suffering in the evolutionary world (theodicy) have little bearing on the orthodox Christian understanding of creation.[13]

A more process-oriented view on creation has been urged by several theologians from both Protestant and Catholic traditions (Haught 2000, McGrath 2011). At the Notre Dame conference, these views were presented in the plenary address that then became a chapter by the late Archbishop Józef Życiński. In this he considers the issues raised by the intelligent design movement, and the debates over chance in evolution. In response, he develops the theme of the need to understand the universe in an evolving and emerging history that is animated by love.

Notre Dame theologian Celia Deane-Drummond then engages both the issue of creation, and the more specific issue of Christology in evolutionary history, dealing both with some of the themes raised by Teilhard de Chardin and other recent process theologians, and then moving away from these by utilizing some of the interpretive framework she draws from the work of theologians Hans Urs von Balthasar and his interpreter, the theologian Ben Quash.

In looking ahead, it is, of course, difficult to make predictions concerning the shape of evolutionary theory as it will appear at the inevitable bicentennial celebration of the *Origin* in 2059. But one does hope for a more interesting and creative dialogue between the sciences and humanities over evolutionary science. In relation to the dialogue of theology and philosophy with mainstream evolutionary theory, it is not the intent of this volume to suggest these issues are resolved by relying on any one particular tradition of evolutionary biology. History has taught well the pitfall of attempting to ground theology, ethics, and other humanistic questions on the deliveries of science. Evo-devo perspectives, for example, may indeed turn out in the end to be assimilated within the framework of an "expanded" Synthesis, which leaves the theoretical "hard core" of assumptions that have governed evolutionary science since the 1930s intact.[14] But the theoretical developments over the last fifty-five years suggest that revisions of the larger framework of evolutionary biology are in process in directions that open up room for dialogue that are closed by the sometimes strident metaphysical naturalism advanced in some common interpretations of evolutionary biology. As this introductory essay has sought to develop, much of the philosophical and theological debate generated by evolutionary theory is a product of history, and this discussion might have developed in more conciliatory ways. It is clear that there is considerable ferment at the present surrounding evolutionary theory, and new lines of conversation have been opened up within the sciences themselves that promise to move us past some of the issues raised over the the last 155 years. Familiar issues surrounding Darwin's achievement, particularly as these bear on evolution-creation debates, continue to play out in the public domain. But the clarification of some important issues within the last half-century, within both Protestant and Roman Catholic thought, has made many of the classic controversies of diminished relevance.

The two reflective papers that end this volume, one by British historian of evolutionary theory Peter Bowler and the other by the French philosopher of biology Jean Gayon, both delivered at the Notre Dame conference, engage us in informed speculations about either what might have been or where we can expect developments in the future. Summarizing an argument developed at greater length in his recent monograph (Bowler 2013), Peter Bowler offers a counter-factual historical reflection on how evolutionary theory might today look if Darwin had simply perished during the HMS *Beagle* voyage rather than written his *Origin of Species*, suggesting that the subsequent history of life science may not have been greatly different. But he also speculates that some of the controversies surrounding evolution since Darwin may have taken very different directions. Jean Gayon looks to the future and offers some predictions of the shape of evolutionary theory as it might appear at the bicentennial in 2059, foreseeing a considerable expansion into the human sciences.

Utilizing both hindsight and foresight, the reflections in this volume display again the perennial richness of Darwin's vision of life, and the way the more general evolutionary picture of life, to which Darwin was the major, if not sole, contributor, has repositioned our understanding of ourselves in the world and cosmos. In this volume are fruitful suggestions of ways in which the future discussions of evolutionary theory and the humanistic disciplines may take on a fresh new character.

NOTES

The author acknowledges with deep appreciation the valuable comments on this introduction from Gerald McKenny, Kathleen Eggleson, David Depew, James Barham, Erik Peterson, M. Katherine Tillman, and the anonymous reviewers of this manuscript.

1. On the role of such commemorations in the creation of consensus and iconography in the history of science, see Abir-Am (1998) and the special issue of *Osiris* 14 (Abir-Am and Elliot 1999). See also Browne (2005). Resolution of the historiographical debate over the "reality" of a neo-Darwinian "synthesis" is not being attempted here. On this debate see Sarkar (2002),

Cain (2009), and Deslisle (2011). The explicit appeal to such a historical "Synthesis" is not common in the presentations of the Chicago symposium (Tax 1960b). I use the term for convenience to denote the achievement that many evolutionary theorists of the 1940s, after the publication of Julian Huxley's *Evolution, the Modern Synthesis* (Huxley 1942), assumed had been achieved with the unification of natural selection theory and Mendelian genetics. It is still described by subsequent historians under this label (e.g., Bowler 2003, chap. 9). It is still commonly referred to in literature on contemporary evolution. See for example Pigliucci and Müller (2010).

2. See on this Ruse (2008) and West-Eberhard (2003, chap. 24). Gould was careful to clarify in subsequent statements that he was not advocating DeVriesian mutationism or rejecting some main aspects of evolutionary gradualism.

3. On several dimensions of the probability revolution see Porter (1986, 2004), Gigerenzer et al. (1990), and Hacking (1990). I am deeply indebted to the forthcoming essay by David Depew, "Contingency, Chance, and Randomness in Ancient and Modern Biology" (forthcoming; personal communication, cited with permission).

4. Fisher was concerned to emphasize causal indeterminacy and statistical probability as a way of defending free will. See Fisher (1934). See also Moore (2007).

5. For a substantial and penetrating examination of the deficiencies in the typical understanding of the whole issue surrounding the "evolution and creation" debate, see Hanby (2013).

6. The common meaning of "teleonomy" in Anglophone discussions, typically indebted to Ernst Mayr's influential definition, builds upon Colin Pittendrigh's initial definition of the concept in 1958 (Pittendrigh 1958). Mayr identifies this with the apparent teleological action produced by the working out of an underlying genetic program established by natural selection (Mayr 1974, 1992). This presumably avoids traditional internal teleology as might be found in Aristotle. It should be evident that Mayr's teleonomy is not the same as that employed in the Auletta, Colagè, and D'Ambrosio chapter in this volume, which draws on some of the wider uses of the concept in physical as well as biological science. For some historical analysis of these concepts in modern biology, see Sloan (2012). The new concern with epigenetic factors in evolutionary change, and the appeal to the role of active organic choice in such things as niche construction (Odling-Smee 2010; Bateson and Gluckman, 2011, 100–103), illustrate in a surprising way the reintroduction of internal teleological principles into recent discussions of evolutionary theory. It is difficult to see how Mayr's concept of teleonomy covers these appeals to organic choice in evolution. For further elaborations on the use of teleological reasoning in recent biology, see Barham (2012a, 2012b) and Walsh (2009).

7. This is developed in the book under preparation, with co-author John Jackson, tentatively entitled *Darwinism, Democracy, and Race: Evolutionary Biology and Boasian Anthropology in the American Century.* I express my appreciation to David Depew for access to a draft chapter dealing with the place of anthropology in the University of Chicago symposium and its background.

8. See the more expansive analysis of these issues in Numbers (2006).

9. "Orthogenetic" here means a directional trend in evolution. This is distinguished from the branching and nondirectional pattern generally associated with Darwin and his principle of divergence. On this see Bowler (1983, chap. 7).

10. See also comments by Joseph Ratzinger (Pope-Emeritus Benedict XVI) in 1986, reprinted in Horn and Wiedenhofer (2008, 8–16).

11. An earlier example is the Notre Dame conference, organized by the late Fr. Ernan McMullin in 1984, to explore the issues of creation and evolution (McMullin 1985). See also the excellent update on these issues in the several papers in Caruana (2009).

12. Mistranslations of a crucial sentence in this statement in several English versions rendered "plus qu'un hypothèse" as "but one hypothesis." The full context of the French original makes it clear that this is saying more: "Aujourd'hui, près d'un demi-siècle après la parution de l'encyclique, et de nouvelles connaissances conduisent à reconnaître dans la théorie de l'evolution plus qu'un hypothèse. Il est en effet remarquable que cette théorie se soit progressivement imposées à l'esprit des chercheurs, à la suite d'une série de découvertes faites dans diverse disciplines du savoir. La convergence, nullement recherchée ou provoquée, de résultats de travaux menés indépendamment les uns des autres, constitue par elle-même un argument significatif en faveur de cette théorie." I am indebted to Professor Kenneth Kemp of St. Thomas University for a copy of the French original.

13. It is important to emphasize that in the Thomistic-Augustinian interpretation of creation, the emphases on undirected process in the neo-selectionist interpretation of evolution are not deeply relevant, since creation *ex nihilo* concerns the *existence* of the world in any order and emerging by any process whatever (Hanby 2013). These issues do seem to have more bearing on process-theology interpretations of evolution.

14. The distinction by Imre Lakatos (1970) between a central irreformable set of assumptions of a research program, and the flexible and reformable surrounding "protective belt" that may change, is useful here. With such an analysis, we can read the contemporary conflicts as a competition between a successful Synthesis program that is now showing signs of degeneration, and a revived developmentalist, morphological program that was overshadowed, but never defeated, by the success of the Synthesis.

REFERENCES

Abir-Am, P. (1998). *La Mise en mémoire de la science: Pour une ethnographie historique des rites commemoratifs.* Amsterdam: Éditions des archives contemporaines.

Abir-Am, P, and C. A. Elliott, eds. (1999). *Commemorative Practices in Science: Historical Perspectives on the Politics of Collective Memory.* Special issue *Osiris* 14.

Allen, G. (1969). "Hugo de Vries and the Reception of the 'Mutation Theory.'" *Journal of the History of Biology* 2:55–87.

Amundson, R. (2007). *The Changing Role of the Embryo in Evolutionary Thought.* Cambridge: Cambridge University Press.

Appleby, S. (1987). "Between Americanism and Modernism: John Zahm and Theistic Evolution." *Church History* 56:474–89.

Artigas, M., T. F. Glick, and R. A. Martinez. (2006). *Negotiating Darwin: The Vatican Confronts Evolution, 1877–1902.* Baltimore: Johns Hopkins University Press.

Auletta, G., M. LeClerc, and R. A. Martinez, eds. (2011). *Biological Evolution: Facts and Theories; A Critical Appraisal 150 Years after "The Origin of Species."* Rome: Gregorian and Biblical Press.

Barham, J. (2012a). *Normativity, Agency, and Life: Teleological Realism in Biology.* Saarbrücken: LAP Lambert.

———. (2012b). "Normativity, Agency and Life." *Studies in History and Philosophy of the Biological and Biomedical Sciences* 43:92–103.

Barkow, J. H. (2006). *Missing the Revolution: Darwinism for Social Scientists.* Oxford: Oxford University Press.

Bateson, P., and P. Gluckman. (2011). *Plasticity, Robustness, Development and Evolution.* Cambridge: Cambridge University Press.

Beurton, P. J., R. Falk, and H.-J. Rheinberger, eds. (2000). *The Concept of the Gene in Development and Evolution: Historical and Epistemological Perspectives.* Cambridge: Cambridge University Press.

Bowler, P. (1983). *The Eclipse of Darwinism.* Baltimore: Johns Hopkins University Press.

———. (1988). *The Non-Darwinian Revolution: Reinterpreting a Historical Myth.* Baltimore: Johns Hopkins University Press.

———. (2003). *Evolution: The History of an Idea.* 3rd ed. Berkeley: University of California Press.

———. (2013). *Darwin Deleted: Imagining a World without Darwin.* Chicago: University of Chicago Press.

Browne, J. (2005). "Commemorating Darwin." *British Journal for the History of Science* 38:251–74.

Buss, D. M. (2004). *Evolutionary Psychology: The New Science of the Mind.* Boston: Pearson.

Cain, J. (2009). "Rethinking the Synthesis Period in Evolutionary Studies." *Journal of the History of Biology* 42:621–48.

Caruana, L., ed. (2009). *Darwin and Catholicism: The Past and Present Dynamics of a Cultural Encounter.* London: Clark.

Clayton, P., and J. Schloss, eds. (2004). *Evolution and Ethics: Human Morality in Biological and Religious Perspective.* Grand Rapids, MI: Eerdmans.

Darwin, C. (1876). *The Variation of Plants and Animals Under Domestication.* 2nd ed. New York: Appleton. Orig. pub. 1868.

———. ([1859] 1964). *On The Origin of Species by Means of Natural Selection.* 1st ed., reprint edition. Cambridge, MA: Harvard University Press.

———. ([1871] 1981). *Descent of Man, and Selection in Relation to Sex.* 1st ed., reprint edition. Princeton: Princeton University Press.

Dawkins, R. (1987). *The Blind Watchmaker: Why Evolution Reveals a Universe Without Design.* New York: Norton.

———. (2008). *The God Delusion.* Boston: Houghton-Mifflin.

Dennett, D. C. (1995). *Darwin's Dangerous Idea: Evolution and the Meanings of Life.* New York: Simon and Schuster.

Depew, D., and B. Weber. (1995). *Darwinism Evolving: Systems Dynamics and the Genealogy of Natural Selection.* Cambridge, MA: MIT Press.

———. (forthcoming). "Contingency, Chance, and Randomness in Ancient and Modern Biology." In Ramsey and Pence forthcoming.

Deslisle, R. G. (2011). "What Was Really Synthesized During the Evolutionary Synthesis? A Historiographic Proposal." *Studies in History and Philosophy of Biological and Biomedical Sciences* 42:50–59.

Desmond, A., and J. R. Moore. (2009). *Darwin's Sacred Cause: How a Hatred of Slavery Shaped Darwin's Views on Human Evolution.* Boston: Houghton Mifflin and Harcourt.

Eldredge, N., and S. J. Gould. (1972). "Punctuated Equilbria: An Alternative to Phyletic Gradualism." In *Models in Paleontology*, ed. T. J. M. Schopf, 82–115. San Francisco: Freeman.

Ewing, J. F. (1960). "Current Roman Catholic Thought on Evolution." In Tax 1960b, 3:19–28.

Falk, R. (2000). "The Gene—A Concept in Tension." In Beurton, Falk, and Rheinberger 2000, 317–48.

Farber, P. L. (1994). *The Temptations of Evolutionary Ethics.* Berkeley: University of California Press.

Fisher, R. A. (1934). "Indeterminism and Natural Selection." *Philosophy of Science* 1:99–117.

Gayon, J. (1998). *Darwinism's Struggle for Survival: Heredity and the Hypothesis of Natural Selection.* Trans. M. Cobb. Cambridge: Cambridge University Press.

Gigerenzer, G., Z. Swijtink, T. Porter, and L. Daston. (1990). *The Empire of Chance: How Probability Changed Science and Everyday Life.* Cambridge: Cambridge University Press.

Gilbert, S. (2000). "Genes Classical and Genes Developmental: The Different Use of Genes in Evolutionary Syntheses." In Beurton, Falk, and Rheinberger 2000, 178–92.

Gilbert, S., J. Opitz, and R. Raff. (1996). "Resynthesizing Evolutionary and Developmental Biology." *Developmental Biology* 173:357–72.

Gingras, Y. (2011). "Revisiting the 'Quiet Debut' of the Double Helix: A Bibliometric and Methodological Note on the 'Impact' of Scientific Publications." *Journal of the History of Biology* 43:159–81.

Gissis, S. B., and E. Jablonka. (2011). *Transformations of Lamarckism: From Subtle Fluids to Molecular Biology.* Cambridge, MA: MIT Press.

Gould, S. J. (1977). *Ontogeny and Phylogeny.* Cambridge, MA: Belknap Press.

———. (1997). "Darwinian Fundamentalism." *The New York Review of Books,* June 12. Available at http://www.nybooks.com/articles/archives/1997/jun/12/darwinian-fundamentalism.

Grene, M. (1967). "Biology and the Problem of Levels of Reality." *New Scholasticism* 41:427–49.

Hacking, I. (1990). *The Taming of Chance.* Cambridge: Cambridge University Press.

Hanby, M. (2013). *No God, No Science? Theology, Cosmology, Biology.* Oxford: Wiley-Blackwell.

Haught, J. (2000). *God after Darwin: A Theology of Evolution.* Boulder, CO: Westview Press.

Hodge, M. J. S. (1992). "Biology and Philosophy (Including Ideology): A Study of Fisher and Wright." In *The Founders of Evolutionary Genetics,* ed. S. Sarkar, 231–93. Dordrecht: Kluwer.

Hodge, M. J. S., and G. Radick, eds. (2009). *The Cambridge Companion to Darwin.* 2nd ed. Cambridge: Cambridge University Press.

Horn, S. O., and S. Wiedenhofer, eds. (2008). *Creation and Evolution: A Conference with Pope Benedict XVI in Castel Gandolfo.* Trans. M. J. Miller. San Francisco: Ignatius Press.

Huxley, J. (1960). "The Evolutionary Vision." In Tax 1960b, 3:249–61.

Huxley, J., ed. (1942). *Evolution, the Modern Synthesis.* New York: Harper and Bros.

John Paul II, Pope. (1996). "Magisterium Is Concerned with the Question of Evolution, for It Involves the Conception of Man." Delivered to the Pontifical Academy of Sciences, 22 October 1996. Originally published in English in *L'Osservatore Romano,* 30 October 1996. Available at http://www.its.caltech.edu/~nmcenter/sci-cp/evolution.html.

Katz, L. D., ed. (2000). *Evolutionary Origins of Morality: Cross-Disciplinary Perspectives.* Bowling Green, OH: Imprint Academic Press.

Keller, E. F. (2000). *The Century of the Gene*. Cambridge, MA: Harvard University Press.

Lakatos, I. (1970). "Falsification and the Methodology of Scientific Research Programmes." In *Criticism and the Growth of Knowledge*, ed. I. Lakatos and A. Musgrave, 91–196. Cambridge: Cambridge University Press.

Larson, E. J. (1998). *Summer for the Gods: The Scopes Trial and America's Continuing Debate over Science and Religion*. New York: Basic Books.

Leland, K., et al. (2014). "Does Evolutionary Theory Need a Rethink?" *Nature* 514, no. 7521:161–64. Online at http://www.nature.com/news/does -evolutionary-theory-need-a-rethink-1.16080.

Mayr, E. (1974). "Teleological and Teleonomic: A New Analysis." In *Methodological and Historical Essays in the Natural and Social Sciences*, ed. R. S. Cohen and M. Wartofsky, 91–117. Boston: Reidel.

———. (1992). "The Idea of Teleology." *Journal of the History of Ideas* 53:117–35.

Mayr, E., and W. Provine, eds. (1980). *The Evolutionary Synthesis: Perspectives on the Unification of Biology*. Cambridge, MA: Harvard University Press.

McGrath, A. E. (2011). *Darwin and the Divine: Evolutionary Thought and Natural Theology*. Oxford: Wiley-Blackwell.

McMullin, E., ed. (1985). *Evolution and Creation*. Notre Dame, IN: University of Notre Dame Press.

Medawar, P., and J. S. Medawar. (1977). *The Life Science: Current Ideas of Biology*. New York: Harper and Row.

Monod, J. (1971). *Chance and Necessity: An Essay on the Natural Philosophy of Modern Biology*. Trans. A. Wainhouse. New York: Knopf.

Moore, J. (2007). "R. A. Fisher: A Faith Fit for Eugenics." *Studies in History and Philosophy of the Biological and Biomedical Sciences* 38:110–35.

Morange, M. (2000). "Developmental Gene Concept: History and Limits." In Beurton, Falk, and Rheinberger 2000, 193–215.

Moss, L. (2003). *What Genes Can't Do*. Boston: MIT Press.

———. (2006). "Redundancy, Plasticity, and Detachment: The Implications of Comparative Genomics for Evolutionary Thinking." *Philosophy of Science* 73:930–46.

Müller, G. B., R. D. Douglas, and B. K. Hall, eds. (2004). *Environment, Development, and Evolution: Toward a Synthesis*. Boston: MIT Press.

Müller, G. B., and S. A. Newman, eds. (2003). *Origination of Organismal Form: Beyond the Gene in Developmental and Evolutionary Biology*. Cambridge, MA: MIT Press.

Neumann-Held, E., and C. Rehmann-Sutter, eds. (2006). *Genes in Development: Rereading the Molecular Paradigm*. Durham, NC: Duke University Press.

Nijhout, H. F. (2002). "The Nature of Robustness in Development." *Bioassays* 24:553–63.

Numbers, R. A. (2006). *The Creationists: From Scientific Creationism to Intelligent Design*. Cambridge, MA: Harvard University Press.

Odling-Smee, J. (2010). "Niche Inheritance." In Pigliucci and Müller 2010, 175–207.

Okasha, S. (2012). "Population Genetics." *The Stanford Encyclopedia of Philosophy* (Fall 2012 edition), ed. Edward N. Zalta. Available at http://plato.stanford.edu/archives/fall2012/entries/population-genetics.

Olby, R. (1987). "Dimensions of Scientific Controversy: The Biometric-Mendelian Debate." *British Journal for the History of Science* 22:299–320.

———. (2003). "Quiet Debut for the Double Helix." *Nature* 421:402–5.

Olson, E. (1960). "Morphology, Paleontology, and Evolution." In Tax 1960b, 1:523–45.

Pelikan, J. (1960). "Creation and Causality in the History of Christian Thought." In Tax 1960b, 3:29–40.

Pence, C. (2011). "Describing Our Whole Experience: The Statistical Philosophies of W. F. R. Weldon and Karl Pearson." *Studies in History and Philosophy of the Biological and Biomedical Sciences* 42:475–85.

Pence, C., and G. Ramsey (2013). "A New Foundation for the Propensity Interpretation of Fitness." *British Journal for the Philosophy of Science*. 64:851–81.

Pigliucci, M., and G. Müller, eds. (2010). *Evolution: The Extended Synthesis*. Cambridge, MA: MIT Press.

Pittendrigh, C. (1958). "Adaptation, Natural Selection, and Behavior." In *Behavior and Evolution*, ed. A. Roe and G. G. Simpson, 391–94. New Haven, CT: Yale University Press.

Porter, T. M. (1986). *The Rise of Statistical Thinking, 1820–1900*. Princeton: Princeton University Press.

———. (2004). *Karl Pearson: The Scientific Life in a Statistical Age*. Princeton: Princeton University Press.

Provine, W. (1971). *The Rise of Theoretical Population Genetics*. Chicago: University of Chicago Press.

Rahner, K. (1966). "Christology within an Evolutionary View." In Karl Rahner, *Theological Investigations*, vol. 5, chap. 5. New York: Seabury.

Ramsey, G., and C. Pence, eds. (forthcoming). *Chance and Evolution*. Chicago: University of Chicago Press.

Rheinberger, H.-J. (2010). *An Epistemology of the Concrete*. Durham, NC: Duke University Press.

Rosenberg, C. (2009). "Darwinism in Moral Philosophy and Social Theory." In Hodge and Radick 2009, 345–67.

Ruse, M. (2008). "Darwinian Evolutionary Theory: Its Structure and Its Mechanism." In *Oxford Handbook of Philosophy of Biology*, ed. M. Ruse, 34–63. New York: Oxford University Press. Available at http://www.oxfordhandbooks.com.

Sarkar, S. (2002). "Evolutionary Theory in the 1920s: The Nature of the Synthesis." *Philosophy of Science* 71:1215–16.

Schrödinger, E. (1974). *What is Life? and Mind and Matter.* Cambridge: Cambridge University Press.

Schwartz, J. (2008). *In Pursuit of the Gene: Darwin to DNA.* Cambridge, MA: Harvard University Press.

Sedley, D. (2007). *Creationism and Its Critics in Antiquity.* Berkeley: University of California Press.

Simpson, G. G. (1944). *Tempo and Mode in Evolution.* New York: Columbia University Press.

———. (1960). "The History of Life." In Tax 1960b, 1:117–80.

Sloan, P. R. (1985). "The Question of Natural Purpose." In McMullin 1985, 121–50.

———. (2009). "Bringing Evolution to Notre Dame: Father John Zahm, C.S.C., and Theistic Evolutionism." *American Midland Naturalist* 161:189–205.

———. (2012). "What Happened to Teleology in Early Molecular Biology?" *Studies in History and Philosophy of the Biological and Biomedical Sciences* 43:140–51.

———. (2014). "Molecularizing Chicago—1945–1965: The Rise, Fall and Rebirth of the Chicago Biophysics Program." *Historical Studies in the Natural Sciences* 44:364–412.

Smocovitis, B. J. (1996). *Unifying Biology: The Evolutionary Synthesis and Evolutionary Biology.* Princeton: Princeton University Press.

———. (1999). "The 1959 Darwin Centennial Celebration in America." *Osiris* (2nd series) 14:274–323.

Sober, E., and D. Sloan Wilson. (1998). *Unto Others: The Evolution and Psychology of Unselfish Behavior.* Cambridge, MA: Harvard University Press.

Stamos, D. (2001). "Quantum Indeterminism and Evolutionary Biology." *Philosophy of Science* 68:164–84.

Strasser, B. J. (2003). "Who Cares About the Double Helix?" *Nature* 422:803–4.

Tax, S. (1960a). "The Celebration: A Personal View." In Tax 1960b, 3:271–82.

Tax, S., ed. (1960b). *Evolution After Darwin.* 3 vols. Chicago: University of Chicago Press.

Waal, F. B. M. (1996). *Good Natured: The Origins of Right and Wrong in Humans and Other Animals.* Cambridge, MA: Harvard University Press.

Waal, F. B. M., ed. (2006). *Primates and Philosophers.* Princeton: Princeton University Press.

Waddington, C. H. (1942). "The Epigenotype." *Endeavour* 1:18–20.

———. (1960). "Evolutionary Adaptation." In Tax 1960b, 1:381–402.

Walsh, D. (2009). "Teleology." In *The Oxford Handbook of Philosophy of Biology*, ed. M. Ruse. New York: Oxford University Press, 113–37.

Waters, C. K. (2013). "Molecular Genetics." *The Stanford Encyclopedia of Philosophy* (Fall 2013 edition), ed. Edward N. Zalta. Available at http:// plato.stanford.edu/archives/fall2013/entries/molecular-genetics.

West-Eberhard, M. J. (2003). *Developmental Plasticity and Evolution*. Oxford: Oxford University Press.

Whye, J., ed. (2009). *University of Cambridge Darwin Website*, http://darwin -online.org.uk.

Wilson, E. O. (1978). *On Human Nature*. Cambridge, MA: Harvard University Press.

———. (1998). *Consilience: The Unity of Knowledge*. New York: Knopf.

PART ONE

Nature

TWO

EVOLUTION THROUGH DEVELOPMENTAL CHANGE
How Alterations in Development Cause
Evolutionary Changes in Anatomy

Scott F. Gilbert

> *Nature interests me because it's beautiful, complex, and robust.*
> *Evolutionary theory interests me because it explains why nature*
> *is beautiful, complex, and robust.*
> —David Quammen (quoted in Valenti 2007)

> *. . . a study of the effects of genes during development is as essential*
> *for an understanding of evolution as are the study of mutation and*
> *that of selection.*
> —Julian Huxley (1942, 8)

THE MODERN SYNTHESIS AND ITS CRITIQUES

For the past half-century, the mechanisms of evolution have been
explained by the fusion of genetics and evolutionary biology called
"the Modern Synthesis." The tenets of the Modern Synthesis have
been generally formulated as such:

1. There is genetic variation within the population.
2. There is competition for survival, with most of the population not reproducing.
3. This leads to the differential survival and reproduction of those organisms with genetic variants that make them more likely to succeed in the particular environment ("survival of the fittest").
4. The offspring have a high likelihood of inheriting those genetic variants that enabled the differential survival of their parents.

This synthesis explains natural selection within a species, and it explains it remarkably well. Moreover, the Modern Synthesis explains evolutionary change in populations throughout the natural kingdoms. For example, drug resistance in bacteria, coat colors in rodents, mimicry in moths and butterflies, carbon metabolism in plants, and malarial resistance in humans have each been shown to be caused by the selection of randomly produced genetic variation by factors within the environment. These genetic variants can be assigned to gene mutations, and the mathematical modeling of the selection on the variations has shown the power of this model (see, for example, Rice 2004; Futuyma 2009; Gilbert and Epel 2009).

Just a few examples will be mentioned here: The ability of mice to blend into the background and avoid predators is provided by a mutation in the *Agouti* gene, which is involved in hair pigmentation (Linnen et al. 2009). The ability of human populations to avoid succumbing to malaria can be acquired by a change in the hemoglobin gene. This genetic change causes the replacement of the amino acid glutamate by valine at the sixth position, and the malarial parasite cannot multiply in red blood cells having this variant of hemoglobin (see Gilbert and Epel 2009). Humans have acted as selective agents on mosquitoes, such that the *Anopheles* mosquito that transmits the malarial parasite is becoming less sensitive to DDT. By widely spraying DDT, we have selected for mosquitoes that have evolved enzymes that are resistant to the pesticide (in that their amino acids do not bind the pesticide) or that actually destroy the pesticides before it can kill the mosquito (Raymond et al. 1998; Donnelly et al. 2009). Similarly, genetic changes in the opsin protein can cause it to be activated by a different wavelength of light, thereby allowing some species to see in the ultraviolet rather than violet light. This mutation enables kestrels to fly high and catch voles by seeing their urine trails, which are visible in the ultraviolet (Viitala et al. 1995).

But while such examples show that evolution can occur within a species, the species itself did not transform into anything else. Malaria-resistant humans and DDT-resistant mosquitoes were still humans and mosquitoes, respectively. The mice with the *Agouti* gene mutation persisted as better camouflaged mice, but they did not evolve into anything un-mousy. Until recently, one could study how the selection of certain genetic variants allowed a particular phenotype to persist and characterize a species; but the differences *between* species (that did not interbreed) could not be studied in this manner.

While there has been excellent evidence for evolution provided by comparative anatomy, biogeography, paleontology, and several other sciences, this inability to study the genetic mechanisms of evolution above the species level provided a space though which critics could claim that evolution had not been proven. Father Stanley Jaki, for instance, wrote that,

> As to the claim . . . that the Darwinian evolutionary mechanism (the interplay of chance mutations with environmental pressure) has solved all basic problems, I hold it to be absurd and bordering at times on the unconscionable. While the mechanism in question provoked much interesting scientific research, it left unanswered the question of transition among genera, families, orders, classes, and phyla where the absence of transitional forms is as near-complete as ever. (quoted in Dembski 1996)

TOOLS FOR ANSWERING THE CRITIQUES: DNA SEQUENCING AND DEVELOPMENTAL GENETICS

This situation changed dramatically, starting in the mid-1970s and continuing into the present day. Here, two major advances contributed to a more complete evolutionary theory that could explain both evolution within a species (microevolution) and evolution between species (macroevolution). The advances were (1) DNA sequencing and (2) developmental genetics. DNA sequencing allows scientists to actually compare the gene sequences between species, and developmental genetics gives scientists a theory of body construction that enables us to see how the changes in the DNA can cause (selectable) changes in anatomy.

Genomics: Comparing Genomes between Species

DNA sequencing and the computer-enhanced science of genomics provided "the ultimate forensic record of evolution" (Carroll 2006). First, they have allowed us to see which animals are grouped together and which are not. DNA contains the documentary record of evolution. Just as a linguist can compare words to show that French, Romanian, Italian, and Spanish descended from a common ancestor (and that Spanish and Portuguese diverged rather recently), so can a biologist compare particular sequences of DNA to show how animals are related to one another (Gilbert 2003; Rice 2004). Descent with modification was shown not only in our bones but also in our genes.

Developmental Genetics: How Cell Communication Controls Cell Determination

The revolution in DNA technology enabled developmental biology to formulate a theory of body construction based on cell-to-cell communications. For decades, it had been known that interactions between vertebrate embryonic cells determined the eventual identities of each cell. Chemical instructions from one cell told the other cell what proteins to make and whether to become a blood cell, a nerve cell, or so forth. However, the actual proteins providing these instructions were not able to be isolated until DNA sequencing enabled scientists to isolate the genes rather than the proteins. The common plan for communicating developmental instructions from one cell to another begins with the inducing cell's secreting a protein called a *paracrine factor*. These factors do not go through the blood (like endocrine factors), but rather they diffuse over a few cell diameters, working on neighboring cells. The responding cell has a receptor for the paracrine factor, and this receptor protein spans the cell membrane. Its outer domain can bind the particular paracrine factor, and when it does so, this changes the structure of the domain of the receptor residing inside the cell. Usually, the binding of a paracrine factor with its receptor gives the inner domain an enzymatic activity that it did not formerly possess. This enzymatic activity can (either directly or indirectly) modify a protein called a *transcription factor*. In many cases, the transcription factor protein is phosphorylated (has a phosphate group added to it) by the new enzymatic activity. The phosphate group activates the transcription factor enabling it to enter the nucleus, bind to a specific sequence of DNA, and activate the gene adjacent to that DNA sequence.

In this manner, one cell type can change gene expression in a neighboring cell, and thereby instruct the cell to become a pancreatic cell, a liver cell, or a neuron. By sequential interactions with other cells, this specification gets finer, such that the cell will be told to be an insulin-secreting beta cell of the pancreas, or an oval cell of the liver, or a dopaminergic neuron of the ventral brain. Cell adhesion and cell migration can also be controlled in this manner.[1] Moreover, changes in gene expression in the embryo can cause different types of anatomical structures to form.

One critical consequence of the combination of DNA sequencing and developmental genetics has been the discovery of "deep homologies" (Shubin, Tabin, and Carroll 1997, 2009). Here, the same gene is responsible for the generation of a particular organ, even if the organ differs greatly between animal groups. The eyes of insects, squids, and vertebrates are very different, and are even formed from different parts of the organism. Evolutionary developmental geneticists such as Walter Gehring (1996, 2005) demonstrated that there is a developmental genetic pathway-specifying photoreceptor development, and that this is common to all animals with eyes. The basic genetic pattern for eye development is specified by homologues of the *Pax6* gene. *Pax6* is one of those transcription factor proteins mentioned above. The pathway initiated by *Pax6* and its homologues probably evolved only once, and the eyes of all the vertebrates and arthropods, flatworms, and snails come from a modification of this central scheme. Indeed, if one activates a mouse *Pax6* gene in those cells of a fly larva destined to become its jaw, that jaw will develop into a fly eye. If one expresses the fly *Pax6* gene in frog skin, the fly *Pax6* initiates eye development in the frog skin and causes it to become an amphibian retina (Halder et al. 1995; R. Chow et al. 1999).

Indeed, several transcription factors have deep homology throughout the animal kingdom. A gene called *tinman* specifies cells to become the heart in both insects and mammals, and another set of genes, called *Hox* genes, specifies the anterior-posterior axis of all animals. Indeed, eyes, hearts, and limbs "arose by the modification of preexisting genetic regulatory circuits established in early metazoans" (Shubin, Tabin, and Carroll 2009).

Deep homology has very important ramifications. First, it is remarkable evidence for the relatedness of all animals. We all came from the same ancestor. Second, it makes obsolete the notion that similarities in animals arose from convergence to a particular state. One can only have convergence when the two lineages were originally divergent. However, we now know that the various lineages have common genes that specify these "convergent traits." Although the notion of convergence has been brought up again by Simon Conway Morris (2003), this idea is based on the disproved notion that homologous genes could not exist between greatly divergent phyla (see Amundson 2005, 217). Rather, DNA sequencing and developmental genetics have shown that there is descent with modification within the genome.

Evolutionary Developmental Biology

Evolutionary developmental biology holds that evolution occurs by changing the trajectory of development. In other words, the major processes in evolution are those involving the changes in the genetic regulatory genes that construct the organism. Rather than studying gene *frequency within* a population, evolutionary developmental biology focuses on the changes in gene *expression between* or within populations. To the evolutionary developmental biologist, evolutionary biology remains incomplete without a theory of body construction and how the construction of bodies can change (Amundson 2005; Gilbert and Epel 2009). Evolutionary explanation must involve not only the "survival of the fittest," but also the "arrival of the fittest." For instance, when confronted with the question of how arthropod body plans arose, Hughes and Kaufman (2002) state:

> To answer this question by invoking natural selection is correct—but insufficient. The fangs of a centipede . . . and the claws of a lobster accord these organisms a fitness advantage. However, the crux of the mystery is this: From what developmental genetic changes did these novelties arise in the first place?

The realization that a theory of body construction was needed to supplement the evolutionary theory of change was proclaimed in a well-publicized 1975 research paper by Mary-Claire King and Alan Wilson. Entitled "Evolution at Two Levels in Humans and Chimpanzees," this study showed that despite the large anatomical differences between chimps and humans, their DNA was almost identical. King and Wilson's solution to this conundrum that bodily change was outpacing gene changes harkened back to the ideas of Richard Goldschmidt, Conrad Waddington, and other developmental biologists who felt that evolutionary change was predicated upon specific changes in gene *regulation*. Specifically, the King and Wilson paper states:

> The organismal differences between chimpanzees and humans would . . . result chiefly from genetic changes in a few regulatory systems, while amino acid substitutions in general would rarely be a key factor in major adaptive shifts. (King and Wilson 1975, 114)

That is to say, the nucleotide substitutions in the DNA sequences that encode protein sequences—which seem to be pretty much the same for chimps and humans—are not what is important. The important differences are where, when, and how much the genes are activated. François Jacob's 1977 paper, "Evolution by Tinkering," suggested a new model, generalizing the work of King and Wilson, and focusing attention on how the same genes can create new types of body plans when their expression pattern is altered during development (Jacob 1977).

Over the next thirty years, it became possible to study gene expression and gene sequence together. Recombinant DNA technology, polymerase chain reaction, *in situ* hybridization, and high-throughput RNA analysis now enable us to look at gene sequences and to compare their expression between species. This ability has revolutionized the way we look at evolutionary processes. The new science of *evolutionary developmental biology* ("evo-devo") is looking at the mechanisms by which changes in gene regulation cause changes in anatomy. It looks at the mechanisms of tinkering.

Wallace Arthur (2004) has catalogued four ways in which Jacob's "tinkering" can take place at the level of gene expression:

- Heterotopy (change in location of gene expression)
- Heterochrony (change in time or duration of gene expression)
- Heterometry (change in amount of gene expression)
- Heterotypy (change in kind of gene being expressed)

These mechanisms are predicated on the modularity of enhancers (see Kirschner and Gerhart 2005; Gilbert and Epel 2009). Enhancers are the genetic sequences that tell a gene where and when (and how much) they are to be transcribed. Enhancers bind the abovementioned transcription factors, thereby allowing their presence or absence to regulate gene expression. One of the most important discoveries of evolutionary developmental biology is that different enhancers can direct the transcription of a gene in different organs. Thus, via mutating enhancers, organisms can express the same gene, but in different cells or for different durations.

This enhancer modularity has major significance in evolution. In humans, the gene for the Duffy protein has enhancers enabling it to become expressed in blood vessels, red blood cells, and cerebellar neurons. In African regions having vivax malaria, the *Plasmodium* parasite can use the Duffy protein as an entry site to get into red blood cells. However, a large percentage of the human population in these regions have become immune to this type of malaria through a mutation in the red blood cell enhancer of this gene. This mutation prevents the expression of this gene in the red blood cell precursors but does not prevent this gene's activity in the blood vessels and neurons (Tournamille et al 1995). Similarly, stickleback fish living in lake regions have mutations in the "pelvic" enhancer of the *Pitx1* gene, preventing the gene from being expressed in the pelvis (causing spines to which predators can bind), while enabling its expression in the thymus and nasal pit (Shapiro et al. 2004). Enhancer modularity thus enables nature to readily experiment, making genes active in different cells, or at different times, or at different amounts.

Heterotopy

Changing the location of a gene's expression (i.e., altering which cells it is expressed in) can result in enormous amounts of selectable variation. For instance, one of the biggest differences between the duck and the chicken is the webbed feet of the duck. This is an obviously important adaptation to swimming (as are the duck's broad beak and the oil glands of its skin). Vertebrate embryonic limbs develop with their fingers and toes surrounded by a web of connective tissue and skin. In chicks (and humans), the cells that make up the webbing between the digits undergo apoptosis (programmed cell death) initiated by the paracrine factor BMP4. This cell death destroys the webbing and frees the digits. It turns out that ducks also express BMP4 in the webbing of their hindlimbs, but this protein is prevented from signaling cell death by the presence of another protein, the BMP inhibitor Gremlin (Laufer et al. 1997; Merino et al. 1999). Chick and mouse limbs also express Gremlin around the cartilaginous skeletal elements of the digits, but not in the webbing. Thus, the webbing in embryonic duck feet remains intact thanks to the heterotopy—the change in place of gene expression—of the Gremlin gene. Indeed, if one adds beads containing the Gremlin protein to the webbing of an embryonic chick limb, the limb will keep its webbing, just as a duck does.

Heterotopy is also seen to be responsible for the formation of bat forelimbs that extend and retain their webbing (Weatherbee et al.

2006), the formation of the carapace of the turtle shell (Cebra-Thomas et al. 2005), and the evolution of the bird feather from the reptilian scale (Harris, Fallon, and Prum 2002). Heterotopy also provides the proximate mechanism for the elimination of snake limbs (Cohn and Tickle 1999; Di-Poï et al. 2010).

Heterochrony

Heterochrony is a shift in the relative timing of two developmental processes from one generation to the next. The elongated fingers in the dolphin flipper appear to be caused by heterochrony, wherein the region responsible for producing the paracrine factors for the developing limb (especially Fgf8) is present longer than in other mammals (Richardson and Oelschläger 2002). The enormous number of ribs formed in embryonic snakes (more than five hundred in some species) is likewise due to heterochrony. In this case, the segmentation reactions cycle nearly four times faster relative to tissue growth in snake embryos than they do in related vertebrate embryos (Gomez et al. 2008).

Some of the most interesting new work on heterochrony concerns the origin of the human brain. Both humans and other apes have rapid brain growth before birth. However, whereas the brain growth slows considerably in other apes, the growth of the human brain continues at the fetal rate for around two years, making over thirty thousand new synapses per second (Rose 1998; Barinaga 2003; Gilbert 2013a.) Some of the genes that differ between humans and other apes are those genes that accelerate human brain growth (McLean et al. 2011; Dennis et al. 2012). Thus, as predicted by King and Wilson (1975), small changes in gene expression during development can cause important alterations in anatomy.

Heterometry

Heterometry is the change in the *amount* of a gene product. One of the best examples of heterometry involves Darwin's celebrated finches, a set of fourteen closely related birds collected by Charles Darwin and his shipmates during his visit to the Galápagos and Cocos Islands in 1835. These birds helped him frame his evolutionary theory of descent with modification, and they still serve as one of the best examples of adaptive radiation and natural selection (see Weiner 1994; Grant 1999; Grant and Grant 2007). Systematists have shown that these finch species evolved in a particular manner, with a major speciation event being the split between the cactus finches and the ground finches. The ground finches evolved deep, broad beaks that enable them to crack seeds open, whereas the cactus finches evolved narrow, pointed beaks that allow them to probe cactus flowers and fruits for insects.

Earlier research (Schneider and Helms 2003) had shown that species differences in the beak pattern were caused by changes in the growth of the neural crest-derived mesenchyme of the frontonasal process (i.e., those cells that form the facial bones). Moreover, they and others (Wu et al. 2004, 2006) showed that birds use BMP4 as a growth factor promoting cell division in their beaks, and that species-specific differences in beaks (such as those between the chick and the duck) were due to the placement of *Bmp4* expression.[2]

Abzhanov and his colleagues (2004) found a remarkable correlation between the beak shape of the finches and the amount of *Bmp4* expression. No other paracrine factor showed such differences. The expression of *Bmp4* in the connective tissue of embryonic ground finch beaks starts earlier and is much greater than the *Bmp4* expression in cactus finch beaks. In all cases, the *Bmp4* expression pattern correlated with the breadth and depth of the beak.

The importance of these expression differences was confirmed experimentally by changing the *Bmp4* expression pattern in chick embryos (Abzhanov et al. 2004; Wu et al. 2004, 2006). When *Bmp4* expression was enhanced in the facial connective tissue, the chick developed a broad beak reminiscent of the beaks of the ground finches. Conversely, when BMP signaling was inhibited in this region (by adding a BMP inhibitor to the developing beak primordium), the beak became narrow and pointed, like those of cactus finches. Thus, enhancers controlling the amount of beak-specific BMP4 synthesis appear to be critically important in the evolution of Darwin's finches.

Heterotypy

In heterochrony, heterotopy, and heterometry, the mutations affect the regulatory regions of the gene. The gene's product—the protein—remains the same, although it may be synthesized in a new place, at a different time, or in different amounts. In heterotypy, the changes affect the protein that binds to these regulatory regions. The changes of heterotypy affect the actual coding region of the gene and thus can change the functional properties of the protein being synthesized.

For instance, insects have six legs, whereas most other arthropod groups (spiders, millipedes, centipedes, and crustaceans) have many more (see Minelli, this volume). How is it that the insects came to form legs only in their three thoracic segments and have no appendages in their abdominal regions? The answer seems to reside in the relationship between Ultrabithorax (Ubx) protein and the *Distal-less* gene.

The *Ubx* gene encodes a transcription factor, Ubx, that is used throughout the arthropods to activate those genes expressed in the lower thorax and abdomen. In most arthropod groups, Ubx does not inhibit the *Distal-less* gene. However, in the insect lineage, a mutation occurred in the *Ubx* gene wherein the original terminus of the protein-coding region was replaced by a group of nucleotides encoding a stretch of about ten alanine amino acids (Galant and Carroll 2002; Ronshaugen, McGinnis, and McGinnis 2002). This alanine-rich region represses *Distal-less* transcription. When a shrimp *Ubx* gene is experimentally modified to encode this polyalanine region, it too represses the *Distal-less* gene. The ability of insect Ubx to inhibit *Distal-less* thus appears to be the result of a gain-of-function mutation that characterizes the insect lineage. One cannot explain insect evolution without a developmental explanation as to why limb development of this lineage has been so constrained.

Maize, commonly known in the U.S. as corn, is an organism whose domestication and subsequent propagation throughout the world may have been effected through a mutation of a transcription factor (Wang et al. 2005). There are many steps postulated between the ancestral teosinte plant and the modern cultivated maize plant *Zea mays*, and one of the most economically important steps in this was the liberation of the kernel from its hard protective envelope (the glume). This mutation exposed the kernel on the cob so that the kernels can be readily harvested for human consumption. This critical event in maize evolution and domestication is controlled by a single gene, *teosinte glume architecture* (*TGA1*). This gene encodes a transcription factor that is active in the development of the influorescence (the "ear") of the plant. When maize *TGA1* is placed into teosinte, the teosinte kernels become corn-like; and when teosinte *TGA1* becomes expressed in maize, the kernels become teosinte-like. These genes are expressed in the respective ears of the plant, and they appear to be expressed in the same amounts.

Within the *TGA1* gene, there is a mutational change resulting in a single amino acid difference between the maize and teosinte proteins the two gene alleles encode: a substitution of lysine in teosinte to asparagine in maize at amino acid position 6. (The lysine at this position is conserved in teosinte, wheat, and rice, suggesting that it is important for protein function or stability.) The mutation to asparagine appears to cause the TGA1 protein to be degraded faster, so perhaps there is a failure to make the glume surface that would normally cover the kernel.

Population genetic analysis shows that there is very little variation in the regions around the *TGA1* gene, suggesting that there was a "sweep" of this allele through the teosinte population. These results demonstrate that some of the complex differences between the teosinte and corn cobs are regulated by a single gene, and the differences are probably the result of a single amino acid substitution.

GENETIC ASSIMILATION

In addition to these four genetic ways of producing variation, there are three other means of developmental change that involve the induction of inheritable change by the environment. Variously called "genetic assimilation" (Waddington 1952), "heterocyberny" (Gilbert and Epel 2009), or the "adaptability driver" (Bateson 2005), the first of these processes denotes changes occurring when a physiological response to the environment becomes incorporated into the genome. In its simplest form, this is a transfer of the trigger for a preexisting phenotype that was originally triggered by the environment to a new initiator within the embryo. Both Waddington (1942, 1952) and Schmalhausen (1949) independently proposed this "genetic assimilation" to explain how some species have evolved rapidly in particular directions. For instance, both scientists were impressed by the calluses of the ostrich. Most mammalian skin has the ability to form calluses on areas that are abraded by the ground or some other surface. The skin cells respond to friction by proliferating. While such examples of environmentally induced callus formation are widespread, the ostrich is born with calluses where it will touch the ground. Waddington and Schmalhausen hypothesized that since the skin cells are already competent to be induced by friction, they could be induced by other signals as well. As ostriches evolved, a mutation (or a particular combination of alleles) appeared that enabled the skin cells to respond to some substance *within* the embryo.

Genetic assimilation has been shown numerous times in the laboratory, and two of the most well-documented laboratory cases include selection for environmentally induced ectopic wings and environmentally induced coloration (see Gibson and Hogness 1996; Suzuki and Nijhout 2006; Gilbert and Epel 2009). In nature, there are numerous phenotypes that could be explained this way. For instance, in regions containing both large and small prey, the tiger snake (*Notechis scutatus*) can develop a large head or a small head, depending on its diet. However, on those islands where there is only large prey, this plasticity is lost, and the tiger snakes are born with large heads (Aubret and Shine 2009).

The advantages of genetic assimilation are: (1) The phenotype is not random. Because the phenotype was originally an adaptive physiological response to an environmental agent, the phenotype has already been tested by natural selection. This would eliminate a long period of testing phenotypes derived by random mutations. (2) This phenotype would have already existed in a major portion of the population.

Genetic assimilation would solve the problem of how a new phenotype gets established in a population. In the Modern Synthesis, one of the problems of having a new phenotype is that the bearers of such a phenotype are "monsters" compared to the wild-type. How would such mutations, perhaps present in one individual or one family, get established and eventually take over a population? With genetic assimilation, this phenotype would have already been seen in a large percentage of the population, whenever the environmental conditions induced it. The phenotype just needs to be genetically stabilized by the selection of modifier genes (that already exist in the population.)

Indeed, West-Eberhard (2005) has proposed that studying gene expression and plasticity will help us understand the genetic divergence that gives rise to speciation. "Contrary to common belief, environmentally initiated novelties may have greater evolutionary potential than mutationally induced ones. Thus, genes are probably more often followers than leaders in evolutionary change." Indeed, new studies strongly suggest that the ability of vertebrates to walk on land was originally an environmentally induced phenomenon that later became genetically incorporated into the genome (Standen, Du, and Larsson 2014). This is not "Lamarckian" in that the inheritance of the variation is not based on physiological use-or-disuse, nor is there any "willing" of the phenotype. Yet, environmentally induced modifications can be selected (at least in the laboratory) to be incorporated into the genetic repertoire of the organism.

Developmental Symbiosis

Another area where developmental biology meets evolution concerns mutualistic symbiosis, where each of the biotic components benefits from association. Recent research has demonstrated that very few, if any animals are truly "individuals" (Gilbert, Sapp, and Tauber 2012; McFall-Ngai et al. 2013). Humans, for instance, are composed of hundreds of bacterial species, whose collective amount outnumbers our "human" cells nearly 9:1 and whose number of genes is about 150-fold that of the human genome (Ley, Peterson, and Gordon 2006; Qin et al. 2010; Gordon 2012). Our physiology is linked into a co-metabolism, such that around 30 percent of the metabolites in our blood have been modified by bacteria (Wikoff et al. 2009; McCutcheon and von Dohlen

2011; Vogel and Moran 2011; Smith et al. 2013). Moreover, the genomes of hosts and symbionts have often become so integrated that one cannot persist without the other. In mice, bacteria provide signals necessary for the normal development of blood vessels, immune system, and brain, and in fish, bacterial signals are required for normal gut stem cell proliferation (Hooper et al. 2001; Rawls, Samuel, and Gordon 2004; Bates et al. 2006; Gilbert, Sapp, and Tauber 2012). In other words, bacteria are needed for normal animal development, and to lack them is like having a loss-of-function mutation. Each "individual" develops as a consortium (McFall-Ngai 2002; Gilbert and Epel 2009; Fraune and Bosch 2010; McFall-Ngai et al. 2013).

Bacteria are also providing numerous animals with selectable variation. In aphids, for instance, variations in color, thermotolerance, and resistance to parasitoid wasp infection are provided by alleles in the symbionts' genes, not by those of the insect (Dale and Moran 2006; Dunbar et al. 2007; Oliver et al. 2009). In mammals, including humans, different bacteria are critical for producing different phenotypes associated with health and disease (see J. Chow et al. 2010).

In addition to being a source of selectable variation, symbionts are also a means of reproductive isolation, one of the other major components of evolution. Symbionts of *Drosophila* (ingested on their larval food) induce mating preferences by altering the cuticular pheromones on the adult fly's body (Sharon et al. 2010). When the symbionts are eliminated from the food (by antibiotics), the mating preferences vanish. In wasps, symbionts have been shown to be responsible for separating species into different groups through cytoplasmic incompatability (Brucker and Bordenstein 2013).

Symbiosis may have played major roles in evolution by rearranging developmental patterns. In addition to the endosymbiotic roles that symbionts have played in forming the eukaryotic cell (see Margulis and Fester 1991), recent studies indicate that the origin of animal multicellularity may have been a symbiosis wherein bacteria induced choanoflagellate protists to divide in a way that generates an integrated multicellular tissue (Dayel et al. 2011; Alegado et al. 2012).

The animal and its symbionts form a complex known as the holobiont (Rohwer et al. 2002; Rosenberg et al. 2007), and one of the first challenges of twenty-first-century evolutionary biology will be to determine how important the holobiont is as a unit of selection. If we are no longer "individuals" by anatomical, physiological, developmental, or genetic criteria, it may be the holobiont that is the target of natural selection (Gilbert et al. 2010; Gilbert 2013b; Guerrero, Margulis, and Berlanga 2013).

Epigenetic Inheritance

The third area where the environment can play an instructive role in development and evolution is the realm of epigenetic inheritance (see Jablonka and Raz 2009; Noble 2013). The silencing or ectopic expression of a gene can be brought about by mutation. But it can also be brought about by chromatin modifications such as DNA methylation (which usually inactivates genes) or histone deacetylation (which usually activates genes). If environmental factors cause chromatin modifications in the germline, they can sometimes be inherited stably over numerous generations. The toadflax variant observed by Linnaeus over 250 years ago is a DNA methylation variant. The DNA between the wild-type and the variant is the same. Only the DNA methylation patterns differ (Cubas, Vincent, and Coen 1999).

Environmental factors have been seen to alter the pattern germline DNA methylation in several animal species, and some of these DNA alterations change gene expression for several generations. This produces particular phenotypes without gene mutation. In mammals, such "epiallelic inheritance" was first documented by studies of the endocrine disruptor vinclozolin, a fungicide used widely on grapes. When injected into pregnant rats during particular days of gestation, vinclozolin will cause testicular anomalies and poor sperm production in the male offspring. Surprisingly, male mice born to those mice that get testicular dysgenesis also get testicular dysgenesis. So do their male offspring and the subsequent generation's male offspring for all the generations (at least five) tested (Anway et al. 2005; Anway and Skinner 2006; Guerrero-Bosagna et al. 2010).

This phenotype is transmitted by DNA methylation, not DNA mutation. Dozens of genes have their methylation pattern changed by vinclozolin, and altered promoter methylation can be seen in the sperm DNA for several subsequent generations (Guerrero-Bosagna et al. 2010; Stouder and Paolini-Giacobino 2010). These genes include those whose products are necessary for cell proliferation, G-proteins, ion channels, and receptors. More common compounds, such as bisphenol-A, also can alter DNA methylation patterns in ways that affect phenotype for at least four generations (Walker and Gore 2011; Wolstenholme et al. 2012).

In roundworms, acquired characteristics can be transmitted for over one hundred generations through the amplification of germline RNAs (Rechavi et. al 2011; Burton et al. 2011), and recently published reports (Cuzin and Rassoulzadegan 2010; Cossetti et al. 2014) strongly suggest that RNA-mediated epigenetic inheritance in mice may be much more prevalent than originally thought. Indeed, Mattick (2012) has suggested that the inherited component of complex traits (which has not been explained by genome-wide association studies) may be due to such epigenetic inheritance.

A NEW EVOLUTIONARY SYNTHESIS:
A QUANTITATIVE OR QUALITATIVE CHANGE?

Evolutionary developmental biology has a program for explaining biodiversity through changes in gene expression during embryonic and juvenile phases. In addition to the "ultimate" causation provided by natural selection, we have a "proximate" causation provided by development. First, new forms can evolve by altering the timing, location, and amount of gene expression during their development. The heterotopy, heterochrony, and heterometry is accomplished by changing the enhancer elements regulating gene expression. Heterotypy can alter the transcription factor that binds to these enhancer sequences, giving these proteins different properties. Second, the environment, in addition to having a selective role in evolution, can also play an instructive role. Genetic assimilation can take a physiological response to the environment and make it part of normal development; and symbionts can provide a second genetic mechanism, paralleling that of the nuclear genes. The alleles of symbionts can provide selectable variation, and they can also effect reproductive isolation. Moreover, epigenetic processes have now been found to transmit altered phenotypes from generation to generation through chromatin modification and DNA methylation.

The question is: How "revolutionary" is this? The "classical" modes of evolutionary developmental biology—heterotypy, heterochrony, heterometry, and heterotopy—augment the Modern Synthesis and represent expansions of the mechanisms of variation. As Carroll (2008) has noted, these findings of evolutionary developmental biology constitute a developmental genetic theory of morphological evolution which can supplement and extend the Modern Synthesis of evolutionary biology and population genetics. But the findings of symbiotic and epigenetic contributions to evolution could be more revolutionary. Developmental symbiosis claims that the holobiont is being selected rather than the genetically uniform animal, and the inheritance of acquired traits through epigenetic means threatens to abolish the Weismannian boundary between soma and germline (Jablonka 2011; Mattick 2012). According to Denis Noble (2013, 1241), "we can now see that the Modern Synthesis is too restrictive and that it has dominated biological science for far too long." The challenge is how to update it in a way that correctly balances and apportions the genetic, epigenetic, environmental, and symbiotic components in a common evolutionary framework.

NOTES

This chapter is substantially revised from an essay by the same title published in Auletta, LeClerc, and Martinez (2011) and reprinted by permission of the Gregorian and Biblical Press.

1. This induction is conveniently ignored by intelligent design proponents. The eye, for instance, develops in such a way that the lens cells produce paracrine factors that induce the retina and the retina makes paracrine factors that induce the lens. In this way, the parts of the eye construct one another. There is no incomprehensible complexity in getting the lens to form in the same organ as the retina. The developing lens will transform adjacent cells into retina and vice versa.

2. The same paracrine factor that induces cell death in the webbing of the birds' limb also induces cell division in the birds' beaks (see Gilbert and Epel 2009). In genetic nomenclature, the gene is usually italicized, while the protein is in Roman script and often capitalized.

REFERENCES

Abzhanov, A., M. Protas, B. R. Grant, P. R. Grant, and C. J. Tabin. (2004). "*Bmp4* and Morphological Variation of Beaks in Darwin's Finches." *Science* 305:1462–65.

Alegado, R. A., L. W. Brown, S. Cao, R. K. Dermenjian, R. Zuzow, S. R. Fairclough, J. Clardy, and N. King. (2012). "A Bacterial Sulfolipid Triggers Multicellular Development in the Closest Living Relatives of Animals." *elife* 1:e00013.

Amundson, R. (2005). *The Changing Role of the Embryo in Evolutionary Thought: Structure and Synthesis.* Cambridge: Cambridge University Press.

Anway, M. D., and M. K. Skinner. (2006). "Epigenetic Transgenerational Actions of Endocrine Disruptors." *Endocrinology* 147:S43–S49.

Anway, M. D., A. S. Cupp, M. Uzumcu, and M. K. Skipper. (2005). "Epigenetic Transgeneration Effects of Endocrine Disruptors and Male Fertility." *Science* 308:1466–69.

Arthur, W. (2004). *Biased Embryos and Evolution.* Cambridge: Cambridge University Press.

Aubret, F., and R. Shine. (2009). "Genetic Assimilation and the Postcolonization Erosion of Phenotypic Plasticity in Island Tiger Snakes." *Current Biology* 19:1932–36.

Auletta, G., M. LeClerc, and R. A. Martinez, eds. (2011). *Biological Evolution: Facts and Theories; A Critical Appraisal 150 Years after "The Origin of Species."* Rome: Gregorian and Biblical Press.

Barinaga, M. (2003). "Newborn Neurons Search for Meaning." *Science* 299:32–34.

Bates, J. M., E. Mittge, J. Kuhlman, K. N. Baden, S. E. Cheesman, and K. Guilemin. (2006). "Distinct Signals from the Microbiota Promote Different Aspects of Zebrafish Gut Differentiation." *Developmental Biology* 297:374–86.

Bateson, P. (2005). "The Return of the Organism." *Journal of Bioscience.* 30:31–39.

Brucker, R. M., and S. R. Bordenstein, Sr. (2013). "The Hologenomic Basis of Speciation: Gut Bacteria Cause Hybrid Lethality in the Genus *Nasonia.*" *Science* 341:667–69.

Burton, N. O., K. B. Burkhart, and S. Kennedy. (2011). "Nuclear RNAi Maintains Heritable Gene Silencing in *Caenorhabditis elegans.*" *Proceedings of the National Academy of Sciences of the United States of America* 108:19683–88.

Carroll, S. B. (2006). *The Making of the Fittest: DNA and the Ultimate Forensic Record of Evolution.* Waltham, MA: Quercus Press.

———. (2008). "Evo-Devo and an Expanding Evolutionary Synthesis: A Genetic Theory of Morphological Evolution." *Cell* 134:25–36.

Cebra-Thomas, J., F. Tan, S. Sistla, E. Estes, G. Bender, C. Kim, and S. F. Gilbert. (2005). "How the Turtle Forms Its Shell: A Paracrine Hypothesis of Carapace Formation." *Journal of Experimental Zoology* 304B:158–69.

Chow, J., S. M. Lee, Y. Shen, A. Khosravi, and S. K. Mazmanian. (2010). "Host-Bacterial Symbiosis in Health and Disease." *Advances in Immunology* 107:243–74.

Chow, R. L., C. R. Altmann, R. A. Lang, and A. Hemmati-Brivanlou. (1999). "Pax6 Induces Ectopic Eyes in a Vertebrate." *Development* 126:4213–22.

Cohn, M. J., and C. Tickle. (1999). "Developmental Basis of Limblessness and Axial Patterning in Snakes." *Nature* 399:474–79.

Conway Morris, S. (2003). *Life's Solution: Inevitable Humans in a Lonely Universe.* Cambridge: Cambridge University Press.

Cossetti, C., L. Lugini, L. Astrologo, I. Saggio, S. Fais, and C. Spadafora. (2014). "Soma-to-Germline Transmission of RNA in Mice Xenografted with Human Tumour Cells: Possible Transport by Exosomes." *PLoS One* 9: e101629.

Cubas, P., C. Vincent, and E. Coen. (1999). "An Epigenetic Mutation Responsible for Natural Variation in Floral Symmetry." *Nature* 401:157–61.

Cuzin, F., and M. Rassoulzadegan (2010). "Non-Mendelian Epigenetic Heredity: Gametic RNAs as Epigenetic Regulators and Transgenerational Signals." *Essays in Biochemistry* 48:101–6.

Dale, C., and N. A. Moran. (2006). "Molecular Interactions Between Bacterial Symbionts and Their Hosts." *Cell* 126:453–65.

Dayel, M. J., R. A. Alegado, S. R. Fairclough, T. C. Levin, S. A. Nichols, K. McDonald, and N. King (2011). "Cell Differentiation and Morphogenesis in the Colony-Forming Choanoflagellate *Salpingoeca rosetta.*" *Developmental Biology* 357:73–82.

Dembski, W. A. (1996). "What Every Theologian Should Know about Creation, Evolution, and Design." Available at http://www.arn.org/docs/dembski/wd_theologn.htm.

Dennis, M. Y., et al. (2012). "Evolution of Human-Specific Neural *SRGAP2* Genes by Incomplete Segmental Duplication." *Cell* 149:912–22.

Di-Poï, N., J. I. Montoya-Burgos, H. Miller, O. Pourquié, M. C. Milinkovitch, and D. Duboule. (2010). "Changes in Hox Gene's Structure and Function During the Evolution of the Squamate Body Plan." *Nature* 464:99–103.

Donnelly, M. J., V. Corbel, D. Weetman, C. S. Wilding, M. S. Williamson, and W. C. Black IV. (2009). "Does *kdr* Genotype Predict Insecticide-Resistance Phenotype in Mosquitoes?" *Trends in Parasitology* 5:213–19.

Dunbar H. E., A. C. C. Wilson, N. R. Ferguson, and N. A. Moran. (2007). "Aphid Thermal Tolerance Is Governed by a Point Mutation in Bacterial Symbionts." *PLoS Biology* 5:e96.

Fraune, S., and T. C. G. Bosch. (2010). "Why Bacteria Matter in Animal Development and Evolution." *Bioessays* 32:571–80.

Futuyma, D. (2009). *Evolution*. 2nd ed. Sunderland, MA: Sinauer Associates.

Galant, R., and S. B. Carroll. (2002). "Evolution of a Transcriptional Repression Domain in an Insect Hox Protein." *Nature* 415:910–13.

Gehring, W. J. (1996). "The Master Control Gene for Morphogenesis and Evolution of the Eye." *Genes and Cells* 1:11–15.

———. (2005). "New Perspectives on Eye Development and the Evolution of Eyes and Photoreceptors." *Journal of Heredity* 96:171–84.

Gibson, G., and D. S. Hogness. (1996). "Effect of Polymorphism in the *Drosophila* Regulatory Gene *Ultrabithorax* on Homeotic Stability." *Science* 271:200–203.

Gilbert, S. F. (2003). "Opening Darwin's Black Box: Teaching Evolution through Developmental Genetics." *Nature Reviews: Genetics* 4:735–41.

———. (2013a). *Developmental Biology*. 10th ed. Sunderland, MA: Sinauer Associates.

———. (2013b). "Symbiosis as a Way of Life: The Dependent Co-Origination of the Body." *Journal of Biosciences* 38:1–9.

Gilbert, S. F., and D. Epel. (2009). *Ecological Developmental Biology: Integrating Epigenetics, Medicine, and Evolution*. Sunderland, MA: Sinauer Associates.

Gilbert, S. F., E. McDonald, N. Boyle, N. Buttino, L. Gyi, M. Mai, N. Prakash, and J. Robinson. (2010). "Symbiosis as a Source of Selectable Epigenetic Variation: Taking the Heat for the Big Guy." *Philosophical Transactions of the Royal Society of London B: Biological Sciences* 365:671–78.

Gilbert, S. F., J. Sapp, and A. I. Tauber. (2012). "A Symbiotic View of Life: We Have Never Been Individuals." *Quarterly Review of Biology* 87:325–41.

Gomez, C., E. M. Özbudak, J. Wunderlich, D. Baumann, and O. Pourquié. (2008). "Control of Segment Number in Vertebrate Embryos." *Nature* 454:335–38.

Gordon, J. I. (2012). "Honor Thy Gut Symbionts Redux." *Science* 336:1251–53.

Grant, P. (1999). *The Ecology and Evolution of Darwin's Finches*. Princeton: Princeton University Press.

Grant, P. R., and B. R. Grant. (2007). *How and Why Species Multiply: The Radiation of Darwin's Finches.* Princeton: Princeton University Press.

Guerrero, R. L. Margulis, and M. Berlanga. (2013). "Symbiogenesis: The Holobiont as a Unit of Evolution." *International Microbiology* 16:133–43.

Guerrero-Bosagna, C., M. Settles, B. Lucker, and M. K. Skinner. (2010). "Epigenetic Transgenerational Actions of Vinclozolin on Promoter Regions of the Sperm Epigenome." *PLoS One* 5(9): e13100.

Halder, G., P. Callaerts, and W. J. Gehring. (1995). "Induction of Ectopic Eyes by Targeted Expression of the Eyeless Gene in Drosophila." *Science* 267:1788–92.

Harris, M. P., J. F. Fallon, and R. O. Prum. (2002). "Shh-Bmp2 Signaling Module and the Evolutionary Origin and Diversification of Feathers." *Journal of Experimental Zoology* 294:160–76.

Hooper, L. V., M. H. Wong, A. Thelin, L. Hansson, P. G. Falk, and J. I. Gordon. (2001). "Molecular Analysis of Commensal Host-Microbial Relationships in the Intestine." *Science* 291:881–84.

Hughes, C. G., and T. C. Kaufman. (2002). "Hox Genes and the Evolution of the Arthropod Body Plan." *Evolution and Development* 4:459–99.

Huxley, J. (1942). *Evolution: The Modern Synthesis.* London: Allen and Unwin.

Jablonka, E. (2011). "Cellular Epigenetic Inheritance in the Twenty-First Century." In *Transformations of Lamarckianism: From Subtle Fluids to Molecular Biology,* ed. S. B. Gissis and E. Jablonka, 215–26. Cambridge, MA: MIT Press.

Jablonka, E., and G. Raz. (2009). "Transgenerational Epigenetic Inheritance: Prevalence, Mechanisms, and Implications for the Study of Heredity and Evolution." *Quarterly Review of Biology* 84:131–76.

Jacob, F. (1977). "Evolution and Tinkering." *Science* 196:1161–66.

King, M.-C., and A. C. Wilson. (1975). "Evolution at Two Levels in Humans and Chimpanzees." *Science* 188:107–16.

Kirschner, M. W., and J. C. Gerhart. (2005). *The Plausibility of Life: Resolving Darwin's Dilemma.* New York: Norton.

Laufer, E., S. Pizette, H. Zou, O. E. Orozco, and L. Niswander. (1997). "BMP Expression in Duck Interdigital Webbing: A Reanalysis." *Science* 278:305.

Ley, R. E., D. A. Peterson, and J. I. Gordon. (2006). "Ecological and Evolutionary Forces Shaping Microbial Diversity in the Human Intestine." *Cell* 124:837–48.

Linnen, C. R., E. P. Kingsley, J. D. Jensen, and H. E. Hoekstra. (2009). "On the Origin and Spread of an Adaptive Allele in Deer Mice." *Science* 325:1095–1108.

Lynch, M. (2005). "Simple Evolutionary Pathways to Complex Proteins." *Protein Science* 14:2217–25.

Margulis, L., and R. Fester. (1991). *Symbiosis as a Source of Evolutionary Innovation*. Cambridge, MA: MIT Press.

Mattick, J. S. (2012). "Rocking the Foundations of Molecular Genetics." *Proceedings of the National Academy of Sciences of the United States of America* 109:16400–16401.

McCutcheon, J. P., and C. D. von Dohlen. (2011). "An Interdependent Metabolic Patchwork in the Nested Symbiosis of Mealybugs." *Current Biology* 21:1366–72.

McFall-Ngai, M. J. (2002). "Unseen Forces: The Influences of Bacteria on Animal Development." *Developmental Biology* 242:1–14.

McFall-Ngai M., M. G. Hadfield, et al. (2013). "Animals in a Bacterial World: A New Imperative for the Life Sciences." *Proceedings of the National Academy of Sciences of the United States of America* 110:3229–36.

McLean, C. Y., et al. (2011). "Human-Specific Loss of Regulatory DNA and the Evolution of Human-Specific Traits." *Nature* 471:216–19.

Merino, R., J. Rodriguez-Leon, D. Macias, Y. Ganan, A. N. Economides, and J. M. Hurle. (1999). "The BMP Antagonist Gremlin Regulates Outgrowth, Chondrogenesis and Programmed Cell Death in the Developing Limb." *Development* 126:5515–22.

Müller, F. (1869). *Facts and Arguments for Darwin*. London: John Murray.

Noble, D. (2013). "Physiology Is Rocking the Foundations of Evolutionary Biology." *Experimental Physiology* 98:1235–43.

Oliver, K. M., P. H. Degnan, M. S. Hunter, and N. A. Moran. (2009). "Bacteriophages Encode Factors Required for Protection in a Symbiotic Mutualism." *Science* 325:992–94.

Qin, J., et al. (2010). "A Human Gut Microbial Gene Catalogue Established by Metagenomic Sequencing." *Nature* 464:59–65.

Rawls, J. F., B. S. Samuel, and J. I. Gordon. (2004). "Gnotobiotic Zebrafish Reveal Evolutionarily Conserved Responses to the Gut Microbiota." *Proceedings of the National Academy of Sciences of the United States of America* 101:4596–4601.

Raymond, M., C. Chevillon, T. Guillemaud, T. Lenormand, and N. Pasteur. (1998). "An Overview of the Evolution of Overproduced Esterases in the Mosquito *Culex pipiens*." *Philosophical Transactions of the Royal Society of London B: Biological Sciences* 353:1707–11.

Rechavi, O., G. Minevich, and O. Hobert. (2011). "Transgeneration Inheritance of an Acquired Small RNA-Based Antiviral Response in *C. elegans*." *Cell* 147:1248–56.

Rice, S. (2004). *Evolutionary Theory*. Sunderland, MA: Sinauer Associates.

Richardson, M. D., and H. H. Oelschläger. (2002). "Time, Pattern and Heterochrony: A Study of Hyperphalangy in the Dolphin Embryo Flipper." *Evolution and Development* 4:435–44.

Rohwer, F., V. Seguritan, F. Azam, and N. Knowlton. (2002). "Diversity and Distribution of Coral-Associated Bacteria." *Marine Ecology Progress Series* 243:1–10.

Ronshaugen, M., N. McGinnis, and W. McGinnis. (2002). "How Protein Mutation and Macroevolution of the Insect Body Plan." *Nature* 415:914–17.

Rose, S. (1998). *Lifelines: Biology beyond Determinism*. Oxford: Oxford University Press.

Rosenberg, E., O. Koren, L. Reshef, R. Efrony, and I. Zilber-Rosenberg. (2007). "The Role of Microorganisms in Coral Health, Disease and Evolution." *Nature Reviews: Microbiology* 5:355–62.

Schmalhausen, I. I. (1949). *Factors of Evolution: The Theory of Stabilizing Selection*. Philadelphia: Blakiston.

Schneider, R. A., and J. A. Helms. (2003). "The Cellular and Molecular Origins of Beak Morphology." *Science* 299:565–68.

Shapiro, M. D., et al. (2004). "Genetic and Developmental Basis of Evolutionary Pelvic Reduction in Threespine Sticklebacks." *Nature* 428:717–23.

Sharon G., D. Segal., J. M. Ringo, A. Hefetz, I. Zilber-Rosenberg, and E. Rosenberg. (2010). "Commensal Bacteria Play a Role in Mating Preference of *Drosophila melanogaster*." *Proceedings of the National Academy of Sciences of the United States of America* 107:20051–56.

Shubin N., C. Tabin, and S. Carroll. (1997). "Fossils, Genes and the Evolution of Animal Limbs." *Nature* 388:639–48.

———. (2009). "Deep Homology and the Origins of Evolutionary Novelty." *Nature* 457:818–23.

Smith, M. I., et. al. (2013). "Gut Microbiomes of Malawian Twin Pairs Discordant for Kwashiorkor." *Science* 339:548–54.

Standen, E. M., T. Y. Du, and H. C. Larsson. (2014). "Developmental Plasticity and the Origin of Tetrapods." *Nature* 513:54–58.

Stouder, C., and A. Paoloni-Giacobino. (2010). "Transgenerational Effects of the Endocrine Disruptor Vinclozolin on the Methylation Pattern of Imprinted Genes in the Mouse Sperm." *Reproduction* 139:373–79.

Sturtevant, A. (1923). "Inheritance of Direction of Coiling in *Limnaea*." *Science* 58:269–70.

Suzuki, Y., and H. F. Nijhout. (2006). "Evolution of a Polyphenism by Genetic Assimilation." *Science* 311:650–52.

Tournamille, C., Y. Colin, J.-P. Cartron, and C. Le Van Kim. (1995). "Disruption of a GATA Motif in the Duffy Gene Promoter Abolishes Erythroid Gene Expression in Duffy-Negative Individuals." *Nature Genetics* 10:224–28.

Valenti, L. (2007). "Wordsmiths of Southwestern Montana: David Quammen." *Outside Bozeman* (Winter 2006/7): 21. Available through http://www.outsidebozeman.com.

Viitala, J., E. Korpimäki, P. Palokangas, and M. Koivula. (1995) "Attraction of Kestrels to Vole Scent Marks Visible in Ultraviolet Light." *Nature* 373:425–27.

Vogel, K. J., and N. A. Moran. (2011). "Sources of Variation in Dietary Requirements in an Obligate Nutritional Symbiosis." *Proceedings of the Royal Society of London B: Biological Sciences* 278:115–21.

Waddington, C. H. (1942). "The Canalization of Development and the Inheritance of Acquired Characters." *Nature* 150:563.

———. (1952). "Selection of the Genetic Basis for an Acquired Character." *Nature* 169:278.

Walker, D. M., and A. C. Gore. (2011). "Transgenerational Neuroendocrine Disruption of Reproduction." *Nature Reviews: Endocrinology* 7:197–207.

Wang, H., et al. (2005). "The Origin of the Naked Grains of Maize." *Nature* 436:714 –19.

Weatherbee, S. D., R. R. Behringer, J. J. Rasweiler, IV, and L. A. Niswander. (2006). "Interdigital Webbing Retention in Bat Wings Illustrates Genetic Changes Underlying Amniote Limb Diversification." *Proceedings of the National Academy of Sciences of the United States of America* 103:15103–7.

Weiner, J. (1994). *The Beak of the Finch: A Story of Evolution in Our Time.* New York: Random House.

West-Eberhard, M. J. (2005). "Developmental Plasticity and the Origin of Species Differences." *Proceedings of the National Academy of Sciences of the United States of America* 102 Suppl 1:6543–49.

Wikoff, W. R., A. T. Anfora, J. Liu, P. G. Schultz, S. A. Lesley, E. C. Peters, and G. Siuzdak. (2009). "Metabolomics Analysis Reveals Large Effects of Gut Microflora on Mammalian Metabolites." *Proceedings of the National Academy of Sciences of the United States of America* 106:3698–3703.

Wolstenholme, J. T., et al. (2012). "Gestational Exposure to Bisphenol A Produces Transgenerational Changes in Behaviors and Gene Expression." *Endocrinology* 153:3828–38.

Wu, P., T. X. Jiang, S. Suksaweang, R. B. Widelitz, and C. M. Chuong. (2004). "Molecular Shaping of the Beak." *Science* 305:1465–66.

Wu, P., T. X. Jiang, J. Y. Shen, R. B. Widelitz, and C. M. Chuong. (2006). "Morphoregulation of Avian Beaks: Comparative Mapping of Growth Zone Activities and Morphological Evolution." *Developmental Dynamics* 235:1400–1412.

THREE

THE EVOLUTION OF EVOLUTIONARY MECHANISMS
A New Perspective

Stuart A. Newman

The Modern Evolutionary Synthesis, based on Charles Darwin's concept of natural selection in conjunction with a genetic theory of inheritance in a population-based framework, has been, for more than six decades, the dominant scientific perspective for explaining the diversity of living organisms. In recent years, however, with the growth in knowledge of the fossil record, the genetic affinities among different life forms, and the roles played by nongenetic determinants of organismal shape and form, there have been challenges to the Synthesis in the realms of both cellular (Margulis and Sagan 2002; Woese and Goldenfeld 2009) and multicellular (Alberch 1989; Newman 1994; Jablonka and Lamb 1995, 2005; Newman and Müller 2000; West-Eberhard 2003) evolution.

According to the eminent evolutionary biologist Ernst Mayr, "Nothing strengthened the theory of natural selection as much as the refutation, one by one, of all the competing theories, such as saltationism, orthogenesis, inheritance of acquired characters, and so forth" (Mayr 1982, 840). As I will argue below, a coherent account of the origination of the morphological motifs of multicellular organisms in fact requires bringing all three of these ideas back into evolutionary theory, though in a fashion that acknowledges their declining efficacy as evolution progresses. Put in other terms: evolutionary mechanisms themselves have evolved.

It is easy to see why the first two of Mayr's "competing theories" are antithetical to the standard picture. *Saltationism* is the idea that organismal phenotypes can change from one generation to the next in a manner that is very large compared to the organism's usual range of phenotypic variation. Saltation is often associated with the concept of the "hopeful monster," a term invented by the geneticist Richard Goldschmidt to refer to novel phenotypes that might arise in a single generation by the mutation of a single broadly acting gene (a "macromutation") (Goldschmidt 1940). Arguments developed by population geneticists such as Ronald Fisher (1930) in the twentieth century contended that such jumps would be exceptionally rare and typically lethal, occurrences not contributing significantly to the origin of species or higher taxa. But even before the Synthesis was formulated, Darwin expressed the incompatibility of saltationism with his own ideas in a statement in the first edition of the *Origin of Species*: "If it could be demonstrated that any complex organ existed, which could not possibly have been formed by numerous, successive, slight modifications, my theory would absolutely break down" (Darwin 1859, 158).

Orthogenesis is the doctrine that organisms change in preferred directions over the course of evolution. Although this is consonant with the fact that all material systems have inherent patterns of organization that may manifest themselves rapidly (as in the transition from waves to vortices in water) or over time (as in the generation of the chemical elements), and despite the purported examples of such inherencies in organismal development and evolution provided by biologists such as William Bateson (1909) and D'Arcy W. Thompson (1942), orthogenesis was forcefully rejected by leading architects of the Synthesis, including George Gaylord Simpson (1944) and Mayr himself (1974).

The most disparaged of all the competing theories listed by Mayr is the third one, the inheritance of acquired characteristics, or Lamarckism. This is ironic, considering that Darwin himself became increasingly receptive to this notion in successive volumes of the *Origin*. Indeed, a role for inheritance of characters acquired during an individual's lifetime was written out of the standard theory not because it conflicted with natural selection, but because it violated the tenet of chromosomal genes as the exclusive medium of inheritance. An added barrier to the operation of Lamarckian mechanisms for animal species first emphasized by the post-Darwinian biologist August Weismann is the sequestering of the germ line from the rest of the body during development. This constraint does not apply to plants, however, making it an inapt hallmark of a general theory of evolution.

It is important to recognize, moreover, that while Darwin and most of his successors concerned themselves with multicellular forms, both plants and animals, they conspicuously avoided providing any account of the origination and innovation of the morphological motifs, such as body plans and organ forms, that are raw material for natural selection. The focus on this missing element by the emerging field of evolutionary-developmental biology (Robert 2004; Müller and Newman 2005; Müller 2007; Callebaut, Müller, and Newman 2007; Moczek 2008; Fusco and Minelli 2008) has highlighted a number of properties of developmental systems that were previously marginal to evolutionary theory. These include developmental and phenotypic plasticity and genotype–phenotype discordances (Newman 1994; Trut, Oskina, and Kharlamova 2009; Pigliucci 2001; West-Eberhard 2003; Badyaev 2005; Badyaev, Foresman, and Young 2005; Salazar-Ciudad 2006; Goodman 2008; Vedel et al. 2008), determination of form by physical and epigenetic factors (Müller and Streicher 1989; Newman and Comper 1990; Newman and Müller 2000), and inheritance systems that extend beyond the gene (Jablonka and Lamb 1995, 2005).

In the remainder of this chapter I will describe how our current inferences about the genetic endowment of the single-celled ancestors of the Metazoa (multicellular animals) combined with knowledge of the physical properties of viscoelastic materials (e.g., cell clusters and tissue primordia) have led to new understanding of the origination and early evolution of animal form that is simultaneously saltational, orthogenic, and Lamarckian. The three erstwhile prohibited concepts converge in this revised explanatory framework precisely because the earliest multicellular animals were subject to external physical forces and effects that would have shaped and reshaped them in often nonlinear and abrupt fashions, and because the morphological outcomes (if not any particular one) emerging under these conditions were to a surprising extent physically inevitable.

We have proposed that this early period of physics-dependent morphogenesis is when much large-scale macroevolution took place (Newman and Müller 2000; Newman 2005; Newman, Forgacs, and Müller 2006; Newman 2012). In particular, it was the era in which the major phyla were established. Only after extensive stabilizing and canalizing selection (Schmalhausen 1949; Waddington 1942) would the descendents of these early-diversifying organisms have settled into the mode described by the Modern Synthesis. In present-day forms, gene mutation generally leads to deleterious or incremental alteration of the phenotype because generation of form has now come to be guided by robust hierarchical programs of gene regulation (Davidson 2006). Any effects on development of the external environment, moreover, are either minimal or specifically incorporated into the generative program of the species (e.g., temperature-dependent sex determination in reptiles; Bowden, Ewert, and Nelson 2000).

DYNAMICAL PATTERNING MODULES VS. CELL STATE SWITCHING MECHANISMS

Our concept is based on the existence of a set of molecules in certain modern unicellular organisms, the Choanozoa, and their inferred presence in the single-celled ancestors of the animals. The molecules in question (products of a subset of the genes of the "developmental-genetic toolkit"; Wilkins 2002; Carroll, Grenier, and Weatherbee 2005) were predisposed (as described below) to assume novel functions in the multicellular state by mobilizing physical effects irrelevant to patterning on the scale of the individual cell. We refer to the joint effect of these ancient molecules and the physical processes they mobilized as "dynamical patterning modules" (DPMs) (Newman and Bhat

2008, 2009). Most fundamentally, the multicellular state itself likely came into being by one class of ancestral molecules, the cadherins, forming a DPM by harnessing the physical effect of *homophilic adhesion*. Because the array of physical effects that apply to chemically and mechanically active matter on the spatial scale of cell aggregates is itself limited, a set of inevitable but restricted morphological motifs arose in animal systems. These include the recurrent appearance of interior body cavities, multiple tissue layers, elongated bodies, segments, and appendages. The emergence of these structures was truly orthogenetic, not requiring adaptive gradualism.

Because physics can act with immediacy to shape and modify form, particularly at the developmental stages of ancient organisms for which canalizing mechanisms were not yet in place, early morphological diversification need not have taken long periods of time. Genetic consolidation therefore likely followed, rather than accompanied, the rapid radiation of body plans referred to as the Cambrian explosion. In addition, while the limited number of physical effects that enter into DPMs can be attributed to relevant laws of nature, the limited set of molecules involved in these modules is just a fortuitous consequence of their having been the only suitable ones present in the ancestral cells. The assertion that the animal phyla emerged early and rapidly by means of the DPMs carries the implication that the associated molecules should be central to the developmental pathways of all extant animals, and this is borne out by the evidence.

The biosynthetic states of all cells are thought to be determined by the dynamics of transcription-factor-mediated gene regulatory networks (GRNs) (Davidson 2006). Such networks, containing feedback and feed-forward loops by which the transcription factors promote and suppress their own and each other's synthesis, exhibit multistability (Forgacs and Newman 2005). The systems can thus switch among discrete states, the number of states always being much smaller than the total number of genes in the organism's genome. Since the genes that specify nontranscription factor proteins and regulatory RNAs are themselves subject to transcriptional control, the alternative stable states of the GRNs contribute to specifying biochemically distinct cell types. Multistable GRNs, in this view, cause all unicellular organisms, be they bacteria, fungi, protists, or algae, to exhibit alternative states of differentiation, both reversible and irreversible, under various conditions (e.g., Blankenship and Mitchell 2006; Vlamakis et al. 2008). If true, this must also have been the case for the single-celled ancestors of the Metazoa, that is, the ancient and modern animals.

The transcription factors contained in the developmental-genetic toolkit are thus organized into GRNs that function in the generation of different cell types during development. But an equally important aspect of development is the arrangement of cells into appropriately coherent spatiotemporal patterns (Salazar-Ciudad, Jernvall, and Newman 2003; Gilbert 2006). Unlike cell switching mechanisms, however, specific mechanisms of developmental pattern formation and tissue morphogenesis cannot have existed before multicellularity. In fact, the processes that generate spatial organization on the multicellular level are of an entirely different character from those that operate in individual cells (Newman and Bhat 2009; Newman, Bhat, and Mezentseva 2009; Hernández-Hernández et al. 2012; Niklas and Newman 2013).

THE EMERGENCE OF THE METAZOA

The evolutionary history of the metazoans was initiated with remarkable rapidity during the late Precambrian and early Cambrian periods and is relatively well described (Rokas, Kruger, and Carroll 2005; Larroux et al. 2008). Development in all the metazoan phyla has been mediated by the same conserved developmental-genetic toolkit regulatory molecules for more than half a billion years (Carroll, Grenier, Weatherbee 2005). The extant metazoans have classically been divided into the Eumetazoa, organisms that exhibit true tissues, epithelia with polarized cells, cell-cell junctions, a well-defined basement membrane, neurons, and muscle cells; and the sponges (Porifera) and Placozoa, which lack these features. *Trichoplax adhaerens,* the single known type of placozoan, contains several cell types and layers, but unlike the sponges (which exhibit gastrulation-like movements during development and complex labyrinthine morphologies) (Larroux et al. 2006), it has a simple, flat body without internal cavities. Surprisingly, on the basis of purely genetic criteria Placozoa may have greater genetic affinity to the Eumetazoa than the earlier-diverging sponges (Srivastava et al. 2008).

The eumetazoans, in turn, are divided into two groups. The *diploblasts,* consisting of the Cnidaria (e.g., hydroids and corals) and, traditionally, the Ctenophora (e.g., comb jellies), have two closely apposed epithelial body layers surrounding a luminal interior space. The triploblasts (echinoderms, arthropods, mollusks, chordates, etc.), in contrast, have a third, mesenchymal, body layer.

Sheetlike and hollow spherical forms (Yin et al. 2007), and budding and segmented tubes (Droser and Gehling 2008), possibly the most ancient metazoans, are seen beginning about 630 million years ago in fossil beds of the Precambrian Ediacaran period. Essentially all the triploblastic metazoan body plans then emerged within the space of no more than 20 million years, beginning about 535 million years ago (Conway Morris 2006), during the well-known Cambrian explosion. It has been suggested that the first Cnidaria (corals, hydroids) may have been holdovers from the Precambrian (Erwin 2008). Modern animals, and perhaps some of the Ediacaran forms, have a common ancestry in the Precambrian along with the Choanozoa, some of whose extant members are conditionally colonial (Wainright et al. 1993; Lang et al. 2002; King, Hittinger, and Carroll 2003; Philippe et al. 2004; Shalchian-Tabrizi et al. 2008; Dayel et al. 2011).

Many of the toolkit genes, including some that have key roles in morphogenesis and pattern formation, are found in the genome of *Monosiga brevicollis,* an exclusively unicellular choanozoan (King et al. 2008). Some additional genes appeared in the toolkit concomitant with the emergence of the sponges, and a few more arose with the cnidarians. The Cambrian explosion followed with few significant additions to the toolkit.

Metazoan complexity was thus achieved in a rapid fashion with a basically unchanging set of ingredients. An essential evolutionary step for multicellularity was the acquisition by single-celled antecedents of the capacity to remain attached to one another after dividing. No new genes or gene products were required to mediate this function. The genome of *M. brevicollis* contains 23 putative cadherin genes, as well as 12 genes for C-type lectins (Abedin and King 2008; King et al. 2008), the protein products of which mediate cell attachment and aggregation in metazoan organisms in the presence of sufficient levels of extracellular calcium ion. Rising oceanic Ca^{2+} levels during the period in which multicellularity was established may have recruited these ancient proteins to new roles (Kazmierczak and Kempe 2004; Fernàndez-Busquets et al. 2009).

THE DEVELOPMENTAL GENETIC TOOLKIT: NEW CONTEXTS, NEW ROLES

Metazoan embryos employ a variety of patterning and shaping processes (Salazar-Ciudad, Jernvall, and Newman 2003), some of which are used in all the animal phyla and others of which are used in some of them. As noted above, the first DPM, designated *ADH* (table 3.1), resulted in the formation of a multicellular cluster. Within such a cluster, any or all of the following can occur: the local coexistence of cells of more than one type, the formation of distinct cell layers, or an internal space or lumen, elongation of the cell cluster, the formation of repeated metameres or segments, the change in state or type of cells in one region of the cluster due to local or long-range signals from another region, the change in stiffness or elasticity of a cell layer, and the dispersal of cells while they continue to remain part of an integral tissue (reviewed in Forgacs and Newman 2005). As the Metazoa emerged, DPMs mediated all of the above transformations by mobilizing physical forces and processes characteristic of viscoelastic, chemically and mechanically active materials on the spatial scale of cell aggregates and tissue primordia (100 μm–1 mm). Such materials are referred to by physicists as being simultaneously "soft matter" (de Gennes 1992) and "excitable media" (Mikhailov 1990).

Detailed descriptions of the properties of the major basic and combined metazoan DPMs and their developmental and proposed evolutionary roles can be found in earlier publications (Newman and Bhat 2008, 2009; Newman, Bhat, and Mezentseva 2009). Here these features will be characterized briefly. Each DPM is given a three-letter designation.

Table 3.1. Names, components, and roles of major dynamical patterning modules (DPMs)

DPM	Characteristic molecules	Physical principle	Morphogenetic role
ADH	cadherins	adhesion	multicellularity
DAD	cadherins	differential adhesion	tissue multilayering
LAT	Notch	lateral inhibition	coexistence of alternative cell types
POL_a	Wnt	cell surface anisotropy	lumen formation
POL_p	Wnt	cell shape anisotropy	tissue elongation
OSC	Wnt + Notch	synchronized biochemical oscillation	morphogenetic fields; segmentation
MOR	TGF-b/BMP; Hh	diffusion	pattern formation
ASM	FGFs	diffusion	induction
TUR	MOR + Wnt + Notch	reaction-diffusion instability	symmetry breaking; periodic patterning
MIT	MAPK	increase in mass	global and localized growth
APO	Bcl-2	decrease in mass	localized cell loss
ECM	collagen; chitin; fibronectin	stiffness; cohesion dispersal	elasticity; exo- and endoskeleton formation; epithelial-mesenchymal transformation

Adhesion and differential adhesion — As mentioned above, the emergence of metazoan multicellularity depended on cadherins and C-type lectins, and possibly other cell surface molecules, of single-cell ancestors taking on the new function of cell-cell adhesion (*ADH*). If, in addition, subsets of cells within an aggregate contain sufficiently different levels of cell adhesion molecules on their surfaces, there will be a sorting into islands of more adhesive cells within lakes of less adhesive ones (Steinberg and Takeichi 1994). This constitutes a second DPM, differential adhesion (*DAD*). Since cells undergo random motion, small islands of like cell types will coalesce, and an interface will be established across which cells will not intermix (Steinberg 2003). This effect, which has the same physical basis as phase separation of two immiscible liquids such as oil and water (reviewed in Forgacs and Newman 2005), leads to the formation of nonmixing cell layers, an early stage of most animal embryogenesis.

Lateral inhibition and choice between alternative cell fates — Morphologically complex organisms always employ lateral inhibition (*LAT*) during embryogenesis, whereby early differentiating cells signal to cells adjacent to them to take on a different fate (Rose 1958; Meinhardt and Gierer 2000). Lateral inhibition in metazoans is generally mediated by the Notch signal transduction pathway, specifically, interaction of the cell surface receptor Notch with members of a class of other integral membrane proteins (Delta, Serrate/Jagged, and Lag2: the DSL proteins) which act as ligands for the receptor and mediators of Notch activity (Ehebauer, Hayward, and Arias 2006). This mechanism does not determine the particular fate of any cell, but only enforces the coexistence of alternative available fates in adjacent cells in the same cluster or aggregate.

The most basal metazoans to contain the Notch receptor are sponges (Nichols, Forgacs, and Müller 2006) though related protein domains probably existed in a choanozoan ancestor (King et al. 2008). Lateral inhibition would have enabled basic cell pattern formation in these organisms and more complex animals.

Apical-basal and planar cell polarity — As noted above, cell aggregates behave like viscoelastic liquid droplets (reviewed in Forgacs and Newman 2005). This means that their default morphology is topologically solid (i.e., having no lumen) and spherical. Animal embryos escape these morphological defaults by employing *cell polarization*. Cells can be polarized in one of two ways. When they become anisotropic along their *surfaces* (referred to as apical–basal [A/B] polarization; Karner, Wharton, and Carroll 2006b), interior spaces or lumens can arise within aggregates. Specifically, when A/B polarization leads cells to have lowered adhesiveness on one portion of their surface, they will preferentially attach to their neighbors on their more adhesive (lateral) portions, leaving the less adhesive (basal) portions adjoining an interior space (Newman 1998). Apical–basal polarity is also important in fostering layered tissue arrangements.

Tissue elongation may occur when cells individually polarize in *shape* (rather than surface properties), a phenomenon called planar cell polarity (PCP; Karner, Wharton, and Carroll 2006a). Planar-polarized cells can intercalate along their long axes, causing the tissue mass to narrow parallel to this direction, and consequently to elongate in the orthogonal direction. This tissue reshaping is known as convergent extension (Keller et al. 2000; Keller 2002).

Both A/B polarity and PCP are mediated by secreted factors of the Wnt family (Karner et al. 2006a, 2006b). Which type of polarization occurs depends on the presence of different accessory proteins. The A/B- and PCP-inducing Wnt pathways are referred to, respectively, as the canonical and noncanonical Wnt pathways. In each case, the structural alterations of individual cells have novel consequences in a multicellular context, permitting multicellular aggregates to overcome the morphological defaults of solidity and sphericity. We designate the DPMs involving the Wnt pathway operating in a multicellular context as POL_a and POL_p (table 3.1).

Some key intracellular components of the Wnt pathway have counterparts in fungi, where they also mediate cell polarity (Mendoza, Redemann, and Brunner 2005). Their role in the shaping of metazoan embryos could only have emerged with the multicellular state, but the appearance of this new function would have been all but automatic. Sponges, which are characterized by many interior spaces, have genes for Wnt proteins and their ligands (Nichols et al. 2006). Such genes are also present in the placozoan *Trichoplax* (Srivastava et al. 2008), which despite containing only four cell types, has them arranged in three distinct layers, which is possible only if the cells are surface polarized.

Small, hollow cell clusters identified in the Precambrian Dou-shantuo Formation in China were first referred to as "embryos" (Chen et al. 2004; Hagadorn et al. 2006; Yin et al. 2007). On the basis of the ideas presented here they were suggested instead to have been the definitive forms of the multicellular organisms of the period (Newman, Forgacs, and Müller 2006), a proposal that is more consistent with newer evidence (Huldtgren et al. 2011). The origination of these hollow forms at the transition between the Ediacaran biota and those of the Cambrian explosion was plausibly based in part on the presence of POL_a. Specifically, the presence of the canonical Wnt pathway in a multicellular context could have readily led to the aggregates developing interior spaces.

Genes specifying components of the noncanonical Wnt pathway do not appear to be present in sponges (Adamska et al. 2010) and placozoans (Srivastava et al. 2008), which correspondingly show no sign of body elongation. They are, however, present in the morphologically more complex cnidarians (Guder et al. 2006), which display elongated body stalks and appendages.

Oscillations in cell state — As noted earlier, cell differentiated states are determined by intracellular transcription factor-based gene regulatory networks (GRNs). Such GRNs are typically multistable, but certain arrangements of positive and negative feedbacks will cause such systems to exhibit temporal oscillations in concentration of gene products (Goldbeter 1996; Reinke and Gatfield 2006).

In a single-celled organism, a periodic recurrence of cell state has no morphological consequences. If the oscillation involves downstream effectors of the Notch pathway (which normally enforces alternative cell fate decisions; see above), the result will be that cells will remain labile and uncommitted (Kageyama et al. 2008). Furthermore, if the oscillations become synchronized, the labile cell state will be coordinated across broad tissue domains, permitting concerted responses to a variety of developmental signals. Such synchronized Notch-associated oscillations have indeed been observed during early vertebrate development (Özbudak and Lewis 2008). Because of the near-ubiquity of oscillatory gene expression (Reinke and Gatfield 2006) and the inevitability of synchronization when oscillators are weakly interacting (as is typical for cells in a common tissue) (Garcia-Ojalvo,

Elowitz, and Strogatz 2004), we have proposed (Newman and Bhat 2009) that the oscillation DPM (*OSC*; table 3.1) is at the basis of the ubiquitous but mechanistically elusive phenomenon of the "morphogenetic field" (Gilbert 2006).

Morphogen gradients and activator-inhibitor systems — While single-celled organisms can change their physiological state in response to molecules secreted into the microenvironment by other such cells (Luporini et al. 2006), this effect has novel developmental consequences when it occurs in a multicellular context and spatial *gradients* can be formed. Secreted molecules that act as patterning signals in metazoan embryos by mediating concentration-dependent responses are termed *morphogens* (*MOR*; table 3.1). Some examples are Wnt, discussed above, Hedgehog, BMP/TGF-□, and FGF (Zhu and Scott 2004). The genome of marine sponges contains genes specifying members of the first two of these categories of morphogens and their receptors (Nichols et al. 2006; Srivastava et al. 2010), whereas Placozoa contains components of the first three (Srivastava et al. 2008) and Cnidaria all four (Holstein, Hobmayer, and Technau 2003; Rentzsch et al. 2008).

The ability of one or a small group of cells to influence other cells via morphogens, either within a common tissue primordium or, in a special case referred to as embryonic induction, asymmetrically across tissue boundaries (the *ASM* DPM; table 3.1), enables the generation of nonuniform cellular patterns. Assuming that the function of morphogens is tied to the physical principle of molecular diffusion, Crick calculated that they would generate patterns over tens of hours on a spatial scale of 100 μ–1 mm, similarly to what is observed in embryos (Crick 1970). Building on this basic mechanism, evolution has often produced transport processes that are formally equivalent to diffusion but that, by using additional cell-dependent modalities, are faster or slower than the simple physical process (Lander 2007).

When morphogens are positively autoregulatory, that is, directly or indirectly stimulatory of their own synthesis in target cells, they tend not to be maintained as gradients, since all cells eventually become morphogen sources. This tendency can be held in check, however, if the positively autoregulatory morphogen elicits a mechanism of lateral inhibition (such as the *LAT* DPM associated with Notch signaling). In this case, a zone will be induced around any peak of morphogen activity within which activation will not spread (Gierer and Meinhardt 1972; Meinhardt and Gierer 2000). Peaks of activation in such systems will form only at distances sufficiently far from one another so that the effects of the inhibitor are attenuated. This arrangement, termed local autoactivation–lateral inhibition (LALI) (Meinhardt and Gierer 2000; Nijhout 2003; Newman and Bhat 2007), can produce regularly spaced spots or stripes of morphogen concentration (*TUR*; table 3.1). Other "reaction-diffusion" mechanisms with somewhat different circuitry can also generate repeating patterns (Harrison 2011; Raspopovic et al. 2014). In contemporary metazoans, the *TUR* DPM (named after the mathematician Alan Turing who first systematically investigated such pattern-forming systems; Turing 1952) has been proposed to underlie pattern formation of the vertebrate limb skeleton (Newman and Frisch 1979; Hentschel et al. 2004; Zhu et al. 2010; Sheth et al. 2012; Raspopovic et al. 2014), the dentition (Salazar-Ciudad and Jernvall 2002), feather germs (Jiang et al. 2004), and hair follicles (Sick et al. 2006).

The *MOR* DPM is used in conjunction with the *OSC* DPM in vertebrate *somitogenesis*, the process by which blocks of tissue, the primordia of vertebrae and associated muscles, form in a progressive spatiotemporal order along the central axis of vertebrate embryos. In the presomitic plate of vertebrate embryos, the expression of certain genes (including mediators of the Notch pathway, as discussed above) undergoes temporal oscillation with a period similar to the formation of the somites (Dequéant and Pourquié 2008). These oscillations then become synchronized across the plate (Giudicelli et al. 2007; Kageyama, Masamizu, and Niwa 2007; Riedel-Kruse, Muller, and Oates 2007). In conjunction with an FGF morphogen gradient with its source at one end of the extended embryo, a subset of the periodically expressed molecules provide the basis for the generation of somites in vertebrate embryos (Dequéant and Pourquié 2008). The *OSC* DPM may have an analogous role in the segmentation of some arthropods (Salazar-Ciudad, Solé, and Newman 2001; Damen, Janssen, and Prpic 2005; Pueyo, Lanfear, and Couso 2008).

Extracellular matrices—The DPMs *ADH* and *DAD* mediate the formation of "epithelioid" tissues and tissue layers, which are composed of cells that are directly attached to each other. In these tissues physical properties such as viscosity, elasticity, and cohesiveness are determined by the strength of cell-cell attachment and the rheology of the cytoplasm. The other major tissue type, "mesenchyme," is composed of cells that are embedded in a secreted macromolecular microenvironment, the extracellular matrix (*ECM*; Comper 1996; table 3.1). In mesenchymal tissues physical properties are largely determined by the ECM, making them subject to a range of physical processes not seen in epithelioid tissues. The ECM molecules and the physics they mobilize thus constitute a unique DPM, designated *ECM*.

Metazoan ECMs consist largely of glycosaminoglycan polysaccharides, which are typically attached to proteins in the form of proteoglycans, and various fibrillar collagens, which occupy the interstitium between mesenchymal cells and the cells of mature connective tissues. Metazoans also produce a network-type collagen, and laminin, components of the basement membrane that attaches epithelial sheets to mesenchymal and connective tissues. Genes specifying a variety of interstitial and basement membrane ECM proteins and cell surface receptors for ECMs are found in the *M. brevicollis* genome (King et al. 2008). It is unclear what function these molecules perform in the single-celled organism, or would have performed in its common ancestor with the Metazoa, but they have clearly been recruited to new roles in the multicellular context.

Sponges contain both epithelial-like and mesenchymal cells, which reside upon and within an ECM called the "mesohyl" (Wimmer et al. 1999). These organisms actively remodel their branched ECM-rich skeletal structures by the continuous movement of their cells (Bond 1992), thus exhibiting environment-dependent morphological plasticity (Uriz et al. 2003), but only a limited array of morphological themes. It is only the triploblasts (arthropods, annelids, echinoderms, mollusks, chordates), which contain true epithelial and mesenchymal tissue types, that collectively exhibit the entire spectrum of DPM-generated motifs (Newman and Bhat 2009; Newman 2012) (see fig. 3.1).

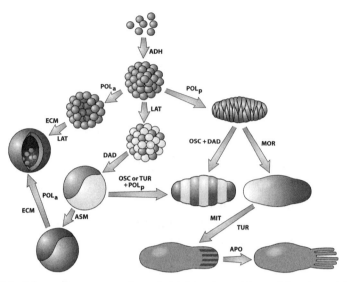

Figure 3.1. Schematic representation of single and combinatorial action of dynamical patterning modules (DPMs) in the generation of potential metazoan forms. Cells are represented individually in the upper tiers of the diagram, while the middle and lower tiers are shown at the scale of tissues. Beginning at the top, single cells form cell aggregates by the action of the *ADH* (adhesion, e.g., cadherins, lectins) module. The *POL* (polarity: Wnt pathway) DPM has two versions, apical-basal (*POL*$_a$) and planar (*POL*$_p$) polarity. *POL*$_a$ causes cells to have different surface properties at their opposite ends, leading to structurally polarized epithelial sheets and lumens within cell aggregates. *POL*$_p$, in contrast, causes cells to elongate and intercalate in the plane, which leads to convergent extension and elongation of the cell mass. The *LAT* (lateral inhibition: Notch pathway) DPM transforms an aggregate of homotypic cells into one in which two or more cell types coexist in the same aggregate, while the expression of *ADH* molecules in different amounts lead to sorting out by the action of the differential adhesion (*DAD*) module. Production of diffusible molecules by cells capable of responding to these same molecules leads to morphogen (*MOR*, e.g., TGF-□/BMP, hedgehog) gradients, whereas morphogens can also act inductively and asymmetrically (*ASM*, e.g., FGFs) by being produced by one type of tissue and affecting a different type. Synchronous biochemical oscillation (*OSC*) of key components of the Notch and Wnt pathways, in conjunction with the *DAD* DPM, can generate segments. Appropriate feedback relationships among activating and inhibitory morphogens can lead to patterns with repetitive elements by Turing-type reaction-diffusion processes (*TUR*). The action of the MAPK signaling pathway in the context of multicellular aggregate containing morphogen gradients leads to nonuniform growth by the mitogenesis (*MIT*) DPM, whereas the apoptosis (*APO*) module leads to differential cell loss. The secretion of extracellular matrix (e.g., collagen, fibronectin) between cells or into tissue spaces creates novel mechanical properties in cell sheets or masses, or new microenvironments for cell translocation, giving rise to the *ECM* DPM. Adapted from Newman and Bhat (2009), which can be referred to for additional details, including DPMs not discussed in the text.

NICHE CONSTRUCTION AND GENETIC ACCOMMODATION

Models based on the ideas of the Modern Synthesis would predict that morphologically aberrant subpopulations brought into being by DPMs would be poorly adapted to the ecological niches inhabited by the originating species. But niches are not preexisting slots in the natural environment passively occupied by organisms that have the right set of characters. They are explored, selected, and in many cases constructed by their inhabitants (Levins and Lewontin 1985; Odling-Smee et al. 2003). When novelties arise in multiple members of a population (as could readily happen with DPM-based innovation mechanisms) there is no requirement for the new forms to remain at the sites of their origin.

If saltation inescapably implied maladaption, we would expect invasive species to be less prevalent than they actually are (Carroll 2007; Stohlgren et al. 2008) and would not anticipate phenomena such as "transgressive segregation" in plants, whereby hybrids which exhibit phenotypes that are extreme or novel relative to the parental lines wind up founding and colonizing new niches (Rieseberg et al. 1999, 2003; Dittrich-Reed and Fitzpatrick 2013). At early stages of a lineage's history, before a high degree of organism-niche coadaptation had evolved, the possibility of organisms "bolting" from their niches and setting up elsewhere would have been even greater than in present-day species.

Since DPMs by definition incorporate physical mechanisms and effects, the results of their action depend on externalities—ionic composition, temperature, pressure, and so on. As long as DPMs were the major determinants of form, consistency of developmental outcome would have required stability of the relevant environmental inputs. The niche in which a novel form would initially be maintained would thus be the one that provided the conditions for its existence. The most effective way a phenotypically plastic organism can assume a morphotype independent of environmental lability, however, is via stabilizing or canalizing genetic or epigenetic change (Schmalhausen 1949; Waddington 1961; West-Eberhard 2003). Selection for persistence of phenotypes can convert body plans and morphological characters that started out as dependent on intrinsic physical properties of tissues and external conditions into products of clade-specific developmental programs (Newman 1994; Newman and Müller 2000; Newman 2011).

DARWINISM AS A LIMITING CASE

I have argued here that evolutionary mechanisms have themselves evolved. This analysis has involved positing roles for features that have been relegated to the margins, or worse, in the standard accounts of evolution associated with the Modern Synthesis: saltation, orthogenesis, and inheritance of acquired characters. These features are in fact all natural consequences of considering biological systems to be material—that is, physical—systems.

To take them in order: the possibility of saltational change in an organism's phenotype arises from the propensity of virtually all complex physical systems to exhibit nonlinear behaviors and changes in state. Examples of abrupt change in a uniform and relatively simple material like water include phase transitions, such as melting or vaporization, or bulk transformations, such as the change from wavelike to vortical motion. Living tissues, being much more complex, have many more latent morphological possibilities.

Orthogenesis is just a reflection of the fact that any physical system will assume a limited set of characteristic forms based on its inherent dynamics or modes of behavior. Liquid water, to use the same example, can be still, or form waves or whirlpools; it can break up into drops or form rivulets on a surface. It cannot form elongated or branched structures in three-dimensions, or enclose hollow spaces.

Inheritance of acquired characters is based on *plasticity*, the propensity of any material entity to assume alternative forms based on its inherent properties and its external conditions. Although animals and plants would have been more polymorphic before evolution rendered them developmentally canalized, no organism, no matter how strongly its form is influenced by the spatiotemporal expression of its genes, is immune from some degree of phenotypic plasticity. That is to say, organisms are not unwavering implementations of rigid genetic programs.

Developmental processes, being functions of physico-genetic DPMs, will necessarily exhibit combinations of the related morphological motifs—multiple tissue layers, lumens, segments, appendages, rod- and nodule-like cell condensations and cell dispersions. These are also the motifs that appeared in the course of metazoan evolution, hence the orthogenetic character of the early stages of this process. The physical nature of the DPMs ensured that development of the earliest multicellular animals was plastic and saltational. Their molecular-genetic nature guaranteed that developmental pathways were inheritable and could evolve. If the quantity and quality of the gene products change (by mutation, for example), the resulting alteration of the material properties of the cell aggregates and tissues can influence which of their inherent structural modes they will assume.

Specifying the boundary and initial conditions for multicellular-stage DPM actions by genenerating nonuniform ooplasms prior to cleavage can restrict, channel, and render increasingly stereotyped the morphological outcomes of developmental processes (Newman 2011). This would suppress the tendency of organisms to deviate dramatically in form from their progenitors, render variation mutation-dependent and incremental, around canalized phenotypic norms with little developmental plasticity. In other words, they will eventually evolve into entities (the fauna and flora of our present-day world) whose further evolution will be largely nonsaltational, nonorthogenic, and non-Lamarckian.

Uniformitarianism is the supposition Darwin adopted from the geologists James Hutton and Charles Lyell, that similar forces drive morphological change at all stages of transformation (Gould 1987). It is clear from the previous discussion that the framework presented here is nonuniformitarian. The uniformitarian Darwinian mode of microevolution, we suggest, is a late product of an evolutionary process that at its earlier, more "physical" stages was generative of macroevolutionary, that is, phylum-scale, transitions.

It should not be surprising that Darwin's mechanism would come to be seen as a special case of a broader theory of organismal change. Darwin (and Alfred Russel Wallace, the cooriginator of natural selection) of necessity based his evolutionary hypothesis on the properties of present-day organisms—Galapagos finches, domesticated animals, orchids—that are hundreds of millions of years removed from their phylogenetic origins. Not only, as often noted by historians of biology, were the mechanisms of inheritance unknown to Darwin and Wallace, but so were the mechanisms of development. Since embryogenesis in animals and most plants leads to stereotypical outcomes, and the only readily observable inherited variations within contemporary populations are small ones, selection based on adaptive advantage exerted over vast periods of time was thus an inspired guess for how large-scale morphological differences could be generated. Darwin's and Wallace's radically materialist theory of evolution was thus formulated as a uniformitarian and incrementalist one.

Our current understanding of the physical-molecular modules that constitute developmental mechanisms, including recognition of their pre-phylum origination and prolific protean dynamics, permits us to take a longer view than was available to Darwin and Wallace, while holding fast to their materialist philosophical perspective (Newman and Linde-Medina 2013). Although present-day organisms are indeed subject to microevolution by natural selection, our new outlook suggests that their prodigious variety can only have been produced in a biological world of the distant past of which only traces remain.

NOTE

I thank Ramray Bhat for very helpful comments on an earlier version of this paper and the National Science Foundation for support. This is a revised from an essay by the same title published in Auletta, LeClerc, and Martinez (2011) and reprinted by permission of the Gregorian and Biblical Press.

REFERENCES

Abedin, M., and King, N. (2008). "The Premetazoan Ancestry of Cadherins." *Science* 319:946–48.

Adamska, M., C. Larroux, M. Adamski, K. Green, E. Lovas, D. Koop, G. S. Richards, C. Zwafink, and B. M. Degnan. (2010). "Structure and Expression of Conserved Wnt Pathway Components in the Demosponge *Amphimedon queenslandica.*" *Evolution and Development* 12:494–518.

Alberch, P. (1989). "The Logic of Monsters: Evidence for Internal Constraint in Development and Evolution." *Geobioscience* 19:21–57.

Auletta, G., M. LeClerc, and R. A. Martinez, eds. (2011). *Biological Evolution: Facts and Theories; A Critical Appraisal 150 Years after "The Origin of Species."* Rome: Gregorian and Biblical Press.

Badyaev, A. V. (2005). "Stress-Induced Variation in Evolution: From Behavioural Plasticity to Genetic Assimilation." *Proceedings of the Royal Society of London B: Biological Sciences* 272:877–86.

Badyaev, A. V., K. R. Foresman, and R. L. Young. (2005). "Evolution of Morphological Integration: Developmental Accommodation of Stress-Induced Variation." *American Naturalist* 166:382–95.

Bateson, W. (1909). "Heredity and Variation in Modern Lights." In *Darwin and Modern Science*, ed. A. C. Seward, 85–101. Cambridge: Cambridge University Press.

Blankenship, J. R., and A. P. Mitchell. (2006). "How to Build a Biofilm: A Fungal Perspective." *Current Opinion in Microbiology* 9:588–94.

Bond, C. (1992). "Continuous Cell Movements Rearrange Anatomical Structures in Intact Sponges." *Journal of Experimental Zoology* 263:284–302.

Bowden, R. M., M. A. Ewert, and C. E. Nelson. (2000). "Environmental Sex Determination in a Reptile Varies Seasonally and with Yolk Hormones." *Proceedings of the Royal Society of London B: Biological Sciences* 267:1745–49.

Callebaut, W., G. B. Müller, and S. A. Newman. (2007). "The Organismic Systems Approach: Streamlining the Naturalistic Agenda." In *Integrating Evolution and Development: From Theory to Practice*, ed. S. R. Brandon and R. N. Brandon, 25–92. Cambridge, MA: MIT Press.

Carroll, S. B., J. K. Grenier, and S. D. Weatherbee. (2005). *From DNA to Diversity: Molecular Genetics and the Evolution of Animal Design.* 2nd ed. Malden, MA: Blackwell Publishing.

Carroll, S. P. (2007). "Brave New World: The Epistatic Foundations of Natives Adapting to Invaders." *Genetica* 129:193–204.

Chen, J. Y., D. J. Bottjer, P. Oliveri, S. Q. Dornbos, F. Gao, S. Ruffins, H. Chi, C. W. Li, and E. H. Davidson. (2004). "Small Bilaterian Fossils from 40 to 55 Million Years before the Cambrian." *Science* 305:218–22.

Comper, W. D. (1996). *Extracellular Matrix.* Amsterdam: Harwood Academic Publishers.

Conway Morris, S. (2006). "Darwin's Dilemma: The Realities of the Cambrian 'Explosion.'" *Philosophical Transactions of the Royal Society of London B: Biological Science* 361:1069–83.

Crick, F. H. C. (1970). "Diffusion in Embryogenesis." *Nature* 225:420–22.

Damen, W. G., R. Janssen, and N. M. Prpic. (2005). "Pair Rule Gene Orthologs in Spider Segmentation." *Evolution and Development* 7:618–28.

Darwin, C. (1859). *On the Origin of Species by Means of Natural Selection, or, The Preservation of Favoured Races in the Struggle for Life*. London: J. Murray.

Davidson, E. H. (2006). *The Regulatory Genome: Gene Regulatory Networks in Development and Evolution*. Amsterdam: Elsevier/Academic Press.

Dayel, M. J., R. A. Alegado, S. R. Fairclough, T. C. Levin, S. A. Nichols, K. McDonald, and N. King. (2011). "Cell Differentiation and Morphogenesis in the Colony-Forming *Choanoflagellate Salpingoeca rosetta*." *Developmental Biology* 357:73–82.

de Gennes, P. G. (1992). "Soft Matter." *Science* 256:495–97.

Dequéant, M. L., and O. Pourquié. (2008). "Segmental Patterning of the Vertebrate Embryonic Axis." *Nature Reviews: Genetics* 9:370–82.

Dittrich-Reed, D. R., and B. M. Fitzpatrick. (2013). "Transgressive Hybrids as Hopeful Monsters." *Evolutionary Biology* 40:310–15.

Droser, M. L., and J. G. Gehling. (2008). "Synchronous Aggregate Growth in an Abundant New Ediacaran Tubular Organism." *Science* 319:1660–62.

Ehebauer, M., P. Hayward, and A. M. Arias. (2006). "Notch, a Universal Arbiter of Cell Fate Decisions." *Science* 314:1414–15.

Erwin, D. H. (2008). "Wonderful Ediacarans, Wonderful Cnidarians?" *Evolution and Development* 10:263–64.

Fernàndez-Busquets, X., A. Kornig, I. Bucior, M. M. Burger, and D. Anselmetti. (2009). "Self-Recognition and Ca^{2+}-Dependent Carbohydrate-Carbohydrate Cell Adhesion Provide Clues to the Cambrian Explosion." *Molecular Biology and Evolution* 26:2551–61.

Fisher, R. A. (1930). *The Genetical Theory of Natural Selection*. Oxford: Clarendon Press.

Forgacs, G., and S. A. Newman. (2005). *Biological Physics of the Developing Embryo*. Cambridge: Cambridge University Press.

Fusco, G., and Minelli, A. (2008). *Evolving Pathways: Key Themes in Evolutionary Developmental Biology*. Cambridge: Cambridge University Press.

Garcia-Ojalvo, J., M. B. Elowitz, and S. H. Strogatz. (2004). "Modeling a Synthetic Multicellular Clock: Repressilators Coupled by Quorum Sensing." *Proceedings of the National Academy of Sciences* 101:10955–60.

Gierer, A., and H. Meinhardt. (1972). "A Theory of Biological Pattern Formation." *Kybernetik* 12:30–39.

Gilbert, S. F. (2006). *Developmental Biology*. 8th ed. Sunderland, MA: Sinauer Associates.

Giudicelli, F., E. M. Özbudak, G. J. Wright, and J. Lewis. (2007). "Setting the Tempo in Development: An Investigation of the Zebrafish Somite Clock Mechanism." *PLoS Biology* 5(6):e150.

Goldbeter, A. (1996). *Biochemical Oscillations and Cellular Rhythms: The Molecular Bases of Periodic and Chaotic Behaviour.* Cambridge: Cambridge University Press.

Goldschmidt, R. B. (1940). *The Material Basis of Evolution.* New Haven, CT: Yale University Press.

Goodman, R. M. (2008). "Latent Effects of Egg Incubation Temperature on Growth in the Lizard *Anolis carolinensis." Journal of Experimental Zoology Part A: Ecology, Genetics, and Physiology* 309:525–33.

Gould, S. J. (1987). *Time's Arrow, Time's Cycle: Myth and Metaphor in the Discovery of Geological Time.* Jerusalem-Harvard Lectures. Cambridge, MA: Harvard University Press.

Guder, C., I. Philipp, T. Lengfeld, H. Watanabe, B. Hobmayer, and T. W. Holstein. (2006). "The Wnt Code: Cnidarians Signal the Way." *Oncogene* 25:7450–60.

Hagadorn, J. W., S. Xiao, P. C. Donoghue, S. Bengtson, N. J. Gostling, M. Pawlowska, E. C. Raff, R. A. Raff, et al. (2006). "Cellular and Subcellular Structure of Neoproterozoic Animal Embryos." *Science* 314:291–94.

Harrison, L. G. (2011). *The Shaping of Life: The Generation of Biological Pattern.* Cambridge: Cambridge University Press.

Hentschel, H. G., T. Glimm, J. A. Glazier, and S. A. Newman. (2004). "Dynamical Mechanisms for Skeletal Pattern Formation in the Vertebrate Limb." *Proceedings of the Royal Society of London B: Biological Sciences* 271:1713–22.

Hernández-Hernández, V., K. J. Niklas, S. A. Newman, and M. Benítez. (2012). "Dynamical Patterning Modules in Plant Development and Evolution." *International Journal of Developmental Biology* 56:661–74.

Holstein, T. W., E. Hobmayer, and U. Technau. (2003). "Cnidarians: An Evolutionarily Conserved Model System for Regeneration?" *Developmental Dynamics* 226:257–67.

Huldtgren, T., J. A. Cunningham, C. Yin, M. Stampanoni, F. Marone, P. C. Donoghue, and S. Bengtson. (2011). "Fossilized Nuclei and Germination Structures Identify Ediacaran 'Animal Embryos' as Encysting Protists." *Science* 334:1696–99.

Jablonka, E., and M. J. Lamb. (1995). *Epigenetic Inheritance and Evolution.* Oxford: Oxford University Press.

———. (2005). *Evolution in Four Dimensions: Genetic, Epigenetic, Behavioral, and Symbolic Variation in the History of Life, Life and Mind.* Cambridge, MA: MIT Press.

Jiang, T. X., R. B. Widelitz, W. M. Shen, P. Will, D. Y. Wu, C. M. Lin, H. S. Jung, and C.-M. Chuong. (2004). "Integument Pattern Formation Involves Genetic and Epigenetic Controls: Feather Arrays Simulated by Digital Hormone Models." *International Journal of Developmental Biology* 2–3:117–35.

Kageyama, R., Y. Masamizu, and Y. Niwa. (2007). "Oscillator Mechanism of Notch Pathway in the Segmentation Clock." *Developmental Dynamics* 236:1403–9.

Kageyama, R., T. Ohtsuka, H. Shimojo, and I. Imayoshi. (2008). "Dynamic Notch Signaling in Neural Progenitor Cells and a Revised View of Lateral Inhibition." *Nature Neuroscience* 11:1247–51.

Karner, C., K. A. Wharton Jr., and T. J. Carroll. (2006a). "Planar Cell Polarity and Vertebrate Organogenesis." *Seminars in Cell and Developmental Biology* 17:194–203.

———. (2006b). "Apical-Basal Polarity, Wnt Signaling and Vertebrate Organogenesis." *Seminars in Cell and Developmental Biology* 17:214–22.

Kazmierczak, J., and S. Kempe. (2004). "Calcium Build-Up in the Precambrian Sea: A Major Promoter in the Evolution of Eukaryotic Life." In *Origins*, ed. J. Seckbach, 329–45. Dordrecht: Kluwer.

Keller, R. (2002). "Shaping the Vertebrate Body Plan by Polarized Embryonic Cell Movements." *Science* 298:1950–54.

Keller, R., L. Davidson, A. Edlund, T. Elul, M. Ezin, D. Shook, and P. Skoglund. (2000). "Mechanisms of Convergence and Extension by Cell Intercalation." *Philosophical Transactions of the Royal Society of London B: Biological Sciences* 355:897–922.

King, N., C. T. Hittinger, and S. B. Carroll (2003). "Evolution of Key Cell Signaling and Adhesion Protein Families Predates Animal Origins." *Science* 301:361–63.

King, N., M. J. Westbrook, S. L. Young, A. Kuo, et al. (2008). "The Genome of the Choanoflagellate *Monosiga brevicollis* and the Origin of Metazoans." *Nature* 451:783–88.

Lander, A. D. (2007). "Morpheus Unbound: Reimagining the Morphogen Gradient." *Cell* 128:245–56.

Lang, B. F., C. O'Kelly, T. Nerad, M. W. Gray, and G. Burger. (2002). "The Closest Unicellular Relatives of Animals." *Currents in Biology* 12:1773–78.

Larroux, C., B. Fahey, D. Liubicich, V. F. Hinman, M. Gauthier, M. Gongora, K. Green, G. Worheide, S. P. Leys, and B. M. Degnan. (2006). "Developmental Expression of Transcription Factor Genes in a Demosponge: Insights into the Origin of Metazoan Multicellularity." *Evolution and Development* 8:150–73.

Larroux, C., G. N. Luke, P. Koopman, D. S. Rokhsar, S. M. Shimeld, and B. M. Degnan. (2008). "Genesis and Expansion of Metazoan Transcription Factor Gene Classes." *Molecular Biology and Evolution* 25:980–96.

Levins, R., and R. C. Lewontin. (1985). *The Dialectical Biologist*. Cambridge, MA: Harvard University Press.

Luporini, P., A. Vallesi, C. Alimenti, and C. Ortenzi. (2006). "The Cell Type–Specific Signal Proteins (Pheromones) of Protozoan Ciliates." *Current Pharmaceutical Design* 12:3015–24.

Margulis, L., and D. Sagan. (2002). *Acquiring Genomes: A Theory of the Origins of Species.* New York: Basic Books.

Mayr, E. (1974). "Teleological and Teleonomic: A New Analysis." *Boston Studies in the Philosophy of Science* 14:91–117.

———. (1982). *The Growth of Biological Thought: Diversity, Evolution, and Inheritance.* Cambridge, MA: Belknap Press.

Meinhardt, H., and A. Gierer. (2000). "Pattern Formation by Local Self-Activation and Lateral Inhibition." *Bioessays* 22:753–60.

Mendoza, M., S. Redemann, and D. Brunner. (2005). "The Fission Yeast MO25 Protein Functions in Polar Growth and Cell Separation." *European Journal of Cell Biology* 84:915–26.

Mikhailov, A. S. (1990). *Foundations of Synergetics I.* Berlin: Springer-Verlag.

Moczek, A. P. (2008). "On the Origins of Novelty in Development and Evolution." *Bioessays* 30:432–47.

Müller, G. B. (2007). "Evo-Devo: Extending the Evolutionary Synthesis." *Nature Reviews: Genetics* 8:943–9.

Müller, G. B., and S. A. Newman, eds. (2003). *Origination of Organismal Form: Beyond the Gene in Developmental and Evolutionary Biology.* Cambridge, MA: MIT Press.

Müller, G. B., and S. A. Newman. (2005). "The Innovation Triad: An Evo-Devo Agenda." *Journal of Experimental Zoology B: Molecular and Developmental Evolution* 304:487–503.

Müller, G. B., and J. Streicher. (1989). "Ontogeny of the Syndesmosis tibiofibularis and the Evolution of the Bird Hindlimb: A Caenogenetic Feature Triggers Phenotypic Novelty." *Anatomy and Embryology* 179:327–39.

Newman, S. A. (1994). "Generic Physical Mechanisms of Tissue Morphogenesis: A Common Basis for Development and Evolution." *Journal of Evolutionary Biology* 7:467–88.

———. (1998). "Epithelial Morphogenesis: A Physico-Evolutionary Interpretation." In *Molecular Basis of Epithelial Appendage Morphogenesis*, ed. C.-M. Chuong, 341–58. Austin, TX: R. G. Landes.

———. (2005). "The Pre-Mendelian, Pre-Darwinian World: Shifting Relations between Genetic and Epigenetic Mechanisms in Early Multicellular Evolution." *Journal of Bioscience* 30:75–85.

———. (2006). "The Developmental-Genetic Toolkit and the Molecular Homology-Analogy Paradox." *Biological Theory* 1:12–16.

———. (2011). "Animal Egg as Evolutionary Innovation: A Solution to the 'Embryonic Hourglass' Puzzle." *Journal of Experimental Zoology B: Molecular and Developmental Evolution* 316:467–83.

———. (2012). "Physico-Genetic Determinants in the Evolution of Development." *Science* 338:217–19.

Newman, S. A., and R. Bhat. (2007). "Activator-Inhibitor Dynamics of Vertebrate Limb Pattern Formation." *Birth Defects Research, Part C: Embryo Today Reviews* 81:305–319.

———. (2008). "Dynamical Patterning Modules: Physico-Genetic Determinants of Morphological Development and Evolution." *Physical Biology* 5:15008.

———. (2009). "Dynamical Patterning Modules: A 'Pattern Language' for Development and Evolution of Multicellular Form." *International Journal of Developmental Biology* 53:693–705.

Newman, S. A., R. Bhat, and N. V. Mezentseva. (2009). "Cell State Switching Networks and Dynamical Patterning Modules: Complementary Mediators of Plasticity in Development and Evolution." *Journal of Bioscience* 34:553–72.

Newman, S. A., and W. D. Comper. (1990). "'Generic' Physical Mechanisms of Morphogenesis and Pattern Formation." *Development* 110:1–18.

Newman, S. A., G. Forgacs, and G. B. Müller. (2006). "Before Programs: The Physical Origination of Multicellular Forms." *International Journal of Developmental Biology* 50:289–99.

Newman, S. A., and H. L. Frisch. (1979). "Dynamics of Skeletal Pattern Formation in Developing Chick Limb." *Science* 205:662–68.

Newman, S. A., and M. Linde-Medina. (2013). "Physical Determinants in the Emergence and Inheritance of Multicellular Form." *Biological Theory* 8:274–85.

Newman, S. A., and G. B. Müller. (2000). "Epigenetic Mechanisms of Character Origination." *Journal of Experimental Zoology B: Molecular and Developmental Evolution* 288:304–17.

Nichols, S. A., W. Dirks, J. S. Pearse, and N. King (2006). "Early Evolution of Animal Cell Signaling and Adhesion Genes." *Proceedings of the National Academy of Sciences* 103:12451–56.

Nijhout, H. F. (2003). "Gradients, Diffusion and Genes in Pattern Formation." In Müller and Newman 2003, 165–81.

Niklas, K. J., and S. A. Newman. (2013). "The Origins of Multicellular Organisms." *Evolution and Development* 15:41–52.

Odling-Smee, F. J., K. N. Laland, and M. W. Feldman. (2003). *Niche Construction: The Neglected Process in Evolution*. Princeton: Princeton University Press.

Özbudak, E. M., and J. Lewis. (2008). "Notch Signalling Synchronizes the Zebrafish Segmentation Clock but Is Not Needed to Create Somite Boundaries." *PLoS: Genetics* 4(2):e15.

Philippe, H., E. A. Snell, E. Bapteste, P. Lopez, P. W. Holland, and D. Casane. (2004). "Phylogenomics of Eukaryotes: Impact of Missing Data on Large Alignments." *Molecular and Biological Evolution* 21:1740–52.

Pigliucci, M. (2001). *Phenotypic Plasticity: Beyond Nature and Nurture*. Baltimore: Johns Hopkins University Press.

Pueyo, J. I., R. R. Lanfear, and J. P. Couso. (2008). "Ancestral Notch-Mediated Segmentation Revealed in the Cockroach *Periplaneta americana*." *Proceedings of the National Academy of Sciences* 105:16614–19.

Raspopovic, J., L. Marcon, L. Russo, and J. Sharpe. (2014). "Modeling Digits: Digit Patterning Is Controlled by a Bmp-Sox9-Wnt Turing Network Modulated by Morphogen Gradients." *Science* 345:566–70.

Reinke, H., and D. Gatfield. (2006). "Genome-Wide Oscillation of Transcription in Yeast." *Trends in Biochemical Science* 31:189–91.

Rentzsch, F., J. H. Fritzenwanker, C. B. Scholz, and U. Technau. (2008). "FGF Signalling Controls Formation of the Apical Sensory Organ in the Cnidarian *Nematostella vectensis*." *Development* 135:1761–69.

Riedel-Kruse, I. H., C. Muller, and A. C. Oates. (2007). "Synchrony Dynamics During Initiation, Failure, and Rescue of the Segmentation Clock." *Science* 317:1911–15.

Rieseberg, L. H., M. A. Archer, and R. K. Wayne. (1999). "Transgressive Segregation, Adaptation and Speciation." *Heredity* 83 (Pt 4): 363–72.

Rieseberg, L. H., A. Widmer, A. M. Arntz, and J. M. Burke. (2003). "The Genetic Architecture Necessary for Transgressive Segregation Is Common in Both Natural and Domesticated Populations." *Philosophical Transactions of the Royal Society of London B: Biological Sciences* 358:1141–47.

Robert, J. S. (2004). *Embryology, Epigenesis, and Evolution: Taking Development Seriously*. Cambridge Studies in Philosophy and Biology. Cambridge: Cambridge University Press.

Rokas, A., D. Kruger, and S. B. Carroll. (2005). "Animal Evolution and the Molecular Signature of Radiations Compressed in Time." *Science* 310:1933–38.

Rose, S. M. (1958). "Feedback in the Differentiation of Cells." *Scientific American* 199:36–41.

Salazar-Ciudad, I. (2006). "On the Origins of Morphological Disparity and Its Diverse Developmental Bases." *BioEssays* 28:1112–22.

Salazar-Ciudad, I., and J. Jernvall. (2002). "A Gene Network Model Accounting for Development and Evolution of Mammalian Teeth." *Proceedings of the National Academy of Sciences* 99:8116–20.

Salazar-Ciudad, I., J. Jernvall, and S. A. Newman. (2003). "Mechanisms of Pattern Formation in Development and Evolution." *Development* 130:2027–37.

Salazar-Ciudad, I., R. Solé, and S. A. Newman. (2001). "Phenotypic and Dynamical Transitions in Model Genetic Networks II: Application to the Evolution of Segmentation Mechanisms." *Evolution and Development* 3:95–103.

Schmalhausen, I. I. (1949). *Factors of Evolution*. Philadelphia: Blakiston.

Shalchian-Tabrizi, K., M. A. Minge, M. Espelund, R. Orr, T. Ruden, K. S. Jakobsen, and T. Cavalier-Smith. (2008). "Multigene Phylogeny of Choanozoa and the Origin of Animals." *PLoS One* 3(5):e2098.

Sheth, R., L. Marcon, M. F. Bastida, M. Junco, L. Quintana, R. Dahn, M. Kmita, J. Sharpe, and M. A. Ros. (2012). "Hox Genes Regulate Digit Patterning by Controlling the Wavelength of a Turing-Type Mechanism." *Science* 338:1476–80.

Sick, S., S. Reinker, J. Timmer, and T. Schlake. (2006). "WNT and DKK Determine Hair Follicle Spacing Through a Reaction-Diffusion Mechanism." *Science* 314:1447–50.

Simpson, G. G. (1944). *Tempo and Mode in Evolution*. Columbia Biological Series 15. New York: Columbia University Press.

Srivastava, M., E. Begovic, J. Chapman, et al. (2008). "The Trichoplax Genome and the Nature of Placozoans." *Nature* 454:955–60.

Srivastava, M., O. Simakov, et. al. (2010). "The *Amphimedon queenslandica* Genome and the Evolution of Animal Complexity." *Nature* 466:720–26.

Steinberg, M. S. (2003). "Cell Adhesive Interactions and Tissue Self-Organization." In Müller and Newman 2003, 137–63.

Steinberg, M. S., and M. Takeichi. (1994). "Experimental Specification of Cell Sorting, Tissue Spreading, and Specific Spatial Patterning by Quantitative Differences in Cadherin Expression." *Proceedings of the National Academy of Sciences* 91:206–9.

Stohlgren, T. J., D. T. Barnett, C. S. Jarnevich, C. Flather, and J. Kartesz. (2008). "The Myth of Plant Species Saturation." *Ecology Letters* 11:313–22.

Thompson, D. A. W. (1942). *On Growth and Form*. 2nd ed. Cambridge: Cambridge University Press.

Trut, L., I. Oskina, and A. Kharlamova. (2009). "Animal Evolution During Domestication: The Domesticated Fox as a Model." *BioEssays* 31:349–60.

Turing, A. M. (1952). "The Chemical Basis of Morphogenesis." *Philosophical Transactions of the Royal Society of Landon B* 237:37–72.

Uriz, M. J., X. Turon, M. A. Becerro, and G. Agell. (2003). "Siliceous Spicules and Skeleton Frameworks in Sponges: Origin, Diversity, Ultrastructural Patterns, and Biological Functions." *Microscopic Research and Technique* 62:279–99.

Vedel, V., A. D. Chipman, M. Akam, and W. Arthur. (2008). "Temperature-Dependent Plasticity of Segment Number in an Arthropod Species: The Centipede *Strigamia maritima*." *Evolution and Development* 10:487–92.

Vlamakis, H., C. Aguilar, R. Losick, and R. Kolter. (2008). "Control of Cell Fate by the Formation of an Architecturally Complex Bacterial Community." *Genes and Development* 22:945–53.

Waddington, C. H. (1942). "Canalization of Development and the Inheritance of Acquired Characters." *Nature* 150:563–565.

———. (1961). "Genetic Assimilation." *Advances in Genetics* 10:257–93.

Wainright, P. O., G. Hinkle, M. L. Sogin, and S. K. Stickel. (1993). "Monophyletic Origins of the Metazoa: An Evolutionary Link with Fungi." *Science* 260:340–42.

Weismann, A. (1892). *Das Keimplasma: Eine Theorie der Vererbung.* Jena: Fischer.

West-Eberhard, M. J. (2003). *Developmental Plasticity and Evolution.* Oxford: Oxford University Press.

Wilkins, A. S. (2002). *The Evolution of Developmental Pathways.* Sunderland, MA: Sinauer Associates.

Wimmer, W., S. Perovic, M. Kruse, H. C. Schroder, A. Krasko, R. Batel, and W. E. Müller. (1999). "Origin of the Integrin-Mediated Signal Transduction: Functional Studies with Cell Cultures from the Sponge *Suberites domuncula.*" *European Journal of Biochemistry* 260:156–65.

Woese, C. R., and N. Goldenfeld. (2009). "How the Microbial World Saved Evolution from the Scylla of Molecular Biology and the Charybdis of the Modern Synthesis." *Microbiology and Molecular Biology Reviews* 73:14–21.

Yin, L., M. Zhu, A. H. Knoll, X. Yuan, J. Zhang, and J. Hu. (2007). "Doushantuo Embryos Preserved Inside Diapause Egg Cysts." *Nature* 446:661–63.

Zhu, A. J., and M. P. Scott. (2004). "Incredible Journey: How Do Developmental Signals Travel Through Tissue?" *Genes and Development* 18:2985–97.

Zhu, J., Y. T. Zhang, M. S. Alber, and S. A. Newman. (2010). "Bare Bones Pattern Formation: A Core Regulatory Network in Varying Geometries Reproduces Major Features of Vertebrate Limb Development and Evolution." *PLoS One* 5(5):e10892.

FOUR

THE EVOLVABILITY OF ORGANIC FORMS
Possible, Likely, and Unlikely Change from the Perspective of Evolutionary Developmental Biology

Alessandro Minelli

Confronted with the extraordinary diversity of animal form, we can ask questions about function and adaptation. How does this animal move? How does it feed? How does it defend itself from its enemies? But we can also ask questions about development, reproduction, and heredity. What mechanisms produce these forms? How are these forms perpetuated throughout generations?

We can also ask more general questions, for example, whether there are intrinsic limits to the diversity of forms we can expect to find among living beings. Limits are indeed there, but the location of these limits is something that the diversity of occasional "monsters" forces us to investigate with care.

A LESSON FROM MONSTERS

What is a monster? In the common usage of this word in biology, a monster is an organism that deviates from what we would regard as the norm to such an extent that its long-term viability, not to say its eventual reproductive success, is seriously impaired. Examples of such deviant organisms are those cattle and sheep with two heads that from time to time make their way into the news, as well as the four-winged fruit-flies that occasionally show up amongst a million normal, two-winged flies.

To be sure, these monsters are unlikely the founders of new lineages within which their monstrosities would be perpetuated. Evolution is not simply a matter of producing new forms, but also a question of their comparative survival and reproductive success in comparison to their conspecifics. At least this is the "orthodox" reading of evolution according to the neo-Darwinian paradigm. However, this is far from being the whole story. Focusing on the survival value of phenotypes exposed to natural selection, our attention is diverted away from another equally important aspect of the question, namely, that the different phenotypes, even among those we would regard as "normal," are not uniformly easy to produce. We can hypothesize the existence of a number of phenotypes only slightly deviating from the "normal" (i.e., existing) ones, with such a structure that we would have no reason to doubt their adaptive success. Nevertheless, these may be impossible to generate. On the other hand, nature is able to produce obviously maladaptive phenotypes, such as fruit-flies with four wings and sheep with two heads.

To summarize evidence for the existence of rules in this apparent chaos of organic form: new phenotypes are never produced from scratch like those freely emerging combinations of heads, legs, bones, and tails that Empedocles imagined to be steadily generated and destroyed by the opposing forces of Love and Hatred (see Depew chapter this volume). Even in the case of "monsters," such as the four-winged *Drosophila*, the differences between these and the "normal" phenotypes, conspicuous as they are, reduce nevertheless to a small number of well circumscribed changes within the expected morphogenetic repertoire of the fruit-fly. Specifically, the four-winged flies differ from their "normal" relatives because of the replacement of the third thoracic segment, and its appendages, with an extra copy of the second thoracic segment and the corresponding appendages. The extra wings of the monster are the wings of a fruit-fly, not those of another kind of insect. Even monsters obey rules (fig. 4.1) to such an extent that we can classify them according to Linnean methods (Geoffroy Saint-Hilaire 1832–37).

Figure 4.1. Comparing the Chimera, a creature of myth, to the real "monsters" occurring in nature can help understanding why the latter can be actually generated, whereas the former cannot. Monsters like a four-winged fruit-fly, or a two-headed lamb, are the result of a mechanistically small deviation from normal development, often traceable to the expression of a gene presenting a point mutation. Selection will eventually sweep out this mutated gene because of the obvious shortcomings suffered by its carriers, but these monsters can nevertheless occur again and again in a population. Very different is the "monstrous" character of the Chimera. Its composite identity, partaking of lion's, goat's, and snake's natures, cannot be obtained by either crossing or grafting, because well-established biological mechanisms like genetic incompatibility and immune reaction would immediately stop any corresponding effort.

Thus, there are preferred pathways constraining even the most conspicuous change of form. This should help explain the widespread occurrence of morphological convergence between representatives of distantly related groups (Conway Morris 2003).

NUMBERS

Adequate knowledge of the genetic and developmental mechanisms underlying the generation of fruit-flies with four wings shows that the effect of a point change in development can be the production of an organism quite different from its closest relatives. On the other

Figure 4.2. The long neck of the giraffe is supported by seven cervical vertebrae, the same number as the vertebrae supporting the neck of all mammals, sloths and manatees excepted. Selection acting in favor of giraffes with longer necks could only act on cervical vertebrae with different degrees of elongation, whereas no variation in number was ever available.

hand, we can imagine an unlimited number of likely adaptive phenotypes that nevertheless do not occur in nature, because current developmental processes are not able to produce them. Think, for example, of the neck of the giraffe. Natural selection (or, perhaps, sexual selection) is arguably responsible for the extraordinary elongation of this neck (Simmons and Scheepers 1996), but this does not say anything about the kind of variation on which selection can operate.

Let us consider, in particular, the neck's skeletal axis (fig. 4.2). In principle, elongation could result either from an increase in the number of cervical vertebrae or from a sizeable elongation of all the elements in the original set of cervical vertebrae, without excluding, of course, the possibility that both mechanisms were in fact cooperating. Eventually a glance at a skeleton shows that the giraffe's neck is supported by seven vertebrae, exactly the same number as in nearly all mammals. This indicates that the number of cervical vertebrae was always strictly invariant in the giraffe lineage, as it likely is, for the indirect reason that this is true of mammals generally (Galis 1999).[1]

Centipedes are also a group of animals showing strong and probably unexpected constraints in the number of parts, in their case, the number of leg-bearing segments. All species in this group have as adults an odd number of leg pairs, even in the cases where leg number varies within the species, and even among the members of the same brood.[2]

The nonexistence of centipedes with an even number of leg pairs in the adult cannot be explained in terms of adaptation, that is, as an effect of natural selection. Notice that many centipedes that have fifteen pairs of legs as adults have juvenile stages with an even number of leg pairs. Thus, if we want to find an explanation for the fact that no centipede species has an even number of leg-bearing segments we must look into their developmental system rather than hypothesize that individuals with odd vs. even number of leg pairs are dramatically different targets of natural selection.

THE ROOTS OF EVO-DEVO

The giraffe's neck and the legs of the centipedes are two conspicuous examples of an increasingly large number of evolutionary traits that we realize we cannot explain without combining together evolutionary biology and developmental biology. Until recently, however, evolutionary biology and developmental biology had very little in common. Different questions were asked in each field, different methods and tools were employed, and different academic journals published their respective results.

Throughout the nineteenth century, developmental biology had been essentially confined to the description of the embryonic and larval development of an increasing diversity of animal species. Early in the twentieth century, developmental biology became an experimental science, focused on mechanically manipulating eggs and embryos. Eventually it supplemented these methods with a biochemical analysis of the corresponding processes. This finally changed, in the last quarter of the century, into developmental genetics. The latter approach has rapidly moved from the identification of the genes involved in the control of the different steps of differentiation and morphogenesis to the study of the spatial and temporal patterns of their expression and the complex relationships among them or their products.

However, with the adoption of increasingly more sophisticated and powerful means of investigation, the range of species to which the new tools were applied was increasingly shrinking to a very small number. Most of the advances were indeed derived from research on one mammal species (the mouse), one bird (the chick), one amphibian (*Xenopus*), one fish (the zebrafish *Danio rerio*), one insect (the fruit-fly *Drosophila melanogaster*), and one nematode or roundworm (*Caenorhabditis elegans*). To be sure, such a tiny number of species was quite inadequate to satisfy the interests of evolutionary biologists, who were therefore not immediately prepared to exploit the incoming information on development to improve their understanding of the evolution of animal form.

However, starting around the mid-'80s of the last century, a new research approach began to emerge, within which problems, concepts, and methods of developmental biology increasingly merged with those of evolutionary biology. This approach is currently known as *evolutionary developmental biology*—an expression first featured as the title of a book by Brian K. Hall (1999)—or in short, *evo-devo*.

Within a few years, this lively branch of biology has rapidly expanded, with the foundation of dedicated academic journals: *Evolution and Development* and a new section—*Molecular and Developmental Evolution*—of the *Journal of Experimental Zoology*. New professional societies were formed, such as the European Society of Evolutionary Developmental Biology, established in 2006. Especially with the publication of monographic volumes (Raff and Kaufman 1983; Arthur 1997, 2004; Minelli 2003, 2009a, 2009b; Carroll 2005; Kirschner and Gerhart 2005), and edited collections (Hall and Olson

2003; Müller and Newman 2003; Minelli and Fusco 2008), evo-devo is progressively defining its autonomy within the current panorama of biological disciplines.

Evolutionary developmental biology has dual roots: on the one side these are found in the comparative method, which, from its old origins in morphology (Cole 1944), has progressively expanded to all other aspects of the living beings, including a comparative analysis of developmental processes (Minelli 2003; Scholtz 2008); and on the other side, in the study of the ways genes control the production of organic forms (Coen 1999; Wilkins 2002; Carroll, Grenier, and Weatherbee 2004).

EVOLVABILITY

The existence of constraints (that is, of preferential paths that the evolution of the form of living organisms seems to follow in spite of the possible better adaptive value of alternative phenotypes that are never, or seldom, generated) has forced evolutionary biology to shift attention from its traditional research target, the *survival of the fittest*, to the logically preceding step of the *arrival of the fittest*.

By realizing that the giraffe has evolved a long neck by the mechanism of elongation of seven cervical vertebrae in the ancestral form rather than by adding more elements to the original set of bones, we open a window into the nature and the amount of variation (in this case, in vertebral number and degree of elongation of the individual vertebrae) available in the population. This variation constrains the possible change the population can undergo under the effects of natural selection, sexual selection, and random drift. In other words, this defines its *evolvability* (Hendrikse, Parson, and Hallgrímsson 2007).

The evolvability of a given trait is thus the probability with which it may change into alternative phenotypes (fig. 4.3). In the case of the neck vertebrae, there is high evolvability in the absolute and relative length of the neck of mammals, as the opposite conditions in the giraffe and the hippo demonstrate, whereas the evolvability in the number of cervical vertebrae, in the same group, is close to zero.

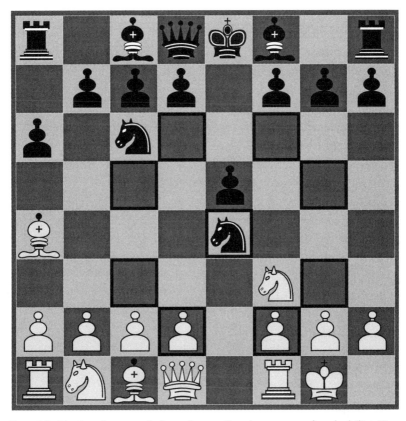

Figure 4.3. Playing chess may help understanding the concept of evolvability. Due to the rules of the game, the range of squares to which a piece can be moved is precisely determined. A choice among those squares will be dictated by "history" (the present distribution of pieces on the checkerboard) and "selection" (the player's strategy). For example, the knight goes two squares in one direction, then one more step ahead at a 90 degree angle. None of the squares to which it can be displaced in one move is thus contiguous to the square occupied by the knight before being moved. In other terms, none of the immediately closest squares belongs to the short-term "evolvability" of a knight's position. Similarly, a seemingly minor morphological change such as acquiring or losing one leg-bearing segment is as much as impossible to a centipede with twenty-one pairs of legs, whereas a morphologically larger change is perhaps easily obtained within the constraints of the animal's current developmental mechanisms.

Evolvability is arguably the single most important conceptual contribution of evo-devo to biology. Its relevance is easily seen when realizing that no selective pressure can operate on phenotypes that development cannot produce, while no developmentally possible "monster" can be regarded as a "hopeful" one (to use Goldschmidt's well-known expression) (Goldschmidt 1940), if the environment does not offer it a chance to survive and reproduce.

Evolution and development are thus so tightly intertwined as to require a combined approach, using the concepts and tools of modern developmental biology together with those of evolutionary biology. It is only within the resulting discipline of evolutionary developmental biology that problems about evolutionary novelties and macroevolution can be seriously addressed.

EVOLUTION BY SALTATION

The success obtained by evolutionary biology operating within the neo-Darwinian paradigm has long left aside, as a problem of dubious scientific legitimacy, the recurrent question as to whether natural selection, as such, is sufficient to generate evolutionary novelties of some importance (Mayr 1960; Maynard Smith 1993). This unease with the dominant views did not extend to questioning the efficacy of natural selection in eliminating the less adaptive minor variations. The problem was whether and by what means could natural selection eventually give rise to the first flower, or the first bird wing, or the first human brain. Were these novelties also produced by the simple accumulation of the effects of selection on minor phenotypic variations produced by blind mutations? Answering this question was hardly possible in the past, because of the very limited knowledge of the correspondence between genes and phenotypes, but still more because of the widespread refusal by leading evolutionary biologists to open the "black box" of development. The possibility of rapid major changes—evolutionary saltation—was simply ruled out in principle rather than rejected on factual evidence. Eventually, however, we have come to recognize actual examples of evolutionary saltation. This has been appreciated, at least, by some evo-devo researchers, as shown by the two cases summarized in the following examples.

The first example involves the evolution of the orchid flowers. Within this family, nothing more than two point mutations has been required to change a flower with 3+3 identical tepals into a flower with two well-distinct groups of 3 tepals and, further, into a flower where one of the 3 internal tepals has become very different from the remaining two. With this second step, a real "leap" has been completed, transforming a flower with radial symmetry into a flower with bilateral symmetry (Theissen 2009).

The other example refers to the animal kingdom, more precisely, to a South American genus of scolopenders (centipedes). One species in this genus (*Scolopendropsis bahiensis*) includes individuals with either 21 or 23 pairs of legs, whereas a very closely allied species described in 2008 (*Sc. duplicata*) includes individuals with either 39 or 43 pairs of legs (Chagas, Edgecombe, and Minelli 2008). This circumstance is very interesting for at least two reasons. First, all contemporary species of scolopenders, of which there are some 700 worldwide, have a fixed number of legs; in the majority of species, this number amounts to 21 pairs, in the remaining ones to 23. Only one species, the just mentioned *Sc. bahiensis*, includes specimens with 21 pairs of legs alongside specimens with 23 pairs. Therefore, *Sc. duplicata* is the only species of scolopenders with a number of leg-bearing segments outside the usual range, and the difference between it and all the other species is quite conspicuous, and no intermediates apparently exist. Second, the two *Scolopendropsis* species are very similar up to the details of the number and position of spines on the legs. Based upon these observations, the divergence between the two species, involving the first appearance of scolopenders with a nearly duplicated set of leg-bearing segments, is very likely a recent event.

The main message we can derive from this example is that the unusual number of trunk segments in *Sc. duplicata* has been very likely obtained through a veritable evolutionary "leap" (Minelli, Chagas, and Edgecombe 2009). However, exactly as originally suggested by Goldschmidt as the mechanism for phenotypic saltation, this major effect is quite probably due to a one-gene mutation in a population originally possessing the segment numbers (21 and 23) still found in *Sc. bahiensis*. This mutation—perhaps through a mechanism involving changes in cell adhesion properties—would have caused a splitting of each original segment into two, excluding perhaps one or two terminal, nonduplicated segments. Considering these two examples—and in the orchid case we have also detailed knowledge at the level of the genes involved in the changes in tepal shape and flower symmetry—it would be difficult to reject outright the idea of evolutionary saltation.

This is not so unorthodox a position in respect to neo-Darwinism as sometimes suggested. To be sure, the orchid and the scolopender story show that major, sudden phenotypic changes are indeed possible, and this seems opposite to what we should expect from a neo-Darwinian point of view, as neo-Darwinism views evolution as strictly the result of the steady accumulation of minor changes over a more or less prolonged time span (J. Huxley 1942; Maynard Smith 1993). However, at the level of the gene, changes are likely minor, even in these putative instances of saltational evolution, and this fits nicely with neo-Darwinism. As a consequence, an evo-devo perspective on these stories of major phenotypic change does not require a truly novel paradigm. What is actually required is not so much a novel view of evolutionary mechanisms, but the willingness, and the ability, to open the black box within which the neo-Darwinian tradition had confined development (Minelli 2010).

WILLISTON'S "LAW" AND THE NEED TO DEVELOP "INERTIAL" MODELS IN BIOLOGY

One of the major tenets of the neo-Darwinian view of evolution is that evolution, as a process, is exclusively microevolution. That is, macroevolutionary trends, such as the reshaping of vertebrate body accompanying the water to land transition, or the impressive novelties represented, for example, by the flower in the world of terrestrial plants, or the wings in the reptilian lineage, can and must be reduced to the long accumulation of many microevolutionary events without a special mechanism for macroevolution.

This basic principle notwithstanding, macroevolutionary theories resurface again and again in the evolutionary literature, especially from the work of palaeobiologists. Besides the singular and problematic examples of long-term evolutionary trends, such as the old textbook example of the horse family, more or less general patterns seem to emerge in macroevolution, for example in terms of histological heterogeneity (Bonner 2004) or of anatomical complexity (McShea 1991, 1996, 2001, 2005). More recently, with the dramatic increase in our knowledge of developmental genetics, another trend has been added to this list, the concept of a hypothetical steady increase in the complexity of the genetic architecture underlying development (Davidson 2006).

The recent evo-devo literature hosts frequent discussions of macroevolutionary trends, such as the so-called Williston's law.[3] According to this principle, the evolution of organisms with serially repeated parts, such as the body segments of earthworms or millipedes, the vertebrae of fishes and snakes, or the sepals/petals/tepals of flowers, should display an evolutionary history moving from an ancestral condition with a high and variable number of identical or nearly identical parts, to a derived condition where the corresponding parts are more diversified and occur in lower and less variable number.

Flowers with a very high number of involucral leaves, especially those where a distinction between sepals and petals is not so clear—a condition found, for example, in water lilies—are thus regarded as closer to the original angiosperm flower compared to flowers with exactly five sepals and five petals, as in wild roses and apple trees. Still more derived would be the condition of mints and peas, where the five petals are differentiated into three different types. Similarly, in the animal world, the marine polychaete *Eunice*, with hundreds of identical segments, would represent a more ancestral type of the Annelida compared to the leeches, with their fixed number of thirty-two body segments, among which five different body regions can be distinguished.

To be sure, fixing the number of serial elements, such as petals and stamens or teeth and body segments, requires a control mechanism additional to those through which an indeterminate number of parts can be generated. This is arguably true of sepals and petals, as it is true of mammal teeth or millipede segments. Regional specializations within a series would require still another form of control in addition to what is required to produce a fixed number of identical elements. Thus, the regionalized body of the leech requires specification, for example, of "sucker" identity for the seven most posterior elements among the series of thirty-two body segments. As a consequence, the open-ended series of elements regarded as primitive in terms of Williston's law represents a kind of default, or "inertial" condition, any deviation from it amounting to evolutionary change. This point is of potentially major relevance for the epistemology of biology. I will return to this point below.

Before leaving the case of Williston's law, however, it is fair to say that its validity is very frequently negated by facts. For example, the phylogenetic relationships among the major groups of centipedes suggest that at the root of the evolutionary radiation of the recent groups there was an animal with exactly fifteen pairs of legs, a condition still found in three major centipede lineages: Scutigeromorpha, Lithobiomorpha, and Craterostigmomorpha. Contrary to expectations based on Williston's law, the higher number of leg-bearing segments found in the two remaining lineages (Scolopendromorpha and Geophilomorpha) are derived from a fifteen-leg pair condition. In addition, the segments in the "basal" lineages with fifteen pairs of legs are less uniform than are the many segments (up to 191!) in the "derived" ones—another feature in blatant conflict with Williston's law. The main point to be made in the face of these facts is that any sensible interpretation of macroevolutionary trends requires an adequate understanding of the developmental mechanisms by which the relevant features are produced, and of their evolvability.

ZERO MODELS

Theoretical biology has largely distanced itself from previous naïve efforts to reduce life sciences to physics, but I do not see why biology should not follow physics in developing some zero models comparable, epistemologically, to the principle of inertia in classical mechanics. Framing questions about evolution or development in terms of changes in respect to a "default" or "inertial" condition would be a healthy move for the life sciences. Framing questions this way would avoid the need to ask questions about origins, which too often imply a finalistic explanation of processes, in developmental biology especially. Although seldom acknowledged (Gayon 1992; Minelli 2011a), "inertial" principles are not completely unheard of in biology. An excellent, effectively working example is the Hardy-Weinberg principle in population genetics. Our interest in the ideal population in Hardy-Weinberg equilibrium begins exactly at the time an actual population moves away from the equilibrium. I want to stress here that *the inertial conditions* (of a material point in classical mechanics or a population in evolutionary biology) *do not represent origins*. They are instead an arbitrarily chosen point of comparison required to study a segment of history. Whatever happened before that point is simply left out of consideration. If we are dealing with evolution, adopting such a "principle of inertia" translates into something like "let's study the modifications affecting a lineage of common descent along a given time interval." Analogously, in developmental biology, we will study deviations from a *local self-perpetuation of cell-level dynamics* (Minelli 2011a, 2011b).

It is indeed reasonable to accept the principle that the "inertial" condition of all living cells is proliferation (Soto and Sonnenschein 2004). This condition is most obviously proper to the early, embryonic phase of development, but adult stem cells also belong here. Other textbook examples of "diffuse inertial multiplicity" are the archaeocytes of sponges and the interstitial cells of hydra. These cells with "inertial" behavior are sometimes very abundant, even in the adult, as in the case of the freshwater flatworms—the planarians—whose stem cells or neoblasts can comprise up to 30 percent of the total cell number in the worm (Ellis and Fausto-Sterling 1997).

Indeterminate growth is a "default" consequence of the prolonged inertial behavior of cells in a developing organism. This condition is indeed widespread in the plant kingdom; it is much less common among animals, but many examples may be given, drawn from several bivalve molluscs, some sea urchins and decapod crustaceans, and even from vertebrates, especially among the bony fishes and the reptiles, plus a few mammals, such as bison, giraffe, and elephant (Karkach 2006).

But cell proliferation is only the most general of the many processes that can run in parallel from within a number of foci within a multicellular system. Evidence of multiple equivalent foci of development is provided by the generation of symmetrical bodies, or by the repetition of segments along an animal's main body axis. The most extreme phenomenon is arguably polyembryony, by which more than one embryo is eventually obtained by multiplying the growing and diferentiating units within the derivatives of a single zygote.

ORIGINS?

Evolutionary developmental biology is often depicted as the branch of biology entitled to address questions about the origin of evolutionary novelties. This may seem a sensible attitude, because novelties (however defined) only occur if they have a non-zero value in the landscape of evolvability of their would-be ancestors, and evolvability is exactly one of the main concepts on which evo-devo focuses attention. There is a problem, however, with this expectation. The problem is not so much in the objective difficulties we meet when we try to define an evolutionary novelty (Müller 1990; Müller and Wagner 1991, idem in Hall and Olson 2003), as it is in the need for caution when adopting the term "origin."

A lesson we can obtain from everyday experience is that all things and all processes last a finite span of time. As a consequence, they must all have had a beginning at a given instant in time and will end at a later instant. But this concept, so deeply rooted in common sense, is often difficult to apply if we want to determine in an unambiguous and sufficiently precise way when a given thing or a given process has actually gotten its start or achieved its end. This happens, for example, when we deal with living organisms, and especially with those belonging to our own species. The ensuing problems are relevant to the topic of this chapter inasmuch as they are in guiding our choices in the face of important questions of ethical, juridical, philosophical, and theological nature. When does a human being's life actually begin? At the very moment the nucleus of a sperm cell unites with the nucleus of an egg, thus first giving rise to a previously nonexistent genome? Or rather at the later time when, following an earlier phase completely under the control of mRNAs and proteins of maternal origin, the new, zygotic genome eventually starts being transcribed and translated? Or, perhaps, even later, when the foetal heart will start beating, or the growing brain will have reached a certain degree of differentiation? Or just at birth?

Answering these questions would be of fundamental ethical relevance, but is there any way a biologist may assist obtaining this answer? The more we have been learning about reproduction and development, the more articulated and precise has become the picture of the events eventually leading to the appearing on the world's scene of a new human being, the more problematic and uncertain the answer to those questions has become. For to be sure, it is not for biology, as a branch of science, to answer questions of an ethical nature. It is, however, its appropriate duty to assist with settling a point of the utmost importance: whether the very question of the existence of a precise point in time at which a new individual's life begins is, or is not, a legitimate and unambiguous question about the natural phenomena this science addresses. I think that the answer must be a negative one. In other words, biology, and science in general, can and must deal with all kind of change occurring in natural systems, but cannot deal with the category of origins, which arguably belongs to contexts other than natural sciences: it pertains to the domain of practical reason, besides its obvious place in the reflections of theology and philosophy.

Currently, biologists do not seem to be cautious enough in respect to framing questions in terms of origins. For example, the most frequent objections raised in respect to the title of Darwin's most famous book deal with the fact that that work does not deal much with species and, especially, with what we currently call speciation. More concern should be directed to the fact that searching for the origin of species, as such, is perhaps not a sensible, or even a legitimate question, especially if one believes, with Darwin, that no clear-cut differences exist between the "true" species and the "mere" variety. Nevertheless, it should not be difficult to realize that along the evolutionary line that eventually led to our species science has not found, and arguably will not find, a unique event of heterogeneous generation, by which a pair of nonhuman apes gave rise to offspring fully worthy of the Linnaean name of *Homo sapiens*. But, under such circumstances, is it still meaningful to speak of origins rather than framing evolutionary questions simply in terms of change?

An excellent example of the dubious utility of the category of origin in addressing questions of biology is provided by developmental biology with regard to the morphogenesis of the fruiting body of *Dictyostelium discoideum*. This is a kind of amoeba usually found in vegetative conditions as a unicellular organism living in the upper layers of the soil, where it feeds on bacteria. When food resources are locally exhausted, these amoebae enter a phase of aggregation. Two or three of them join together, then with other small cell clusters, until a multicellular mass quite similar to a minuscule slug is formed. This moves for a while and eventually stops, to give rise to a kind of tiny mushroom's fruiting body, within which some cells differentiate into spores and eventually disperse (Bonner 1967; Loomis 1982; Spudich 1987).

Of this unique story of aggregation and morphogenesis, the moment most relevant to our argument is exactly the beginning. How does this aggregation behavior originate? Whole research groups have been keenly looking into the molecular singularities that turn an ordinary, free-living amoeba into the founder of an aggregation. Which genes does that cell uniquely express? What aspects of its metabolism single it out among the many amoebae in the now fully exploited pasture? It turned out that all those research efforts were fated to fail, simply because in *Dictyostelium* there are no founder cells at all. All cells in this kind of amoebae produce cAMP (a cyclic form of adenosine monophosphate) and all of them are positively attracted by this substance. But bacteria produce greater amounts of this same molecule. Therefore the *Dictyostelium* amoebae are usually attracted by their food items rather than by their relatives. However, as soon as the last bacterium in the close surroundings has been phagocytized, the only local sources of cAMP will be other *Dictyostelium* amoebae. As a consequence, aggregation will start where one of these cells happens to be close enough to another cell to be stimulated by the cAMP released by it, and vice versa. In other words, aggregating cells are ordinary amoebae that only happen to come under the influence of the attractive molecules surrounding another cell, or a group of cells, of their own species. It is all a matter of the relative distance among them: the most closely clustered amoebae have the highest chance to aggregate. But there is no founding cell at all. In other words, there is no "origin" of the multicellular phase in the life cycle of *Dictyostelium discoideum*, if by this term we mean a unique event due to a distinct cause or with a distinct developmental mechanism.

THE ATTRACTION OF THE PRESENT

Palaeontologists have often remarked on the negative consequences of reconstructing history backwards, that is, starting from living species and looking back to their ancestors. For example, if we obey such an "attraction of the present," we risk losing sight of those innumerable lineages that have failed to leave descendants up to present time, and those that have escaped extinction until now. We are all too easily inclined to focus only on the putative ancestors of recent forms. Eventually, evolutionary biologists have realized the need to look at the organisms of the past not simply, or preferably, as the ancestors of the species living around us. That attitude would represent an unjustified and unproductive concession to finalism.

Things are different in developmental biology, within which the adult occupies, by definition, a unique position, as if it were the only genuine goal of development, whereas all other phases, starting from the egg (or the seed, or the spore) would be nothing else than preparatory stages through which the organism is somehow forced to proceed in its way to adulthood.

This erroneous perspective has arguably several causes. The most important is perhaps the fact that the adult is the developmental phase through which reproduction is accomplished. Therefore this is a phase that cannot be erased from an organism's life cycle, unless we want to bring about its immediate extinction. But this is the same as to say that continuity along a necklace is only provided by the clasp by which the two open ends can be firmly joined, thus ignoring the continuity provided by the thread passing all along the pearls. Another cause of the failure to appreciate the relevance of life cycle phases, other than the adult, is the usually higher structural complexity that an organism exhibits as an adult in comparison to other phases of its life cycle. In the end, however, the continuity of the species is nothing less than the perpetuation of the complete, cyclical succession of developmental stages, rather than the mere periodic occurrence of the adult stage, generation after generation.

EVOLVABILITY AND CATEGORIES

The notion of evolvability does not apply to phenotypes only. It also applies to the rules by which phenotypes evolve (this could be called *meta-evolvability*). As a consequence, the very existence of phenotypes corresponding to even the apparently most obvious categories cannot be taken for granted.

Consider, for example, the individual. This category is usually well suited to articulate a description of human beings and the other vertebrates, and also of many other animals and even plants. But this is not true of all living organisms generally. More important, from an evolutionary perspective, individuality is itself a product of evolution, along different lineages of descent, and also a condition that has often been lost and perhaps regained.

For example, how are we to apply the notion of individuality to a coral? Will we use it to describe one of the white flower-like polyps emerging from the red, branched structure of the coral, which in this case would qualify as a colony? Or will we rather say that the whole, tree-like structure is an individual, and the polyps are thus parts of it, more or less in the sense that a brain, a liver or a hand are parts of a human being?

There are also problems, although of quite different nature, in applying the notion of individual to organisms with complex life cycles. Think for example of a fly, from whose egg hatches a maggot that will eventually turn into an immobile pupa, from whose broken skin an adult fly will emerge in its turn. The question is whether we can simply assume that the maggot is the same individual as the fly.

To be sure, the problem is an old one. In antiquity, authors would have solved it by saying that two different animals are involved in this story, the pupa being the larger egg within which the fly is formed, but this large egg is not different in kind from the small egg within which the maggot is formed. Today, however, all biologists will probably follow T. H. Huxley's concept: "The individual animal is the sum of the phenomena presented by a single life; in other words, it is all those animal forms which proceed from a single egg taken together" (T. Huxley 1852, 188). This is indeed a reasonable position, but on closer inspection, it is more a circular statement than a true definition. It also does not apply to those many animals whose life cycle does not include an egg stage. At any rate, we cannot deny the material continuity through which the egg, the maggot, the pupa and the adult are linked in a sequence regularly repeated along the generations. Thus, rather than recognizing these stages as distinct individuals, we prefer to describe the structural changes along the sequence as the metamorphosis of one and the same individual.

This apparently safe solution leaves indeed many aspects unaddressed (Minelli 2009a). Too many things may happen during metamorphosis. In particular, many larval structures are frequently demolished or abandoned. Histolytic processes, for example, are really dramatic in the case of a fly, whose larval organs are largely destroyed during the pupal stage. Similarly conspicuous events occur during the metamorphosis of many marine invertebrates, for example in sea urchins and sea stars. In these animals, only a small fraction of the larval cells contribute to producing the adult body. This is clearly more than the one-cell contribution provided—through the egg or the sperm cell—by a sexually reproducing animal to generate each of its descendants, but there are examples of reproduction where the parent's contribution is more abundant, possibly not that different from the amount of larval cells contributing to forming an adult sea urchin. This is the case of the hydra and of other animals reproducing asexually by budding.

Thus, we should seriously question why we describe the metamorphoses of a tiny jelly pluteus larva into an adult sea urchin as a series of changes in the life of one individual, whilst the life cycles of many cnidarians (jellyfish), where a polyp gives rise asexually to a medusa and the medusa produces eggs from which new polyps will develop, are regarded instead as comprising two distinct generations.

The evolvability of individuality has likely passed through alternative times of narrow and broad possibilities. The distinct individuality of the unicellular ancestors of the animal lineage was largely lost when the multicellular, loose organization of sponges was obtained, but new forms of individuality evolved later, especially with the advent of bilaterian animals (the lineage to which vertebrates belong, together with insects, molluscs, and many other lineages). But even the emergence of the "new" multicellular organism has not remained unchallenged in evolutionary history, as demonstrated by the evolution of colonial forms, as the sea squirts or ascidians, and also by the nine-banded armadillo and the other species normally reproducing by polyembryony, including, most disquietingly, human Siamese twins.

THE EVOLUTION OF EVOLUTIONARY THEORY

The Darwinian bicentenary in 2009 has fueled often heated discussion about the status and future prospects of evolutionary theory in general, neo-Darwinism in particular. In this context of conceptual debate, evolutionary developmental biology has often played a central role. According to several authors, evo-devo represents a real alternative to the neo-Darwinism, in so far as it downplays the importance of natural selection as a constructive force in evolution (e.g., Stoltzfus 1999; Yampolsky and Stoltzfus 2001; Stoltzfus and Yampolsky 2009; Laubichler 2010). Others regard it as one among several novel approaches to evolution that demand a revised, expanded synthesis (Müller and Pigliucci 2010).

Evo-devo does not negate the central tenets of neo-Darwinism (Minelli 2010), unless we regard only this theory's extreme form, as represented for example by Richard Dawkins's writings, as the legitimate version of the theory (Dawkins 1986, 1996). We should not ignore indeed that, like any other theory, neo-Darwinism has shown a plurality of different facets throughout the time and in the views of its many representatives. It would be easy indeed to accommodate evo-devo, with its interest in "developmental genes" and its sensible pluralism in respect to the way the genotype maps onto the phenotype, within the early neo-Darwinism of a Julian Huxley (1942), or a Gavin de Beer (1930, 1940).

The most myopic attitude an evolutionary biologist may adopt would be to work with semantically frozen concepts in the framework of a dogmatically fixed theory. One of the major lessons of evo-devo is that what changes in living nature are not simply species and organismal traits, but also the rules according to which the traits are produced and changed, as well as the ways species split into daughter species. Change affects even those forms of biological organization we are accustomed to treat as given and virtually "out of the time," while these are in fact products of evolution. Many living forms are not obviously articulated into species, or distinctly recognizable as individuals. Species and individuals are products of history, done, undone, and redone as many times as historical contingency and developmental and ecological constraints permit. In face of this sea of flux, we must bravely address a wide revisitation of concepts and approaches if we want to establish on better footing our understanding of life.

NOTES

1. There are a few exceptions, such as the sloths and the sirenians (manatees and dugong). The range of variation in the number of cervical vertebrae remains nevertheless limited to the six to nine range; Minelli (2009b).

2. A single "monstrous" specimen with eighty pairs of legs has been recently found amongst other defective specimens in a Polish population of the geophilomorph centipede *Stigmatogaster gracilis*; Leśniewska et al. (2009).

3. This principle is usually credited to the paleontologist and entomologist Samuel Wendell Williston (1851–1918), but this is historically incorrect; Minelli (2003).

REFERENCES

Arthur, W. (1997). *The Origin of Animal Body Plans: A Study in Evolutionary Developmental Biology*. Cambridge: Cambridge University Press.
———. (2004). *Biased Embryos and Evolution*. Cambridge: Cambridge University Press.
Ayala, F., J. Ayala, and R. Arp, eds. (2010). *Contemporary Debates in Philosophy of Biology*. Malden, MA: Wiley-Blackwell.
Bonner, J. T. (1967). *The Cellular Slime Molds*. Princeton: Princeton University Press.
———. (2004). "The Size-Complexity Rule." *Evolution* 58:1883–90.
Carroll, S. B. (2005). *Endless Forms Most Beautiful: The New Science of Evo Devo and the Making of the Animal Kingdom*. New York: Norton.
Carroll, S. B., J. K. Grenier, and S. D. Weatherbee. (2004). *From DNA to Diversity: Molecular Genetics and the Evolution of Animal Design*. 2nd ed. New York: Norton.
Chagas, A., Jr., G. D. Edgecombe, and A. Minelli. (2008). "Variability in Trunk Segmentation in the Centipede Order Scolopendromorpha: A Remarkable New Species of *Scolopendropsis* Brandt (Chilopoda: Scolopendridae) from Brazil." *Zootaxa* 1888:36–46.
Coen, E. (1999). *The Art of Genes: How Organisms Make Themselves*. Oxford: Oxford University Press.
Cole, F. J. (1944). *A History of Comparative Anatomy: From Aristotle to the Eighteenth Century*. London: Macmillan.
Conway Morris, S. (2003). *Life's Solution: Inevitable Humans in a Lonely Universe*. Cambridge: Cambridge University Press.
Davidson, E. H. (2006). *The Regulatory Genome*. Amsterdam: Academic Press-Elsevier.
Dawkins, R. (1986). *The Blind Watchmaker*. New York: Norton.
———. (1996). *Climbing Mount Improbable*. Harmondsworth: Penguin.
de Beer, G. (1930). *Embryology and Evolution*. Oxford: Clarendon Press.
———. (1940). "Embryology and Taxonomy." In *The New Systematics*, ed. J. Huxley, 365–93. Oxford: Oxford University Press.
Ellis, C. H., and A. Fausto-Sterling. (1997). "Platyhelminths, the Flatworms." In *Embryology: Constructing the Organism*, ed. S. F. Gilbert and A. M. Raunio, 115–30. Sunderland, MA: Sinauer Associates.
Galis, F. (1999). "Why Do Almost All Mammals Have Seven Cervical Vertebrae? Developmental Constraints, Hox Genes and Cancer." *Journal of Experimental Zoology (Molecular and Developmental Evolution)* 285:19–26.
Gayon, J. (1992). *Darwin et l'après-Darwin: Une histoire de l'hypothèse de sélection naturelle*. Paris: Editions Kimé. English translation by M. Cobb, *Darwinism's Struggle for Survival: Heredity and the Hypothesis of Natural Selection*. Cambridge: Cambridge University Press, 1998.

Geoffroy Saint-Hilaire, I. (1832–37). *Histoire générale et particulière des anomalies de l'organisation chez l'homme et les animaux.* 4 vols. Paris: Baillière.

Gerhart, J., and M. Kirschner. (1997). *Cells, Embryos and Evolution.* Oxford: Blackwell Science.

Goldschmidt, R. (1940). *The Material Basis of Evolution.* New Haven, CT: Yale University Press.

Hall, B. K. (1999). *Evolutionary Developmental Biology.* 2nd ed. Boston: Kluwer.

Hall, B. K., and W. M. Olson, eds. (2003). *Keywords and Concepts in Evolutionary Developmental Biology.* Cambridge, MA: Harvard University Press.

Hendrikse, J. L., T. E. Parson, and B. Hallgrímsson. (2007). "Evolvability as the Proper Focus of Evolutionary Developmental Biology." *Evolution and Development* 9:393–401.

Huxley, J. S. (1942). *Evolution: The Modern Synthesis.* Repr. Cambridge, MA: MIT Press, 2010.

Huxley, T. H. (1852). "Upon Animal Individuality." *Proceedings of the Royal Institution* 1:184–89.

Karkach, A. S. (2006). "Trajectories and Models of Individual Growth." *Demographic Research* 15:347–400.

Kirschner, M., and J. Gerhart. (2005). *The Plausibility of Life: Resolving Darwin's Dilemma.* New Haven, CT: Yale University Press.

Laubichler, M. (2010). "Evolutionary Developmental Biology Offers a Significant Challenge to the Neo-Darwinian Paradigm." In Ayala, Ayala, and Arp 2010, 199–212.

Leśniewska, M., L. Bonato, A. Minelli, and G. Fusco. (2009). "Trunk Anomalies in the Centipede *Stigmatogaster subterranea* Provide Insight into Late-Embryonic Segmentation." *Arthropod Structure and Development* 38:417–26.

Loomis, W. F., ed. (1982). *The Development of* Dictyostelium discoideum. New York: Academic Press.

Maynard Smith, J. (1993). *The Theory of Evolution.* Cambridge: Cambridge University Press.

Mayr, E. (1960). "The Emergence of Evolutionary Novelties." In *Evolution After Darwin*, 3 vols., ed. S. Tax, 1:349–80. Chicago: University of Chicago Press.

McShea, D. W. (1991). "Complexity and Evolution: What Everybody Knows." *Biology and Philosophy* 6:303–24.

———. (1996). "Metazoan Complexity and Evolution: Is There a Trend?" *Evolution* 50:477–92.

———. (2001). "Parts and Integration: The Consequences of Hierarchy." In *Evolutionary Patterns: Growth, Form, and Tempo in the Fossil Record,* ed. J. Jackson, S. Lidgard, and F. K. McKinney, 27–60. Chicago: University of Chicago Press.

———. (2005). "A Universal Generative Tendency towards Increased Organismal Complexity." In *Variation: A Central Concept in Biology*, ed. B. Hallgrímsson and B. K. Hall, 435–53. Amsterdam: Elsevier.

Minelli, A. (2003). *The Development of Animal Form: Ontogeny, Morphology, and Evolution*. Cambridge: Cambridge University Press.

———. (2009a). *Perspectives in Animal Phylogeny and Evolution*. Oxford: Oxford University Press.

———. (2009b). *Forms of Becoming*. Princeton: Princeton University Press.

———. (2010). "Evolutionary Developmental Biology Does Not Offer a Significant Challenge to the Neo-Darwinian Paradigm." In Ayala, Ayala, and Arp 2010, 213–26.

———. (2011a). "A Principle of Developmental Inertia." In *Epigenetics: Linking Genotype and Phenotype in Development and Evolution*, ed. B. Hallgrímsson and B. K. Hall, 116–33. San Francisco: University of California Press.

———. (2011b). "Development, an Open-Ended Segment of Life." *Biological Theory* 6:4–15.

Minelli, A., A. Chagas, Jr., and G. D. Edgecombe. (2009). "Saltational Evolution of Trunk Segment Number in Centipedes." *Evolution and Development* 11:318–22.

Minelli, A., and G. Fusco, eds. (2008). *Evolving Pathways: Key Themes in Evolutionary Developmental Biology*. Cambridge: Cambridge University Press.

Müller, G. B. (1990). "Developmental Mechanisms at the Origin of Morphological Novelty: A Side-Effect Hypothesis." In *Evolutionary Innovations*, ed. M. H. Nitwecki, 99–130. Chicago: University of Chicago Press.

Müller, G. B., and S. A. Newman, eds. (2003). *Origination of Organismal Form*. Cambridge, MA: MIT Press.

Müller, G. B., and M. Pigliucci, eds. (2010). *Evolution: The Extended Synthesis*. Cambridge, MA: MIT Press.

Müller, G. B., and G. P. Wagner. (1991). "Novelty in Evolution: Restructuring the Concept." *Annual Reviews in Ecology and Systematics* 22:229–56

———. (2003). "Innovation." In Hall and Olson 2003, 218–27.

Raff, R. A., and T. C. Kaufman. (1983). *Embryos, Genes, and Evolution: The Developmental-Genetic Basis of Evolutionary Change*. Bloomington: Indiana University Press.

Scholz, G. (2008). "On Comparisons and Causes in Evolutionary Developmental Biology." In Minelli and Fusco 2008, 144–59.

Simmons, R. E., and L. Scheepers. (1996). "Winning by a Neck: Sexual Selection in the Evolution of Giraffe." *American Naturalist* 148:771–86.

Soto, A., and C. Sonnenschein. (2004). "The Somatic Mutation Theory of Cancer: Growing Problems with the Paradigm?" *BioEssays* 26:1097–1107.

Spudich, J. A., ed. (1987). Dictyostelium discoideum: *Molecular Approaches to Cell Biology.* Methods in Cell Biology 28. Orlando: Academic Press.

Steenstrup, J. J. S. (1845). *On the Alternation of Generation or the Propagation and Development of Animals through Alternate Generations.* London: Ray Society.

Stoltzfus, A. (1999). "On the Possibility of Constructive Neutral Evolution." *Journal of Molecular Evolution* 49:169–81.

Stoltzfus, A., and L. Yampolsky. (2009). "Climbing Mount Probable: Mutation as a Cause of Non-Randomness in Evolution." *Journal of Heredity* 100:637–47.

Theissen, G. (2009). "Saltational Evolution: Hopeful Monsters Are Here to Stay." *Theory in Biosciences* 128:43–51.

Wilkins, A. S. (2002). *The Evolution of Developmental Pathways.* Sunderland, MA: Sinauer Associates.

Yampolsky, L., and A. Stoltzfus. (2001). "Bias in the Introduction of Variation as an Orienting Factor in Evolution." *Evolution and Development* 3:73–83.

FIVE

ACCIDENT, ADAPTATION, AND TELEOLOGY IN ARISTOTLE AND DARWINISM

David J. Depew

THE ROLE OF CHANCE IN DARWINISM

Charles Darwin framed the *Origin of Species* to meet criteria for inductive science set out by John Herschel in his *Preliminary Discourse on the Study of Natural Philosophy* (Herschel 1830; Hodge 1977). Accordingly, he was distraught when he learned that Herschel, to whom he had sent a copy of his newly published book, was not persuaded. "I have heard by a round about channel," Darwin wrote to Charles Lyell, whose methodology in the science of geology Herschel had praised and that he took himself to be extending to biology, that "Herschel says my book 'is the law of higgledly-pigglety.' What exactly this means I do not know, but it is evidently very contemptuous. If true, this is a great blow and discouragement" (Darwin to Lyell, 10 December 1859, in Burkhardt et al. 1985–, 7:423).

It got worse. Misconstruing Darwin's greatly expanded sense of time, an incoherent interpretation of the *Origin* soon took hold in quarters hostile to evolution according to which natural selection is no better than the pounding of a large number of monkeys on a large number of typewriters over a period of time long enough to produce, by pure chance, *Hamlet, King Lear, Othello,* or the *Principia*

Mathematica of Newton. Not very likely! When Francis Bowen, a professor of natural theology at Harvard University, took this line, Darwin protested in a letter to his botanist correspondent Asa Gray, who was managing his case in America:

> It is monstrous . . . that [Bowen] should argue against the possibility of accumulative variation and actually leave out entirely selection. . . . The chance that an . . . improved pouter-pigeon should be produced by accumulative variation without man's selection is almost infinity to nothing; and so with natural species without selection. (Darwin to Gray, 26 November 1860, in Burkhardt et al. 1985–, 8:496)

Darwin had reason to feel discouraged. His theory did not equate natural selection with the mere persistence of coincidentally useful variations. It postulated the gradual shaping of heritable variations into adaptations by a process extending across many generations. At first, these variations merely happen to elevate the probability of survival and reproduction. But as they spread through an interbreeding population they do so precisely *because* they amplify and elevate these probabilities. They *become* adaptations.

To see how Darwinism construes adaptations, we should distinguish between the *process* of adaptation, *states* of (relative) adaptedness that make a trait *adaptive* in a certain population, and *traits* that are adaptations. These distinctions became clearer and more tractable when they were mathematically reformulated in terms of the probability revolution that was only beginning when Darwin wrote but subsequently transformed twentieth-century science (Hodge 1987; Gayon 1992; Depew and Weber 1995). Probabilistic reframing has allowed us to see that fitness is not reducible to discrete traits, mostly mean-spirited ones, that cause differential deaths under strict Malthusian competitive conditions. The overall fitness of organisms is a dispositional property, like the solubility of sugar in water. It is undergirded by myriad slight differences in the developmental process that in a particular selective environment enhance chances of reproductive success. It is identical to the relative adaptedness of closely related populations. It exists even when the carrying capacity of environments has not been reached (Mills and Beatty 1979; Brandon 1981; Sober 1984).

Darwin himself was not in a position to enjoy the ripe fruits of the probability revolution. Still, he was aware that there is an element of chance in whether variations that would be useful in a particular situation will be forthcoming and whether, even if they are available, they will actually result in enhanced survival and reproduction (C. Darwin 1859, 82; Hodge 1987). What he meant by calling variation "chance" in relation to natural selection, however, is that variation, whether its own causes are deterministic or chancy, must originally arise independently of its utility if it is subsequently to be shaped into an adaptation by the transgenerational selective process of adaptation (Beatty 2013). It was on this point that Darwin took himself to differ from Lamarck, especially Lamarck as interpreted by Lyell (Lyell 1830–33, vol. 2).

Unfortunately, formulations of Darwin's theory in which chance concurrences are presented more as causes than as conditions of organic structure and function persist to this day. They are especially common in two sorts of critics who are otherwise often at odds with each other: defenders of intelligent design and some mathematical physicists. Intelligent design advocates have an inclination to exaggerate the element of chance in Darwinism because intentional design is especially persuasive when it is presented as the only reasonable alternative to a succession of highly improbable coincidences. That was Bowen's tactic. It was not, however, the source of Herschel's complaint. He objected because he was a physicist. He could not recognize in the process Darwin described any laws of nature of the sort with which he was professionally familiar. The laws of physics apply to homogeneously described quantities, not wildly heterogeneous ones. That is what Herschel meant by "higgledy-pigglety."

There remains, however, another more challenging factor at work in characterizations of Darwinism as a theory of preserved accidents. Even when we have acquired a more accurate understanding of natural selection, Darwinism's appeal to chance denies that the history of life "from monad to man," as the phrase has it, consists of a process in which earlier stages unfold for the sake of later stages, with its usual implication that properties of lower organisms come to be for the sake of the uses to which higher organisms put them. Among the growing variety of evolutionary ideas that had been advanced since about 1800, the *Origin* was novel in proposing that environmental exigencies that impinge on developing organisms in what Darwin called "the struggle for existence" are strong enough to belie an intrinsic causal bond between ontogeny (the development of individual organisms) and phylogeny (the successive emergence of higher forms of life). In contrast to his evolutionary rivals, Darwin held that internal directional tendencies, on the basis of a strong analogy with the end-oriented character of individual development, play no role in explaining adaptive fit:

> It is preposterous . . . to say that, after a certain unknown number of generations, some bird had given birth to a woodpecker and some plant to the mistletoe. . . . This assumption seems to me to be no explanation, for it leaves the case of the co-adaptations of organic beings to each other and to their physical conditions of life untouched and unexplained. (C. Darwin 1859, 3-4)

To be sure, unlike his twentieth-century successors, Darwin thought that spasms of undirected variation are set off when an interbreeding population meets difficulties (Darwin, 1859, 43, 82; Hodge 1987). But which, if any, of these variations will be amplified in a concerted way by the process of adaptive natural selection was for him antecedently unpredictable and initially unrelated to whatever good effects result. Darwin did not deny a parallel between ontogeny and phylogeny. He recognized almost the same progressive phylogenetic order that other advocates of what the British called "the doctrine of development" observed (Richards 1992; Nyhart 2009). But he thought that this order is an *ex post facto* result of the fact that as life ramifies under the imperatives of struggle the difficulty of accessing new resources and so of opening up new niches will bias evolution's direction toward kinds that possess increasingly powerful means of exerting agency on their environments and, in supporting these abilities, are more structurally complex.

In Darwin's day, it was difficult even for his converts, not to speak of his opponents, to separate the etiology of adapted traits from the larger issue of evolutionary direction and purpose. In the absence of such a *telos,* the history of life seemed to many random, mindless, blind, and accidental. So it still seems to many today. Even when they are trying to be fair, contemporary intelligent design advocates, following the lead of Darwinians who causally privilege what happens at the molecular level, typically characterize Darwinism as "random variation plus natural selection," with the added implication that randomness prevails over selection in the evolutionary process to such an extent that according to Darwin and Darwinism we humans are nothing more than historical accidents.[1]

This is better than Bowen. It gives selection a role. But it is still a misrepresentation. Genetic variation may ultimately rest on unpredictable and seemingly causeless copying errors in DNA, but it is easy to confuse and conflate (1) randomness in this sense, and even more the quantum randomness that lurks below; (2) chance fixation of genotypes in populations by genetic drift; and (3) the chance that what turns out to be an adaptive variation will arise and present itself to natural selection when needed. As we have seen, the variation of which Darwin speaks in connection with natural selection is chancy in only the last of these three senses. The first two phenomena, being fruits of the probability revolution, were unknown to him. Since chance variation in Darwin's sense alone is relevant to the process of natural selection—not only as Darwin conceived it, but also as contemporary biologists and philosophers of biology do—formulations that speak of "random variation plus natural selection" invite subtle equivocations with the other senses. In so doing they invite us to view evolution as a whole as the result of a series of preserved accidents, with the implication that the human species itself is just such an accident. Theists who appeal to this supposed implication to register an objection to Darwinism can find support in a proclamation of so respected a physicist as Steven Weinberg, who remarks that evolutionary biology differs from physics insofar as it "depends on historical accidents" (Weinberg 1992, 42) rather than on laws of nature of the sort with which he, like Herschel, claims familiarity: universal laws that deductively subsume a homogeneously described array of instances and are in principle reducible to the laws of more basic sciences. Weinberg parts company with intelligent design proponents only in accepting that human beings are, in fact, accidents of history (245). Both, however, misunderstand Darwinian chance in its most basic meaning.

Why do they misunderstand it? My suspicion is that interpretations of Darwinism that speak of "random variation plus natural selection" project the element of contingency that is to one degree or another present in the overall course of evolution back into the core of the theory of natural selection itself. The notion supposedly found there that natural selection ranges over a sea of purely chance events in the sense of random errors in DNA is then reprojected outward onto evolution as a whole. This has the effect of exaggerating the accidental quality of evolution as a whole.

In what follows, I will attempt to redress this exaggeration. I will do so by speaking about biological teleology. First, I will identify a limited, but objective degree of what Kant called natural purposiveness in the process by which natural selection maintains adaptedness and brings forth adaptations. I will then suggest how this kind of natural teleology affects large-scale teleology. To make my points clearer I will at times compare Darwin to Aristotle, the father of biological teleology.

HUXLEY CONFOUNDED

Thomas Henry Huxley was devoted to spreading the scientific worldview through society. He championed Darwin to that end. In consequence, he opposed "fallacies . . . loudly expressed in the early days of the Darwinian controversy . . . that charge Mr. Darwin with having attempted to reinstate the old pagan goddess Chance" (Huxley 1887, 552–53). We have already encountered the interpretive tendencies to which he was referring. Huxley responded by taking a 180-degree turn. For Darwinism, he says, chance is merely ignorance of deterministic causes. As science progresses, it will be found that there is not a curve of the waves, not a note in the howling chorus, not a rainbow-glint on a bubble, which is other than a necessary consequence of the ascertained laws of nature; and that with a sufficient knowledge of the conditions competent physico-mathematical skill could account for, and indeed, predict every one of these so-called "chance" events. (Huxley 1887, 554–55)

In saying this, Huxley was reacting so strongly to exaggerations of the role of chance in natural selection that his Laplacean determinism undermined the concept of natural selection as badly as Bowen's appeal to the goddess Chance. Since Darwin, too, had an ignorance-based interpretation of chance, one might think that he agreed with Huxley that natural selection reduces to microphysics. In a letter to his pen pal and confidant Gray, Darwin guessed that God had set up laws governing matter and then left the rest to "what *we may call* chance" (Darwin to Gray, 22 May 1860, in Burkhardt et al. 1985–, 8:224, my italics). But with characteristic and admirable clarity about what was unclear to him Darwin never inferred from his implicit theory of probability or from his claim about initially designed laws that govern fundamental particles and forces what Huxley did, namely, that "molecules of . . . the primitive nebulosity of the universe" contain *in nuce* everything that subsequently evolved (Huxley 1887, 555). Darwin did not infer, that is to say, that the chance element in natural selection is ultimately non-chancy. On the contrary, he was probably groping toward something like the open, chancy, but still law-governed universe that Charles Sanders Peirce, upon reading the *Origin,* saw in it (without, however, Peirce's sense that the laws themselves evolve).

In our own time, the deterministic dream has been revived by molecular geneticists, who sometimes believe that in the genetic code they have at last found ways to "compute" the organism in ways comparable to physicists' ability to compute orbits (Gilbert 1993). Reframing Darwinism in terms offered by the probability revolution allows us to see why this dream is illusory. As the philosopher of biology Elliott Sober has shown, the properties over which natural selection and other determinants of evolutionary trajectories range, such as fitness or relative adaptedness, belong to, or at the very least are only visible in, organisms and genotypes viewed as members of populations whose reproductive rates are being compared. A Laplacean or Maxwellian demon whose eyes were trained only on the computable paths of individual atoms, molecules, genotypes, or even single organisms would fail to notice the causes that figure in evolutionary explanations, and even most of the phenomena to be explained (Sober 1984, 118–28). Even philosophers of biology who subscribe to reductionistic ideals in science have given up on the idea of reducing evolutionary to molecular biology in this way (Rosenberg 2007). Huxley was wrong to predict that the course of science would ultimately reveal that what his contemporaries thought of as chancy is actually fully determined.

Dubious, too, is Huxley's concession that theists might "without risk of contradiction" believe that an intelligence that computes the deterministic history of the universe might also have had an overall purpose in mind (Huxley 1887, 554). The more mechanist in the Laplacean sense one is, Huxley argues, the more one is also "at the mercy of the teleologist," who is free to say that a divine intelligence (presumably the same intelligence that laid down the law for the initial molecules that were so pregnant with evolutionary consequence) did so for a purpose. Who can say he didn't? On this view Darwinism is "agnostic"—a term Huxley invented—between theism and atheism (Huxley 1887, 555–56).

Huxley's clever, table-turning point is directed against the infinite monkey and typewriters argument. If natural selection reduces to pure chance, as some of Darwin's early misinterpreters took him to imply, Huxley's "wider teleology," as he calls it, would be impossible. The history of life would be too much of a jumble to make much sense at all. Determinism, by making the process intelligible in principle, removes this impediment. So theists should embrace it. The problem is that Huxley's determinism is so strong that it renders the teleology of which he speaks too weak to count as teleological at all. Admittedly, teleology has various meanings. However, the anti-Darwinian teleologists whom Huxley was seeking to mollify took it to be an explanatory, if not a fully causal, notion, not a mere mode of interpreting a cosmic process that *ex hypothesi* moves along entirely under the steam of deterministic mechanical causes. So should we. Teleology cites ends, purposes, goals, functions, intentions, and other such items as answers to questions about *why* a sequence of events unfolds as it does (Brandon, 1981, 1990). As such, teleology is more closely linked to causality than Huxley admits.

Huxley commends his "wider teleology" as better than the "old" teleology that "supposed that the eye . . . was made with the precise structure it exhibits for the purpose of enabling the animal which possesses it to see." He says that the old teleology "has undoubtedly received its death blow" at Darwin's hands (Huxley 1887, 554). My own view is that while a sound account of adaptation by natural selection will avoid exaggerating the role of chance, it will also evade reductionistic determinism enough for a Darwinian to affirm without equivocation that in point of fact the eye *has* come to be "for the purpose of enabling the animal which possesses it to see." Allow me support this claim by putting it in historical context.

BIOLOGICAL TELEOLOGY IN EMPEDOCLES AND ARISTOTLE: NATURAL SELECTION AS PROPERLY FINAL CAUSALITY

It was probably Huxley's widely circulated deterministic and reductionistic argument that accounts at least in part for Darwinism's reputation among Oxford and Cambridge intellectuals of the 1870s and 1880s as having revived ancient Greek materialism. In sending the aged Darwin a copy of his translation of Aristotle's *De Partibus Animalium*, for example, the Oxford classicist William Ogle playfully imagined a meeting between the author of the *Origin* and Aristotle. "I can fancy the old teleologist," Ogle wrote, "much astounded to find that . . . [the atomist] Democritus, whom he thought to have effectually and everlastingly squashed, had come alive again in the man he saw before him" (Ogle to Darwin, 17 January 1882; Cambridge University Library, Darwin Papers, DAR 173:10; see Gotthelf 1999).

Ogle might more aptly have cited Empedocles than Democritus as a suitable challenger to Aristotle. Empedocles was no less a materialist than Democritus, but he introduced more objectivity into our inescapable impression of the natural end-directedness of organic traits. By allowing spontaneity (*to automaton*) into the universe, he was able to regard claims that organic traits "save" (*esothê*) the organisms that happen to have them as objectively true, even if only accidentally caused, rather than as subjective projections onto an objectively deterministic universe, as Democritus (and Huxley) held. The story Empedocles tells is this. When cosmic history was more under the influence of love than of strife, various body parts were independently formed by the chemical affinities of the elements—so much air, so much water, so much fire, so much earth. This much is the work of necessity (*anankê*). These parts then spontaneously (*automatou*) conjoined and remain conjoined whenever in combination they just happened to be stable in, and so fit for (*epitêdeios*), the environments in which they sprang up. All the organisms we observe, Empedocles claims, are coalescences of parts "saved" in this chancy way. Other possible combinations never got off the ground or quickly died out (Empedocles, *Fragment* 57 DK; Aristotle, *Physics* II.8.198b29–32).[2]

This is indeed a selection process that involves but does not reduce to sheer chance. To that extent, Empedocles is closer than Democritus to Darwin. Beyond this similarity, however, there are large differences between Darwinian natural selection and Empedocles' appeal to what his critic Aristotle called incidental (*kata symbêbêkos, per accidens*) final (*hou heneka*) causality: causality in which an event that arises independently of a good outcome just happens to bring about that outcome. One key difference is that there is no appeal in Empedocles to differential reproductive success within kinds. This rules out the transgenerational etiology required for Darwinian adaptations, leaving Empedocles with the "frozen accident" view of the selection process from which I have from the outset of this chapter been distancing Darwin and Darwinism. On any neo-Empedoclean reconstruction of Darwinism, the first, accidental instance of a variant that is causally responsible for comparative reproductive success already counts as an adaptation. We have already seen, however, that on any coherent formulation of Darwinian theory adaptations lie at the end of a process of adaptation or *pari passu* along the way. In fact, the necessarily gradual emergence of adaptations, which enhance the capacities of populations of interbreeding organisms for metabolizing, growing, and reproducing, would count in Aristotle's terminology as proper (*kata auta, per se*), not as incidental (*kata symbêkêkos, per accidens*), final causality: as biologically teleological causality that conforms to criteria Aristotle himself championed, although it is not Aristotle's particular account (Lennox 1993, 1994; Depew 2008; Lennox and Kampourakis 2013).

Proper final causality is causality that runs through a process whose constituent moments, to the extent that something does not interfere, emerge as they do *because* they have a good effect—as in the case of the eye. To be sure, Aristotle would not agree with Darwin that ontogenetic orientation toward a developmental endpoint finds a counterpart in the evolutionary emergence of phylogenetically new traits. He did not think that large-scale, or perhaps even small-scale, evolution is even possible. Still, in Aristotle's technical terms Darwinian adaptations do have properly final causes. They reliably have certain effects and they come to be precisely because they have these good effects (Depew 2008; Lennox and Kampourakis 2013).[3] So Ogle's analogy between Darwinism and Democritus is unsound for the same reason Huxley's deterministic interpretation of natural selection is unsound. It would be only a little less misplaced if Ogle had imagined Empedocles instead of Democritus as reborn in Victorian England in the person of Charles Darwin.

DESIGN-WITHOUT-A-DESIGNER DARWINISM VS. CONSEQUENCE TELEOLOGY

Those who see natural selection as doing naturally what designers of machines and other artifacts do on purpose usually decline to regard "design without a designer" as teleological (Ghiselin 1984; Dennett 1995; Ruse 2003). For them teleology is by definition modeled on intentional, means-end acts. An Aristotle reborn into Victorian England would have taken exception to this restriction of teleology. For him there are two forms of intrinsically final causation: natural and intentional (Aristotle, *Physics* II.5.196b17–19). Biological teleology exhibits the first, not the second. Even if we accept for the sake of argument the antiteleological protestations of its partisans, however, the design-without-a-designer way of framing the process and products of natural selection inevitably gets the delicate relationship between chance variation, environmentally influenced selection, and end-directed adaptedness just wrong enough to invite objections from intelligent design advocates and their naturalistic enemies alike. Why is that?

The design-without-a-designer metaphor tacitly presumes that Darwinism picks out and explains the same kinds of phenomena to which intelligent design theorists refer. Both approaches are assumed to cover the same *explananda*. This explains why appeals to intelligent design and design-without-a-designer adaptationism tend to rise to prominence at the same time, most recently the 1980s and 1990s. In their contestation they ratchet each other up. But this contest has some bad side effects on naturalized design. For one thing, the analogy causes us to miss adaptive phenomena that are only visible through an evolutionary lens, such as mechanisms that preserve as well as use genetic variation, such as diploid chromosomes.[4] Lineages with these capacities are prevalent in evolutionary history because they enhance the adaptability and evolvability of populations (Dobzhansky 1970). Without evolution they would be invisible or unintelligible (Depew 2010).

For another thing, natural selection framed as design without a designer inclines its advocates to fix reference to a trait as an adaptation far too early in the process of inquiry. A designed trait is designed from its first appearance. Accordingly, Darwinians who work in terms of the naturalized design metaphor tend to identify traits, both morphological and behavioral, as adaptations by using the same criteria as natural theologians like Paley. Only afterward do they look for genetic changes in the composition of populations as causes of these presumed adaptations. Even when such correlations are found, it is usually difficult to tell whether a trait has arisen because of these genetic changes or as side effects of other changes or indeed by chance. It is a matter of presumption and burden of proof. Slighting causes other than directional adaptive natural selection is a proclivity that springs from crypto-creationist ways of establishing reference to phenomena to be explained. This premature way of identifying a trait as an adaptation has provoked critics to see in the so-called "adaptationist research programme" methodological biases and premature conclusions (Gould and Lewontin 1979; Lewens 2004, 29; Buller 2005; Richardson 2007). It may be replied that these critics are reading too much into the design metaphor. But this is a perilous rebuttal. It amounts to tacitly confessing that design-without-a-designer advocates do not have an analysis of adaptation at all, but, like Huxley, merely a way of commenting on it.

We have already identified the source of the difficulty. The first instance of an intelligently designed trait exists for the sake of its intended task as much as, or even more than, later instances. But this is decidedly not the case for any coherent version of Darwinism, which makes it a conceptual point that adaptations must result from a transgenerational process. Accordingly, seeing traits through the eyes of the metaphor of design without a designer, especially when coupled with an inclination to identify adaptations in the same way natural theologians do, encourages a subtle disposition to think of the first occurrence of a variation that just happens to be reproductively advantageous as already implicitly performing a functional role that in principle can be acquired only in later generations.[5]

In *Darwin's Dangerous Idea*, for example, Daniel Dennett, seeking a way to get selection going in a world whose natural state is random in the strong sense that I criticized in the first section, describes the first instance of a behavior that over multigenerational time is amplified into an adaptation as having "a limited amount of 'look-ahead'" (Dennett 1995, 8). To compensate for the unwanted teleological implications that wobbles of this sort invite—wobbles that give intelligent design advocates an opportunity to seize their chance—partisans of naturalized design usually go out of their way to stress, and sometimes to overstress, either the deterministic or the accidental aspects of evolution. Perhaps "forced moves" result in "arms races" that give the history of life an apparently progressive character (Dawkins 1986). Or perhaps we humans are accidents after all. Either way the correct balance among the three conceptual components of natural selection—chance, teleology, and the demands placed on organisms by environments—is lost.

For these and others reasons, most analytic philosophers of biology have opted for what is called the consequence-etiological, historical, or selected-effects analysis of what counts as an adaptation (Ayala 1970; Wright 1976; Brandon 1981, 1990; Neander 1991; Sterelny and Griffiths 1999; Lennox and Kampourakis 2013).[6] We have seen that for Darwinism the effects of initially chance variations gradually increase the relative adaptedness of populations that possess, transmit, and amplify these variations. The consequence-etiological analysis of adaptations that become over time entrenched in the ontogeny of organisms holds that their proper final cause, in Aristotle's sense, is a concerted series of past effects, all driven by ordinary efficient causality, that propel reproduction-enhancing traits through populations. In accounts of this sort—they do vary in details—the pull of the future that is overtly present in design teleology, and that reappears in mystified form in design-without-a-designer versions of natural selection, is entirely absent. The analysis is purely historical. Absent, too, is any appeal to intentions. For this reason, the selected effects or consequence-etiological analysis of adaptive natural selection neither exaggerates nor underestimates the interplay between chance, environmental pressure, and differential retention of variants in Darwinism. It gets them just right.[7]

Francisco Ayala may have been the first, but he was not the last, to argue that consequence-etiological explanations of adaptation by natural selection also count as teleological explanations in the sense of proper final causality that Aristotle urged against Empedocles (Ayala 1970; Dobzhansky 1970; Dobzhansky et al. 1977; Wright 1976; Brandon 1981; Lennox 1993, 1994; Depew 2008; Lennox and Kampourakis 2013). Consequence teleology has genuine historical-causal import without entangling itself in talk about intentions. Adaptations not only have certain reliable effects in the efficient conduct of organic life, but also come to be just because they have these effects. Having evolved in and through the selective process for the sake of realizing the fitness-enhancing effects that they bring into existence and sustain, they are teleologically caused and explained without having much, if anything, to do with the overall teleology that confused the issue in the elite culture in the nineteenth century and still confuses it in the popular culture of our day.

From a philosophical point of view, the consequence-etiological approach to biological teleology relieves the antinomy that Kant, who could think of natural teleology only on the model of intentional design, thought he saw in natural purposes. Specifically, it alleviates the problem of reverse causality. Kant could readily imagine how conscious representations of intended or designed future events could lead to their achievement. But he could not imagine how the end-states of the embryological process, for example, could influence the sequence of earlier events that bring functional traits into existence by natural, not intentional, means. The consequence-etiological view enables us to see that this problem is not as intractable as Kant thought. What has changed the situation is cybernetics. Adaptations come to be from selected past effects that feed forward into new generations by a combination of positive and negative feedback loops. This makes it possible to say that adapted traits come to be and to exist for the sake of their adaptive effects with as much conviction as Aristotle, who based his claim on a non-evolutionary metaphysics.

Reductionists, who tie explanation to underlying efficient-material causes, will reject this solution to the problem of biological teleology. They cannot imagine that the term "teleology" means anything other than intentional design and overall directedness. But we have already seen that evolutionary biology at its most scientific is resistant to their view because reductionism cannot achieve anything close to the right balance between chance, end-directedness, and environmental utility in analyzing the process of natural selection. "Design without a designer," as we have also seen, doesn't do much better.

BUT IS ETIOLOGICAL CAUSALITY REALLY TELEOLOGICAL?

I hear an insistent objection. Consequence teleology "isn't really teleology" (Walsh 2006). Whether this is true depends on what one thinks teleology "really" is and how it is related to final causes. The term itself goes back only to about 1730. It was devised by the philosopher Christian Wolff to refer to the providential nature of the pervasive system of mutual uses that connects various parts of the *scala naturae* (Wolff 1732). It originally referred, that is, to sallies of the sort satirized by Voltaire in *Candide*, such as that Providence has providentially provided us with nose bridges in order to support eyeglasses (Gould and Lewontin 1979). Treatises on natural theology, which reached their most elaborate development in eighteenth- and early nineteenth-century Great Britain, are full of arguments of this "prevision and provision" sort (Hodge 1982). We learn, for example, that coal was put into the secondary strata so that much later we would have fuel to burn (Buckland 1836, 497–98). This was originally called "the doctrine of final causes" because Wolffians ascribed other aspects of biology, including reproduction, to mechanical efficient causes. The point of teleology was to show that the hierarchical system of mutual uses among parts of the *scala naturae*, culminating in the way inanimate stuffs and lower kinds serve the higher purposes of human beings, is so finely tuned that it must be referred directly to a benevolent designing intelligence. In this way "final causes" and "teleology" became associated with intelligent design on a cosmic scale. It was evidenced only in phenomena that resisted efficient causal explanation. It was a "god-of-the-gaps" theory. This is, I believe, what teleology connotes even today. It is the unfortunate source of biology's entanglement with theology. So the objection that consequence teleology isn't real teleology is not to be dismissed lightly.

There is, however, more to be said. Wolff's privileging of intentionally designed external use over the end-orientation of the ontogenetic or development process bothered Kant. He responded not by rejecting the doctrine of teleology, but by turning it upside down. By adopting a finally caused epigenetic rather than an efficiently caused preformationist view of reproduction, Kant, in his *Critique of Judgment*, made end-directed organic self-formation or ontogeny the primary locus of natural teleology, reduced mutual uses to mere effects of efficient causal chains, and treated the pervasive, mutually supportive web of linkages in the system of nature as a whole as a matter of reflective, not determinate, judgments. By this he meant that describing ecological webs as if they were purposive is the best, perhaps the only way, of identifying and beginning to untangle the chain of material and efficient causes that undergird them (Kant 1790).

Darwin, too, flatly rejected Wolffian teleology of the sort that was pervasive in the natural theology of his day. "If it could be proved that any part of [an organism's] structure had been formed for the exclusive good of another species," we read in the *Origin*, "it would annihilate my theory" (C. Darwin 1859, 201). Still, Darwin himself sometimes traced the etiology of adaptations to what he called final causes (Lennox 1993, 1994, 1998). Moreover, on at least one occasion he was willing to approve using the word "teleology" to refer to explanations of finally caused adaptations. In rebutting his Harvard colleague Bowen's accident-based misinterpretation, Darwin's American friend Gray claimed that "Darwin's great service to natural science is bringing back to it teleology; so that instead of morphology versus teleology, we have morphology wedded to teleology" (Gray 1963, 237).[8] Darwin replied, "What you say about teleology pleases me especially" (Darwin to Gray, 5 June 1874, in F. Darwin 1887, 2:267).

How was this possible? The answer is that on this occasion Darwin was using the term "teleology" not in the Wolffian-Voltairean-Paleyean sense, but as a way of referring to the pervasive internal purposiveness that Aristotle, Kant, and the Kant-inspired comparative anatomist Georges Cuvier saw in the relation between each species' integrated structure and its unique "conditions of existence" (Reiss 2009).[9] This "internal-developmental" sense of teleology had been brought to Britain by William Whewell, whom Gray was actually quoting (without citation) when he spoke of Darwin as showing how "morphology [is] wedded to teleology."[10] If William Ogle, the classicist whom we earlier glimpsed sending Darwin his translation of *De Partibus Animalium,* had known as much as we now know about Aristotle's biological treatises he would have seen that "the father of teleology" himself was not much less insistent than Darwin that species cannot acquire traits oriented to the good of other species by any form of properly final, that is teleological, causality. Aristotle certainly recognized that members of each species play systematically supportive roles in one another's flourishing. But for him this system of mutual uses is an effect, not a cause, of each organism's coming to be for the sake of realizing its distinctive range of species-specific psychic capacities (Lennox 2001; Falcon 2005, *contra* Sedley 2007).

To be sure, Aristotle thought this system is good for all concerned. This good is all the more realizable and all the more explainable because it unfolds against a metaphysical backdrop in which Being is inherently a plentitude that stretches from pure mind to pure matter. For Aristotle in consequence, emergence is unnecessary for the same reason that reductionism is impossible. Even so, Aristotle (if not his English, or even his Latin, translators) distinguishes between coming to be for the sake of a good proper to oneself (*heneka* + genitive) and, having done so, serving the good of another (*heneka* + dative) (Kullmann 1998). It would seem that by Wolff's standards Aristotle, far from being the "father of teleology," was not a teleologist at all!

What had happened is that by Darwin's time Wolff's sense of the term no longer monopolized its meaning. Indeed, except where the argument from design continued to play an ideological role in public discourse, as in Great Britain and America—and as a foil in the *Origin of Species*, which was addressed to this very public (Depew 2009)— teleology of the Aristotelian-Kantian-Cuvierian sort had become the dominant technical meaning of the term among experts, including Darwin. It referred to the internal final causality involved in epigenetic self-formation, which Darwin thought of as modified by natural selection and other processes that affect ontogeny. That is why for Darwin embryology was an important test for and source of evidence for his theory. It is probably true that Darwin used "teleology" and "final cause" in several senses. Still, design-without-a-designer Darwinians have overlooked the important fact that he sometimes used them in Cuvier's and Whewell's sense. Writing in opposition to direct descendants of British natural theology, design-without-a-designer Darwinians treat intelligent design as the fixed, unproblematic meaning of teleology and in consequence as Darwinism's whipping boy (Ghiselin 1984; Ruse 2003). But by treating teleology in this univocal way they ignore highly relevant historical aspects of the term.

These historical reflections on the meanings of "teleology" are relevant to our discussion not only because Darwin himself sometimes thought this way, but because the epigenetic view of development and evolution is once again coming to the fore in evolutionary science. Natural selection properly so-called is a process that has adaptive effects only in reproductively linked developmental systems (Oyama, Griffiths, and Gray 2001; Weber and Depew 2001; Depew 2010). Adaptive natural selection is not what Dennett calls "substrate neutral" (Dennett 1995, 82). It cannot occur in silicon-based computers as well as in hydrocarbon-based organisms. It is not an algorithm. The variation on which natural selection depends consists in environmentally induced changes in rate, placement, and intensity in the developmental process (Gilbert and Epel 2009; Pigliucci and Müller 2010). The retention of some of these variations in the selective process affects the life histories of organisms considered as wholes rather than as collections of separately aggregated and adaptively optimized parts. Organisms so construed are deeply immersed in dynamic ecologies, where they serve as nodal points for the transfer of matter, energy, and information (Gilbert and Epel 2009; Ulanowicz 2000).

Let me draw the threads of this chapter together by asserting that design-without-a-designer versions of Darwinism inevitably open themselves up to their theological shadow selves just because they aren't really natural enough to capture the nature of organisms as ecologically embedded developmental processes. Intentional teleology shadows any version of Darwinism in which organisms are thought of as assemblies of machine-like modules. This conceit carries with it an aggregative conception of organic composition that echoes both the design paradigm against which it protests too much and Empedocles' account of the agglomeration of already individuated hands, legs, heads, and other body parts. The latter, as we saw earlier, cannot correctly represent the logic of Darwinian explanations at all. By contrast, the consequence etiological analysis of adaptation is a fully naturally teleological process in virtue of its entirely *ex post facto* but at the same time end-oriented character. Moreover, this way of analyzing biological teleology is at its most explanatorily powerful when it ranges over organisms conceived of as epigenetically developmental processes in the way Aristotle, Kant, Cuvier, and Darwin conceived of them rather than as analogues of designed objects.

What remains, of course, is to ask whether consequence teleology so conceived can or should be extended to the course of evolution as a whole. The objection that consequence etiology is not really teleological rests squarely on the presumption that "real" teleology is progressive and directional on a cosmic scale rather than on the issue of whether specific phenomena are in some sense teleological if they are adaptations.

In addressing this question, we must, if we are to remain true to the purely *ex post facto* character of consequence teleology, avoid the old, nineteenth-century view of progressive evolution to which many theologians who embrace evolution still cling. As Peter Bowler has shown, post-Darwinian elites, including many clerics, made themselves comfortable with macroevolution by convincing themselves that the evolutionary process preserves much of the premodern, anthropocentric *scala naturae* (Bowler 1988, 2013). They simply regarded the notion that lower beings emerge for the sake of human beings as an unfolding diachronic, rather than an eternally synchronic, affair. Upon closer inspection, however, this representation actually retains the external Wolffian teleology to which its advocates were already accustomed. It simply disguises use teleology by dressing up phylogeny as internal teleological ontogeny on a grand scale.[11] I agree with Bowler that this picture of teleology is inherently non-Darwinian. Contrary to core Darwinian principles, the picture assumes inner tendencies that pull ontogeny toward greater complexity, autonomy, mentality, or other such valued good things. Its advocates too readily comfort themselves either by misinterpreting natural selection or by downplaying its role in the evolutionary process. Pierre Teilhard de Chardin, for example, retains rather large doses of Bergsonian metaphysics, Lamarckian inheritance, and orthogenesis.

Small wonder that twentieth-century Darwinians have been hostile to this approach—so hostile that at times they have overreacted to it by abandoning the developmental locus of evolutionary dynamics altogether and by asserting that population-genetic evolution exhibits no inherent directional tendencies at all. It is possible, however, to be genuinely Darwinian while thinking of living systems as energy-using, entropy-dissipating, partially self-organizing developmental processes that, as they evolve by consequence-teleological feedback, are bounded from the environments on which they are nonetheless dependent. The trick is to see the biosphere, indeed the whole earth, in as processive and probabilistic a way as Peirce, for one, saw it and, in addition, to cease thinking, as perhaps Peirce was not yet able to do, about entropy in the depressing and as it turns out false way that took hold in the later nineteenth century. Order in our part of the universe is no less probable than disorder and, as Stuart Kauffman has said, some of it comes for free (Kauffman 1993). Having made this perceptual shift, one will no longer view evolutionary change as a tacitly nonnatural uphill battle, thereby rendering it unnecessary to invoke either self-interested competition ("arms races") or benevolent "prevision and provision." In viewing evolution this way it may be wise to restrict natural teleology, analyzed in terms of consequences, to adaptive natural selection working on genetic and epigenetic variation. But it will be no less wise to recognize that natural selection so construed takes place in a world full of other feedback-driven processes that are just as irreversible, directional, and organization-generating.

It will also become easier in this light to discern the foundations of what Darwin called "the co-adaptation of organic beings to each other and to their physical conditions of life" (C. Darwin 1859, 3–4). Darwin, we recall, saw a great vision of ecological mutualism precisely when as a young man he let go of inner drives toward phylogenetic diversification. The interpretation of Darwinism toward which I am pointing has the added advantage of the anthropocentrism that lies at the heart of the defective view of evolution that gave rise to modern racism with an ethic of care in which, with Darwin, we can discern and appreciate the kinship of all living things and the unity of our species (Desmond and Moore 2009). It allows us to contemplate the emergent biosphere with far greater theological equanimity than we can look on the barren Laplacean scene conjured up by Huxley. Theologians do not have to add teleology to nature. A good deal of it is already there.

NOTES

This chapter is a slightly amended version of Depew 2011 and is reprinted with permission of the Gregorian and Biblical Press. I have changed the wording here and there and added some recent references. My thanks especially to Phillip Sloan for facilitating the original paper, arranging for its reappearance in the present volume, and helping me edit the text of the present version. My thanks, too, to my dialogue partner and frequent co-author Bruce H. Weber and to James Lennox for helping me develop these ideas. The mistakes are my own.

1. From its origin the version of creationism that resulted by the 1990s in what is called the theory of intelligent design carried the implication that "all the adaptive design of life has resulted from a blind random process" and bears the marks of that randomness (Denton 1986, 351). The formula "random design plus natural selection" is still routinely used by writers associated with the Discovery Institute, which promotes intelligent design, to suggest that in the absence of overt design human life must be a cosmic accident devoid of meaning. One aim of this chapter is to suggest that this underestimation of the telic quality of natural selection's products and of the geophysical processes from which selection arises results from a question-begging assumption that biological teleology depends for its very meaning on intentional design, which comprises not only the uses that organisms make of their own parts but also that species make of each other. The idea that humans are at the top of the *scala naturae* depends on the latter inference.

2. For Empedocles, apt combinations of body parts are inherited in male and female seeds, each preserving only part of an apt organic structure. The separate halves combine in reproduction to yield a whole organism. Empedocles probably regarded organisms as ontologically identical with these seeds, which expand in the course of growth, contract in the generative cycle, and then expand again. In addition to holding that even this hypothetical process, let alone the real one, would be too marvelous to be an accident, Aristotle objected that Empedocles' doctrine denied that organisms actually come to be at all; as opposed to the elements, they are not, *ex hypothesi*, independent substances for Empedocles, or indeed Democritus (Aristotle, *Physics* 2.8.198b29–32; Depew 2010, 94). For Aristotle, on the other hand, organisms are paradigmatic substances. See Sauvé Meyer 1992.

3. Aristotle retains the merely incidentally final bond between the origin of a trait and its life-enhancing effect that he ascribes to Empedocles in the etiology of simple organic kinds. He calls them spontaneously generated (*automatou*) because they exhibit something like Empedocles's meaning of that term as accidentally final (Depew 2010).

4. The diploid chromosome at one and the same time randomizes genetic variability, preserves it for future evolvability, and reliably passes on to the next generation genetic combinations that confer enhanced fitness or reproductive success on their possessors. In saying that "Nothing in biology makes sense except in the light of evolution," Theodosius Dobzhansky might well have cited the chromosome as an *explanandum* that doesn't make sense except in relation to evolution. A designer god could scarcely have created it for any other purpose. Such a god must be an instigator of evolution, not an alternative to it.

5. Lewens (2004, 47–50) too traces the difficulty of premature identification of adaptations to problems in the analogy between organism and artifacts, although in general I seem to be harder on the comparison. The attraction of the analogy, and so, too, the reason its weaknesses were long obscured, may historically have something to do with the fact that, in the later nineteenth century, natural selection was often conceived by those who called themselves Darwinians as eliminating all organisms except the ones that were, from the start, the fittest or most adapted. This misunderstanding of the role Darwin assigned to natural selection was a result of Spencer's cooptation of Darwinism by way of their shared appeal to Malthusian population pressure as the force that drives selection. On Spencer's interpretation the fittest are fit from the first generation; time merely makes this plain by trimming away all the other "types." After August Weismann's restriction of heredity to the germ line was confirmed by Mendelian genetics in the early twentieth century, the conviction that mutation, rather than selection or, as Spencer thought, use-inheritance, is the "creative factor" in evolution allowed the assumption that natural selection is a merely eliminative force, and with it the presumption that fit organisms are fit or unfit from the outset, to persist even when Malthusianism was no longer as powerful a thread in the argument. The error was not corrected until the work of R. A. Fisher in the late 1920s and 1930s; even the pioneer geneticist Thomas Hunt Morgan shared it (Morgan 1932). The claim of the Modern Evolutionary Synthesis to be restoring and defending Darwin's original insight into the creativity of natural selection depends on the pervasive availability of variation in natural populations needed to gradually shape adaptations. That is why this issue has for decades been central to debates about evolutionary theory.

6. Adaptations are often viewed as biological functions played in an evolutionary key. This hasty identification has been a comfort to those who reject the consequence etiological analysis of adaptations. Didn't physicians like William Harvey know how to recognize the biological function of the heart without knowing about evolution? The identification of adaptations and functions is also one of the sources of the temptation to think that the *explananda* of evolutionary biology are the same as those of natural theology. How to analyze biological functions seems to me a different issue than the

consequence etiological analysis of adaptations. My point in this chapter is restricted to the latter.

7. To be fair, consequence etiology generates some problems of its own. One of them is the difficulty of finding the information needed to identify and confirm them, much of which is lost in time. I don't regard the difficulty as so insurmountable that it renders the analysis of adaptation suspect. No less insistent is a related question, namely, whether only traits currently or recently maintained by selection should be called adaptations or whether the term should be extended, or even restricted, to traits that were maintained by selection in the past but are no longer adaptive in some or all environments. See Sterelny and Griffiths 1999, 217–20, for an excellent discussion of the issue.

8. Gray understood natural selection so well that he was forced to add *ad hoc* stipulations in order to remain faithful to his theological commitment to evolutionary progressivism. The most well-known of these expedients is his speculation that if adaptive natural selection cannot be intelligently directed perhaps God directs the potentially useful stream of variations from which selection evolves adaptations (Gray 1963, 122, 308–10). Darwin appreciated Gray's grasp of adaptation by natural selection. But he rejected directed variation (Darwin to Gray, 5 June 1861, in Burkhardt et al. 1985–, 9:162). I suspect that his decades-long argument with Gray led Darwin to his ever more intense appreciation of evolution's contingencies and toward his confessed withdrawal from a religious view of the world.

9. Reiss (2009) shows that Darwin accepted but watered down Cuvier's concept of "conditions of existence" by treating these conditions as external, environmental facts more than as internal relations among an organism's parts. Cuvier's main idea remained sufficiently intact, however, to inspire Whewell's remark about teleology; the paragraph in which he makes the remark is about Cuvier and Geoffroy. Whether Gray was using the term "teleology" in Cuvier's sense is open to doubt. Still, echoes of it exist in the main claim of his later essay on "Evolutionary Teleology," namely, that Darwinism distributes part-whole support across whole ecosystems rather than to organisms (Gray 1963).

10. Huxley, in the deeply sophistical argument about teleology that I criticized earlier (Huxley 1887), tried to interpret Whewell's remark about "morphology wedded to teleology," of which Darwin approved, as entangling the great man in his own Laplacean fantasy. There is no evidence whatsoever that Darwin had this view in mind.

11. This way of thinking goes back not to Aristotle but to the Stoics, whose sense of the world as a single living substance denied the distinction between internal and external teleology. For Stoicism all uses are as internal as they are external. Judeo-Christian-Islamic creationism supervened on this picture, undermining internal with explicitly external design teleology.

REFERENCES

Ayala, F. (1970). "Teleological Explanations in Biology." *Philosophy of Science* 37:1–15.

Beatty, J. (2013). "Chance and Design." In *The Cambridge Encyclopedia of Darwin and Evolutionary Thought,* ed. Michael Ruse, 146–51. Cambridge: Cambridge University Press.

Behe, M. (1996). *Darwin's Black Box: The Biochemical Challenge to Evolution.* New York: Free Press.

Bowler, P. (1988). *The Non-Darwinian Revolution.* Baltimore: Johns Hopkins University Press.

———. (2013). *Darwin Deleted: Imagining a World Without Darwin.* Chicago: University of Chicago Press.

Brandon, R. (1981). "Biological Teleology: Questions and Explanations." *Studies in the History and Philosophy of Science* 12:91–105.

———. (1990). *Adaptation and Environment.* Princeton: Princeton University Press.

Buckland, W. (1836). *Geology and Mineralogy with Reference to Natural Theology.* London: Pickering.

Buller, D. J. (2005). *Adapting Minds: Evolutionary Psychology and the Persistent Quest for Human Nature.* Cambridge, MA: MIT Press.

Burkhardt, F., S. Smith, et al., eds. (1985–). *The Correspondence of Charles Darwin.* Cambridge: Cambridge University Press.

Darwin, C. (1859). *On the Origin of Species.* 1st ed. London: John Murray.

Darwin, F., ed. (1887). *Life and Letters of Charles Darwin.* Amer. ed. in 2 vols. New York: Appleton and Co.

Dawkins, R. (1986). *The Blind Watchmaker.* New York: W. W. Norton

Dennett, D. (1995). *Darwin's Dangerous Idea.* New York: Simon & Schuster.

Denton, M. (1986). *Evolution: A Theory in Crisis.* Bethesda, MD: Adler and Adler.

Depew, D. (2008). "Consequence Etiology and Biological Teleology in Aristotle and Darwin." *Studies in the History and Philosophy of the Biological and Biomedical Sciences* 39:379–90.

———. (2009). "The Rhetoric of Darwin's *Origin of Species.*" In *The Cambridge Companion to "The Origin of Species,"* ed. M. Ruse and R. Richards, 237–55. Cambridge: Cambridge University Press.

———. (2010). "Adaptation as Process: The Future of Darwinism and the Legacy of Theodosius Dobzhansky." *Studies in the History and Philosophy of Biology and the Biomedical Sciences* C 42:89–98.

———. (2011). "Accident, Adaptation, and Teleology in Aristotle, Empedocles, and Darwinism." In *Biological Evolution: Facts and Theories; A Critical Appraisal 150 Years after "The Origin of Species,"* ed. G. Auletta, M. LeClerc, and R. A. Martinez, 461–78. Rome: Gregorian and Biblical Press.

Depew, D., and B. Weber. (1995). *Darwinism Evolving: Systems Dynamics and the Genealogy of Natural Selection.* Cambridge, MA: MIT Press.

Desmond, A., and J. Moore. (2009). *Darwin's Sacred Cause.* Boston: Houghton Mifflin Harcourt.

Dobzhansky, T. (1970). *Genetics of the Evolutionary Process.* New York: Columbia University Press.

Dobzhansky, T., F. Ayala, G. L. Stebbins, and J. Valentine. (1977). *Evolution.* San Francisco: Freeman.

Falcon, A. (2005). *Aristotle and the Science of Nature.* Cambridge: Cambridge University Press.

Gayon, J. (1992). *Darwin et l'apres Darwin: Une histoire de l'hypothèse de selection naturelle.* Paris: Editions Kimé. English translation by M. Cobb. *Darwinism's Struggle for Survival.* Cambridge: Cambridge University Press, 1998.

Ghiselin, M. (1984). "Introduction." In C. Darwin, *The Various Contrivances by Which Orchids Are Fertilized by Insects.* Chicago: University of Chicago Press.

Gilbert, W. (1993). "A Vision of the Grail." In *The Code of Codes,* ed. D. Kevles and L. Hood, 83–97. Cambridge, MA: Harvard University Press.

Gilbert, S., and D. Epel. (2009). *Ecological Developmental Biology.* Sunderland, MA: Sinauer Associates.

Gotthelf, A. (1999). "Darwin on Aristotle." *Journal of the History of Biology* 32:3–30.

Gould, S. J., and R. C. Lewontin. (1979). "The Spandrels of San Marco and the Panglossian Paradigm: A Critique of the Adaptationist Programme." *Proceedings of the Royal Society of London,* Series B, 205, no. 1161:581–85.

Gray, A. (1963). *Darwiniana.* Cambridge, MA: Harvard University Press. Orig. pub. New York: Appleton, 1876.

Herschel, J. (1830). *Preliminary Discourse on the Study of Natural Philosophy.* London: Longmans, Rees, Orme, Brown and Green.

Hodge, M. J. S. (1977). "The Structure and Strategy of Darwin's 'Long Argument.'" *British Journal for the History of Science* 10:2437–46.

———. (1982). "Darwin and the Laws of the Animate Part of the Terrestrial System (1835–1837): On the Lyellian Origins of His Zoonomical Explanatory Program." *Studies in the History of Biology* 6:1–106.

———. (1987). "Natural Selection as a Causal, Empirical, and Probabilistic Theory." In *The Probabilistic Revolution,* ed. L. Kruger, G. Gigerenzer, and M. Morgan, 2:233–70. Cambridge: Cambridge University Press.

Hull, D., and M. Ruse, eds. (2007). *The Cambridge Companion to the Philosophy of Biology*. Cambridge: Cambridge University Press.

Huxley, T. (1887). "On the Reception of the *Origin of Species*." In F. Darwin 1887, 1:533–58.

Kant, I. (1790). *Kritik der Urteilskraft*. In *Kants gesammelte Schriften*, ed. Deutschen Akademie der Wissenschaften, 5:165–485. Berlin: Reimer, 1908–13.

Kauffman, S. (1993). *The Origins of Order: Self-Organization and Selection in Evolution*. New York: Oxford University Press.

Kullmann, W. (1998). *Aristoteles und die Moderne Wissenschaft*. Stuttgart: Franz Steiner Verlag.

Lennox, J. (1993). "Darwin *Was* a Teleologist." *Biology and Philosophy* 8:405–21.

———. (1994). "Teleology by Another Name: A Reply to Ghiselin." *Biology and Philosophy* 9:493–95.

———. (1998). "Teleology." In *Keywords in Evolutionary Biology*, ed. E. F. Keller and E. Lloyd, 324–33. Cambridge, MA: Harvard University Press.

———. (2001). *Aristotle's Philosophy of Biology*. Cambridge: Cambridge University Press.

Lennox, J., and K. Kampourakis. (2013). "Biological Teleology: The Need for History." In *Philosophy of Biology: A Companion for Educators*, ed. K. Kampourakis, 421–54.

Lewens, T. (2004). *Organisms and Artifacts*. Cambridge, MA: MIT Press.

Lyell, C. (1830–33). *Principles of Geology*. 3 vols. London: John Murray.

Matthen, M., and A. Ariew. (2002). "Two Ways of Thinking about Fitness and Natural Selection." *Journal of Philosophy* 99:55–83.

Mills, S., and J. Beatty. (1979). "The Propensity Interpretation of Fitness." *Philosophy of Science* 46:263–86.

Morgan, T. H. (1932). *The Scientific Basis of Evolution*. New York: W. W. Norton.

Neander, K. (1991). "Functions as Selected Effects." *Philosophy of Science* 58:168–84.

Nyhart, L. (2009). "Embryology and Morphology." In Ruse and Richards 2009, 194–215.

Oyama, S., P. Griffiths, and R. Gray, eds. (2001). *Cycles of Contingency: Developmental Systems and Evolution*. Cambridge, MA: MIT Press.

Pigliucci, M., and G. Müller, eds. (2010). *Evolution: The Extended Synthesis*. Cambridge, MA: MIT Press.

Reiss, J. (2009). *Retiring Darwin's Watchmaker*. Berkeley: University of California Press.

Richards, R. (1992). *The Meaning of Evolution*. Chicago: University of Chicago Press.

Richardson, R. (2007). *Evolutionary Psychology as Maladapted Psychology*. Cambridge, MA: MIT Press.

Rosenberg, A. (2007). "Reductionism (and Anti-reductionism) in Biology." In Hull and Ruse 2007, 120–38.

Ruse, M. (2003). *Darwin and Design*. Cambridge, MA: Harvard University Press.

Ruse, M., and R. Richards. (2009). *The Cambridge Companion to the "Origin of Species."* Cambridge: Cambridge University Press.

Sauvé Meyer, S. 1992. "Aristotle, Teleology and Reduction." *Philosophical Review* 101:791–825.

Sedley, D. (2007). *Creationism and Its Critics in Antiquity*. Berkeley: University of California Press.

Sober, E. (1984). *The Nature of Selection*. Cambridge, MA: MIT Press.

Sterelny, K., and P. Griffiths. (1999). *Sex and Death: An Introduction to Philosophy of Biology*. Chicago: University of Chicago Press.

Ulanowicz, R. (2000). *Growth and Development: Ecosystems Phenomenology*. San Jose: To Excel Press.

Walsh, D. M. (2006). "Organisms as Natural Purposes: The Contemporary Evolutionary Perspective." *Studies in the History and Philosophy of the Biological and Biomedical Sciences* 37:771–91.

Weber, B., and D. Depew. (2001). "Developmental Systems, Darwinian Evolution, and the Unity of Science." In Oyama, Griffiths, and Gray 2001, 239–54.

Weinberg, S. (1992). *Dreams of a Final Theory*. New York: Pantheon.

Wolff, C. (1732). *Philosophia rationalis sive logica, pars I*. In *Gesamelte Werke*, ed. J. Ecole, Band I. Repr. Hildesheim: Georg Olms Verlag, 1983.

Wright, L. (1976). *Teleological Explanations*. Berkeley: University of California Press.

SIX

THE GAME OF LIFE IMPLIES BOTH TELEONOMY AND TELEOLOGY

Gennaro Auletta, Ivan Colagè, and Paolo D'Ambrosio

NEW WAYS TO COPE WITH LASTING PROBLEMS

The present contribution is mainly aimed at suggesting the importance of teleonomy and teleology as explanatory mechanisms in biology in the light of recent achievements in the field, and at showing that they play an actual and relevant role in the realm of life.

The issue of finality in biology still provokes lively debates in the twenty-first century, especially as far as the dialogue between science, philosophy, and theology is concerned. Often, the question is raised about the extent to which the evolutionary process is to be considered as somehow necessarily determined to bring about the life-forms that we know, especially with reference to the appearance of human beings, or as ruled by nothing else than pure chance and contingency so that any form of finality should be seen as merely apparent. Another recurring relevant problem concerns the finality that may be ascribed to living beings in their ontogenetic activity as to their self-production, reproduction, organization, and behavior. Even if some goal-directed processes are to be acknowledged, many wonder whether the goal is to be conceived as intrinsic or extrinsic with respect to the organism or to the process itself. Recently, the view of an extrinsic goal has been proposed in a particular sense by the supporters of the so-called intelligent design, who maintain that the inherent complexity of organisms as such should be "explained" by turning to direct and specific divine interventions according to goals that are obviously supernatural. This conception may be seen as a rough reprise of the design argument typical of English natural theology dating back to the seventeenth century (Paley 1802; McGrath 2001–3, 1:241–48; Tanzella-Nitti 2009, 77).

Without focusing here on theological matters, we shall address the philosophical questions concerning the finalistic dimensions of living beings within a purely naturalistic framework and in accordance with some lessons derived from contemporary biological sciences. Recent developments have gone in the direction of radical revision of the traditional neo-Darwinian approach to evolution. Nevertheless, it holds true that fundamental aspects of the Darwinian view turn out to remain fundamental also for further progress in biology. In agreement with Darwinian tenets, we do see evolution as a non-predetermined natural process characterized by an actual and irreducible component of contingency. As a matter of fact, Darwin himself, in his observations and theoretical efforts that brought him to the conception of the *Origin of Species*, did not reject finalistic explanations, especially with regards to the fulfillment of biological functions; his true concern was rather to rule out divine intervention as the explanation of natural adaptation (Lennox 1992).

In the *Origin*, the solution represented by natural selection was meant to supply the mechanism that, in the long run and without recourse to final causes in phylogeny, eventually led to the evolution and preservation of complex functional traits (Ruse 2006). Obviously, the problem remains whether or not we can speak of finality for other domains or aspects of biology. As a matter of fact, more recently, after the establishment of the neo-Darwinian synthesis and the genetic and molecular revolutions in biology, some prominent scientists who contributed to the current understanding of evolution have reconsidered the issue of finality as a characteristic of living beings as real as any other property (Ayala 1970; Monod 1970, chap. 1; Ayala 1998). We propose to develop these suggestions according to further scientific research, particularly within the framework of a possible paradigmatic turn within biology (Auletta 2010). As we shall see, such a framework is mainly based on concepts like information control and top-down causation that are not usually employed within mechanist-centered biological explanations. Rather, these concepts are proposed as conceptual foundations of a new general research strategy able to address urgent and significant matters in the field.

In general, many (if not all) activities performed by an organism may appear to be directed toward certain states corresponding to the satisfaction of the needs of the organism itself, from the microscopic level of DNA expression and protein synthesis up to the macroscopic level of phenotypic behavior. Moreover, it could appear that the formation of phenotypic characters and of species that are selected and preserved in accord with their adaptation to different environments is achieved precisely *for the sake* of their (relative) stability and fitness. Therefore, one could be led to the conclusion that life processes are "finalized." However, the crucial point is, in our view, to *understand the specificity of those states and the mechanisms employed to achieve them* (Auletta 2008). First of all, we distinguish between teleology and teleonomy (Ayala 1970, 1998). In general terms, teleonomy may be ascribed to all biological processes implying forms of co-adaptation but not organismal built-in goals. Teleology, instead, concerns processes that are regulated by built-in goals nested in the constitution of an organism. Accordingly, teleological processes are to be ascribed exclusively to *individual* organisms, as they are essential in fulfilling vital functions and maintaining homeostasis (Rosenblueth, Weiner, and Bigelow 1943; Auletta, Ellis, and Jaeger 2008). However, we point out from the start that an organism is characterized by both teleonomic and teleological aspects (Auletta 2011a, chap. 8), whose interpolation will be clarified in the following. Another issue is whether or not there are forms of behavior that are regulated not merely by built-in goals, but by ends that are consciously perceived and elaborated. It is currently under discussion whether this kind of behavior might pertain to higher mammals like primates (Auletta 2007; 2011a, 3rd part; see also Changeux 2006; 2008, 130; Lagercrantz and Changeux 2009). This topic is, however, outside of the scope of the present paper.

Therefore, in the following we shall focus on the basic notions mentioned above supporting the new trends in biological research.

TELEONOMY IN EPIGENY

Epigeny is here understood as the process that brings an organism from biological conception to the stage of maturation (Auletta 2011a, chap. 11). It should be regarded as involving both teleonomic and teleological aspects. When considered in this context, teleonomy can be understood as a process in which the internal (genetic) program initiates a process in which environmental signals are used as cues for giving rise to structures and functions that are adaptive. In other words, teleonomy is a process of co-adaptation. This means that, although natural selection could be conceived in its immediate effect as acting in a point-like way, its final result is the selection of a whole *process of co-adaptation* between an organism and its environment. Obviously, teleonomy strongly relies on feedback circuits through which such a co-adaptation is established, and for this reason no goal is required as far as teleonomic processes are concerned. As a consequence of the previous considerations, the genetic program is only an initiator and not a determiner of the epigenetic process.

As a matter of fact, the epigenetic process eventually leads through further development to the formation of an adult organism that matches the general and distinguishing characteristics of the species-specific final steady state. Remarkably similar outcomes are produced in each of the various individuals of a species through a process that is somehow forced to proceed along different (developmental) pathways leading to a stable result, eventually coherent with the initial set of genetic instructions as well as with the developmental program.

Indeed, it is distinctive of teleonomic systems to be *robust* with respect to the final state but not relative to the initial conditions or the details of the singular (and mostly contingent) path followed by the system: the final outcome turns out to be like an attractor with respect to the different possible trajectories that the system may follow.

But how might the final state be achieved, given the insufficiency of information (the fact that organism relies on environmental inputs) and absence of goals in the whole process?

The genome does not contain all the instructions necessary for the building of the adult organism. Indeed, neither does any embryonic stage of development contain all the information needed to accomplish the final state. Development is rather an actual process of progressively building the organism, starting from what precedes, thus being truly "epi-genetic" in the proper (Greek) sense of the word (Jablonka and Lamb 2002; Griesemer 2002). The whole process is indeed made possible and accomplished through complex feedback circuits involving the genome, the developing phenotype, and the environment, in which context-dependence plays a crucial role. Therefore, the environment turns out to be an essential component for the establishment of the morphological configuration of the mature phenotype that will be produced (i.e., the final state of the process). In other words, the genetic endowment not only provides the initial set of instructions to start the developmental process, but also represents a source of variety, that is, a resource pool that is differently exploited along the various stages. As is well known, differential gene expression involved in cell differentiation and in the formation of tissues and organs often occurs thanks to regulatory mechanisms that do not directly affect the DNA sequence, yet influence the *activation* of coding sequences. Such regulatory mechanisms are in turn affected by several kinds of inputs, including environmental cues. Variations in these mechanisms also turn out to be inheritable (epigenetic inheritance). For instance, physical permanent factors or partially predictable parameters (like gravity and temperature, respectively) are integrated in the developmental process, so that an alteration in these factors may cause different gene expression and therefore remarkable morphological differences (Gilbert 2006, chap. 22). It is also widely acknowledged that the production of phenotypic variation and novelties (also possibly relevant for evolution) strongly depends on changes in the environmental conditions during developmental processes (Newman and Müller 2000; Müller and Newman 2003; more recently, see Petronis 2010). This does not mean at all that the environment *instructs* the developmental process, since it only provides a negative (selecting) feedback (Auletta 2011a, chaps. 8, 9, 11) to which the organism has to respond by means of its own resources as far as it can. Indeed, organisms show a certain degree of *plasticity* in response to environmental stresses that in certain situations allows them to switch developmental pathways so that new phenotypic solutions arise able to preserve an integrated unity of multiple functionalities (Waddington 1953, 1957; Reik and Dean 2002; West-Eberhard 2003). In other words, we may say that environmental signals, which are in principle a continuous threat to homeostasis, are (partially) integrated during development, thus enabling further specific developmental pathways. Therefore, the organism, by means of its internal network of regulatory mechanisms, is capable of a partial canalization of external selective influences or pressures, so that an unceasing teleonomic process of organism-environment co-*adaptation* takes place (Auletta 2011a, sec. 8.2).

To summarize, the epigenetic process paradigmatically shows that one can no longer maintain a strict genetic determinism (Macagno, Levinthal, and Lopresti 1973; Levinthal, Macagno, and Levinthal 1976); however, it additionally shows that even a strong environmental determinism or "instructionalism" is not tenable. This point will be better clarified by taking into account the other finalistic component of epigeny, that is, teleology that, in contrast to teleonomy, concerns goal-oriented activities. Indeed, teleological mechanisms are part of the epigenetic and developmental process in that they concur in the achievement of the species-specific steady state. Interestingly, the maturation that the organisms achieve in this way also coincides with the attainment of a full capability to perform teleological behaviors.

TELEOLOGY IN EPIGENY AND ONTOGENY

Having clarified what we mean by teleonomical processes, we would like to point out that if the game of life were to be ruled solely by teleonomy, organisms would comply passively with environmental pressures. Instead, organisms clearly display an *active* engagement with their environmental surroundings. This is the hallmark of teleo-logical processes concurring to selection for adaptive solutions (Ayala 1998). Teleology may be generally defined as the ability of an individual organism to informationally control another system, for instance its environment, according to the goal of its survival, that is, for preserving and maintaining its metabolic activity. It should be regarded as a top-down causal mechanism since it is related to endogenous processes that the organism carries out during epigeny and ontogeny, in order to satisfy vital needs.

More specifically, teleology always involves (Auletta, Ellis, and Jaeger 2008):

(a) Goal-directed actions, in the sense that they are actually aimed at reaching more or less well-defined outcomes;
(b) an information control system, whose general aim is that of lowering as much as possible the gap existing between the actual state of the system and the required or "to-be-implemented" state, in other words, the goal; this in turn implies
(c) the semiotic capability of organisms to refer to other systems in relation to the proper goal. A semiotic relation consists in something standing for something else under a certain perspective or in a certain context. An information control system is in fact able to treat external signals (the something) as signs about the states of other systems (the something else) and their significance for its vital needs (the context).

It is important to stress briefly the relevance of all these features. Beginning with the last one, we have to acknowledge that, in the natural world, only organisms are able to take physical patterns or signals as signs representing the *function* that an object may have for the organisms themselves. In other words, in the presence of a stimulus or a signal, an organism "guesses" about the survival-related significance of the source (whether noxious or not, whether survival-promoting or not) and acts accordingly. Even organisms as simple as the bacterium *E. coli* display this characteristic, as can be shown, for example, in the case of chemotaxis (Jurica and Stoddard 1998; Auletta 2011a, sec. 8.3.1; 2011b; 2011c; see also: Alon 2007; Weiner 2002).

Information control may be simply seen as the fundamental way in which organisms actively keep themselves alive, and it is instantiated in a feedback control loop with reference to a goal to be attained. The performing of an operation (understood as a pathway of physical-chemical interactions) that is needed to fulfill a vital function has to be checked step by step by the organism with regards to the desired outcome. The organism can do this by means of regulative circuits capable of measuring the gap between the current outcome and the "expected" one, which is relative to the goal to be accomplished. The presence of a relevant gap between these two states is taken as a sign of something to be somehow corrected (again a semiotic relation), so that the organism is induced to draw upon the genetic system, the metabolic one, or even the external environment in order to find resources for filling the gap and restore the functional efficiency of the operation. This kind of goal-oriented strategy may be thought of as being deployed both when some problems arise within the internal constitution of the organism (even at the microscopic level of pathways representing a molecular operation), and when unexpected challenges coming from the external environment are to be faced.

At a general and basic level of consideration, we may say that teleological mechanisms are in play as organisms perform activities and behaviors aimed at maintaining or restoring *homeostasis* with respect to a number of vital functions, but also in building up more efficient ways to fulfill the latter. Therefore, it may be now clear how teleology is also a crucial component of epigenesis. During development, not only may organisms, within a certain tolerance window, alter regulatory responses when internal or external factors cause dysfunction, they also *actively* search for environmental cues and resources that are indispensable for the accomplishment of the epigenetic process. For instance, the larvae of many insects feed themselves by eating other animals eventually provided by the mother. In the case of post-natal growth, we may recall here the function of play in mammals, which turns out to be crucial for their brain and cognitive development and is, moreover, a spontaneous and *endogenously* ignited behavior.

The realization of the complete organism's teleological capability is, to a wide extent, the result of the developmental process. To build an adult individual means to build an organism able to behave properly in its environment eventually to ensure, during the rest of its lifespan, reproductive success within its population and the preservation of its lineage. *Ontogeny* may be defined as the whole biography of an individual, from conception to death; in this sense, it includes also epigenesis as its first stage. We may say that the passage from development to maturity is marked by an increase of the control that organisms exert on the environment according to their needs and that is already displayed in epigenesis to a minor extent. In particular, during maturity the organism shows itself to be more shielded against external environmental fluctuations, and its stability is by far less sensitive to potentially disruptive impinging factors. This is due to the fact that the teleological systems that the previous epigenetic stage has produced are at their maximal efficacy in the mature stage.

A useful way to distinguish the developmental and the mature ontogenetic stages is to resort to the dichotomy between *assimilation* of the environment and *accommodation* to it (Auletta 2011a, secs. 8.4 and 10.4). Accommodation refers to the capability of the organism to become adjusted to its environment and is rooted in teleonomic processes. Assimilation, on the contrary, is the capability of the organism to carve out and transform the external environment according to its needs. As we have seen, during development, the organism is more prone to suffer environmental changes, up to the point that overwhelming pressures may determine a switch in the developmental pathway itself. Accommodation is therefore preponderant in development, especially in the early stages (though still bearing in mind that the epigenetic process should be understood as a combination of teleonomic and teleological components). In the mature stage of life, the organism is, instead, widely committed in controlling its immediate environment in order to submit it to its needs, thus showing a higher degree of assimilation. Assimilation requires and finds its root in teleology, as it presupposes the capability to exert informational control on the environment.

One of the most important expressions of the control exerted by the organism is the construction of suitable environmental niches (Auletta 2011a, sec. 10.3). In this case, organisms can exploit environmental resources or ameliorate unfavorable external conditions and strong selective pressures. The construction of a niche represents the way to shield the organism against the invasive or even potentially catastrophic pressures coming from the universal environment (understood as the general and wide environment that also contains the niche) by diminishing some of its features and enhancing other ones (Lewontin 2000, 60). From this point of view, niche construction can be regarded as a "filter" that organisms endeavor to build (in a teleological way) between themselves and the universal environment. Cumulative behaviors and actions undertaken by the organism in order to carve out the environment in this way influence the interactions within and between species and ultimately impact the external conditions in which future generations will live and reproduce. Therefore, the construction of environmental niches may have, in the long run, *indirect* yet remarkable consequences in the evolution of the species (Laland, Odling-Smee, and Feldman 1996, 1999, 2001; Odling-Smee, Laland, and Feldman 2003).

TELEONOMY IN PHYLOGENY: EVOLVABILITY AND CONVERGENCES

Evolution, generally understood as the phylogenetic accumulation of variations and interdependences with the environment (Auletta 2011a, sec. 9.1), is, in our view, a teleonomic process, thus essentially involving the accommodation of organisms to the environment. As mentioned above, teleonomic processes are not goal-oriented; therefore, we regard phylogeny as non-targeted in advance. Even if, as we have just seen, organisms can have an indirect influence on evolutionary dynamics by means of niche construction, they cannot control their own evolution since this would be possible only if a *complete* control on the environment were allowed, other species included. At an individual level, the cumulative actions and behaviors of organisms turn out to bring about an increasing assimilation of the environment. However, as already seen, such assimilation can never be completely accomplished. Therefore, whenever environmental conditions become extremely challenging or not manageable, the only alternative is to *accommodate* to the (new) environment. In other words, the epigenetic and developmental processes eventually bring about the formation of adult organisms that are maximally able to control their environment by means of teleological mechanisms. But the inability to have complete control over the environment can also result in the failure of adult forms to cope with environmental changes. When this is the case, to change crucial aspects of the epigenetic process may *possibly* give rise to new adult forms with a renewed capability to assimilate the environment, in other words, to face successfully the environmental challenges. Here, we see one of the most important lessons coming from the emerging research program called "evolutionary developmental biology" (evo-devo): changes in the developmental pathways of organisms have an essential relevance for phylogeny.

Nevertheless, the search for new suitable solutions is a random one, not only in the sense that the genetic source of modifications, at the basis of this process, has genuine chance components, but also because the appropriate phenotypic solution *relative* to particularly problematic environmental changes is not set in advance or chosen according to a stable criterion. It is rather selected *only once* it is implemented, and because of its successfully dealing with the relative problematic situation; consequently, the population will drift toward this new relative phenotypic optimum and a new equilibrium in fitness is eventually reached. We think that such an issue remarkably supports the teleonomic view of the phylogenetic process.

Recently, some evo-devo paths of research have focused on the concept of "evolvability," in which an interesting connection is established between genetic and phenotypic variability. Indeed, heritable genetic variability can provide the phenotypic variation that may turn out to be suitable for dealing with challenging environmental conditions. In other words, the issue of evolvability is regarded as concerned with the ways in which genetic variance is used or exploited to bring about phenotypic adaptive solutions. This is, in our view, the conceptual core common to many remarkable recent studies that unfortunately do not share the same definition (Wagner and Altenberg 1996; Kirschner and Gerard 1998; Hansen and Houle 2004; Schlosser and Wagner 2004; Wagner 2005, 2008). The kind of genetic variability we are speaking about in this context does not deal with genetic point mutations, but rather with changes in larger portions of the genome. There is evidence showing that in the course of the evolution of the species, an extremely relevant role has been played by mutations involving large portions of the genome, from hundreds to tens of thousands of base-pairs, through such mechanisms as transpositions, homologous recombination, insertion and deletion, and so on (McClintock 1984; Gerhart and Kirschner 1997, 218–27; Rosa et al. 1999; Ayala and Coluzzi 2005; Jablonka and Lamb 2005, 68–70. See also Alberts et al. 2008, 305–26). In other words, we draw our attention here to large-scale genome rearrangement or reshuffling. In principle, indeed, if only point-mutations were to trigger evolution, there would be likely far less probability to achieve new and stable phenotypes capable of surviving and reproducing. Moreover, it is widely proven that point-mutations accumulate substantially less frequently in significant DNA sequences than in sequences devoid of functional import. Consequently, we think that large-scale genome rearrangements are important in assuring the evolvability of organisms, especially considering that the rearranged segments might be regarded as minimal *modules*, in that they are partially independent and capable of being activated in order to contribute to the fulfillment of biological functions (e.g., they may code for a protein that, per se, is able to fold properly and catalyze a metabolic reaction, provided, of course, appropriate general conditions). Since what matters here is the different and variable distribution of these modules without changing their internal constitution (i.e., the specific base-sequence), we may suggest that evolvability stems from variability on the *architecture* in which such modules are arranged rather than in variability *within* them.

The rearrangement of these minimal modules can well assume relevance in the epigenetic developmental pathways. Indeed, in consequence of such changes, some minimal modules may be put under the control of a different regulatory gene, or framed in different regulatory networks. This may eventually lead (i) to the capability of deploying a function in a new or more efficient way, (ii) to the emergence of a new functionality, and, in general, (iii) to sensibly different, and possibly adaptive, phenotypic configurations. Of course, we are not suggesting that such a way to evolve is somehow a built-in strategy that makes the organisms "opt" for it rather than for a point-mutation-based evolution. Our point is that the probability of achieving reliable and stable solutions by this means is higher than in the other case: this may have also caused the molecular mechanisms responsible for the large-scale rearrangement (and the successive different regulative "recruitment") to be selected and preserved (Macagno, Levinthal, and Lopresti 1973; Levinthal, Macagno, and Levinthal 1976). Again, this speaks in favor of the teleonomic, rather than teleological, character of phylogenetic evolution, as we see that the latter is not targeted toward specific and anticipated solutions; rather it stems from a search for stable and adaptive ones, which are attainable through alternative possible paths.

Another way to appreciate the teleonomic character of evolution may be found starting from a different point of view, that is, taking into account true *evolutionary convergences* (Conway Morris 2003, 2008). The central idea is that starting from highly different biological structures, belonging to distant branches of the phylogenetic tree, and based sometimes even on significantly divergent genetic material, very similar solutions appear for the *same* function that, given internal and external constraints and environmental challenges, must be fulfilled for survival. Convergences do not take place only at the level of complex phenotypical traits (as, for example, the camera-eye structure, the wings, the fins/flippers, or the insects' halteres), but also at the level of behavior, especially in higher life-forms: think, for example, of some forms of parental care, cooperative hunting, and social play. Moreover, interesting examples of convergences at the molecular level are increasingly acknowledged. A particularly striking example is provided by the yeast a2 and the *Drosophila* engrailed DNA binding proteins: though being separated by billions of years of evolution and sharing only seventeen of sixty amino acid residues, the proteins have an extremely similar three-dimensional structure and the same regulatory function.

Obviously, many convergences can be understood as independent expressions, in different species, of homologous genes. Even in those cases, this parallel expression could not happen at all without (i) suitable developmental and environmental constraints and solicitations and (ii) developmental networks in which it could be embedded. Therefore, once again, the issue of genetic variability or expression must be always understood in the wider context of the teleonomic feedback circuits established between organisms and environment. Moreover, in most cases, genes of different species share common ancestors and are involved in similar epigenetic processes, but are not truly identical. In our view, these considerations fully justify the concept of convergence even in the cases of genetic homology.

A useful way to look at the mechanisms possibly involved in convergent evolution is to consider *exaptations*, that is, characters previously evolved for other usages (or for no function at all) and later coopted for their current role (Gould and Vrba 1982). In other words, organisms may accumulate, in the course of their phylogenetic evolution, a series of traits that are not of immediate utility but that can turn out to be useful to face new challenges or to increase fitness. This is made possible by the *degeneracy* characterizing biological structures. Degeneracy designates, indeed, the possibility for biological structures involving different elements to yield the same or different functions depending on the context (Auletta 2011a, sec. 8.2.5; Edelman and Gally 2001).

On this basis, it may be said to have convergences when such a process (1) happens in organisms belonging to two different species evolving along divergent evolutionary trajectories (that have already significantly diverged), and (2) brings about the capability of fulfilling the same function thanks to similar structural motifs, which however rely on different genetic-epigenetic endowments. Although there is never a one-to-one correspondence between the biological structures and the functions fulfilled through them, it remains true that certain structural motifs (or hubs) at the molecular or the morphological level are to be maintained whenever a particular function is concerned. From this point of view, it is possible to regard convergences as the result of successive steps of independent exaptations in different phylogenetic branches bringing about structures having the motifs required to fulfill the concerned function (Auletta 2011a, sec. 9.5.3). The requirement of common motifs for a certain function may be regarded as the manifestation of the fact (often recalled in what precedes) that the solutions adopted by life must be *stable*, so that such stability significantly constrains and limits the space of the possible *viable* solutions. Moreover, the involvement of exaptation in the process of generating convergences allows us to explain why phylogeny, which is an essentially divergent and therefore teleonomic process, may also display convergent solutions. It is indeed evident that exaptations cannot emerge *because of* the need to fulfill the concerned function since, by definition, they are prior to evolutionary convergences, although they significantly contribute to the satisfaction of the needs to which those convergences are solutions.

FINALISM, CONTINGENCY, AND NECESSITY

In this chapter, we have tried to clarify our general view on teleonomic and teleological processes in the three different (but interrelated) contexts of epigeny, ontogeny, and phylogeny. We have also dealt with these delicate issues by taking into account some recent advancements and promising perspectives in biology. This is the reason why the epigenetic process has received particular emphasis: it is the keystone on which the convergence of the hemi-arcs of ontogeny and phylogeny are constructed. The attention to epigeny should be regarded at the same time as a chance to look at teleonomy and teleology in a new way and in light of significant new achievements. Epigenetics finds teleology and teleonomy useful, if not indispensable, explanatory tools.

In conclusion, we would like to restate, in the light of the above considerations, the philosophical questions mentioned in the introduction, namely, (1) whether evolution should be regarded as necessarily driven or as merely the product of blind chance, and (2) what kind of finalism is to be acknowledged in an organism's self-production, re-production, regulation, and behavior.

As to the latter issue, in dealing with ontogeny and epigeny we have seen that the two processes are mainly characterized by teleological mechanisms and by a trade-off of teleonomic and teleological mechanisms, respectively. In the case of teleonomy, we have canalization toward stable solutions in the production of which the environment (or whatever external influence) also plays a crucial role. In other words, the stable solution reached does not depend on goals somehow "inscribed" within the system, but on a genuine trade-off between internal drives and variable, unpredictable external inputs. Therefore, teleonomic mechanisms cannot be considered as aiming at a definite final state. In the case of teleological goal-directed behaviors, on the contrary, the goal is to be seen as intrinsic to the organism and it has to be attained within relatively narrow tolerance windows. This confers on teleological processes a genuine finalistic character.

As to the first question above, we have argued that phylogeny is a teleonomic process and not a goal-directed one. In this case, it is true that evolution gives rise to life-forms that, as a matter of fact, are stable and sufficiently adapted to their environments. However, it would be a mistake, in our view, to consider this result as goal-driven evolution and therefore targeted in advance. The specific adaptive forms, indeed, cannot be predetermined because, as we have seen: (a) they have to adapt to an environment that is continuously and unpredictably changing (also as a consequence of the actions that other organisms perform on it), and (b) the variability triggering evolution has irreducible random components.

Consequently, we stress the undeniable contingency in evolution and life processes in general. We suggest that "necessity," in this domain, is rather the *a posteriori* result of dynamic (exaptation) processes in which random elements are integrated and constrained in view of the capability to fulfill vital functions. Thus, the process brings about final results that display a certain necessity. Therefore, necessity should not be regarded as something imposing conditions in a mechanical way from the start.

NOTE

This chapter was delivered at the 2009 Notre Dame conference "Darwin in the Twenty-First Century: Nature, Humanity, and God," and has been previously printed under the same title in Auletta and Pons 2013, 267–84. Republished with permission of the Gregorian and Biblical Press.

REFERENCES

Alberts, B., A. Johnson, J. Lewis, M. Raff, K. Roberts, and P. Walter. (2008). *The Molecular Biology of the Cell.* 5th ed. New York: Garland Press.

Alon, U. (2007). "Simplicity in Biology." *Nature* 446:497.

Auletta, G. (2007). "Information, Semiotics, and Symbolic Systems." *Semiotica* 166:359–76.

———. (2008). "How Many Causes Are There?" *21mo secolo. Scienza e tecnologia* 5:41–48.

———. (2010). "A Paradigm Shift in Biology?" *Information* 1:28–59.

———. (2011a). *Biology and Cognition: Biological Systems Dealing with Information.* Oxford: Oxford University Press.

———. (2011b). "Teleonomy: The Feedback Circuit Involving Information and Thermodynamic Processes." *Journal of Modern Physics* 2 (3): 136–45.

———. (2011c). "A Mathematical Model of Complexity and Its Application to Chemotaxis." *British Journal of Mathematics and Computer Science* 1:204–27.

Auletta, G., G. Ellis, and L. Jaeger. (2008). "Top-Down Causation by Information Control: From a Philosophical Problem to a Scientific Research Program." *Journal of the Royal Society Interface* 5:1159–72.

Auletta, G., and J. S. Pons, eds. (2013). *Si può parlare oggi di una finalità dell'evoluzione? Riflessioni filosofiche e teologiche alla luce della scienza cintemporanea.* Rome: Gregorian University Press.

Ayala, F. J. (1970). "Teleological Explanations in Evolutionary Biology." *Philosophy of Science* 37:1–15.

———. (1998). "Teleological Explanations versus Teleology." *History and Philosophy of the Life Sciences* 20:41–50.

Ayala, F. J., and M. Coluzzi. (2005). "Chromosome Speciation: Humans, Drosophila, and Mosquitoes." *Proceedings of the National Academy of Sciences U.S.A.* 102:6535–42.

Caporale, L. H. (2003a). "Natural Selection and the Emerence of a Mutation Phenotype: An Update of the Evolutionary Synthesis Considering Mechanisms That Affect Genome Variation." *Annual Review of Microbiology* 57:467–85.

———. (2003b). "Foresight in Genome Evolution." *American Scientist* 91:234–41.

Changeux, J-P. (2006). "The Ferrier Lecture 1998: The Molecular Biology of Consciousness Investigated with Genetically Modified Mice." *Philosophical Transactions of the Royal Society of London B* 361:2239–59.

———. (2008). *The Physiology of Truth: Neuroscience and Human Knowledge.* Cambridge, MA: Harvard University Press.

Conway Morris, S. (2003). *Life's Solution: Inevitable Humans in a Lonely Universe*. Cambridge: Cambridge University Press.

———. (2008). "Evolution and Convergence." In *The Deep Structure of Biology*, 46–47. West Conshohocken, PA: Templeton Foundation Press.

Edelman, G. M., and J. A. Gally (2001). "Degeneracy and Complexity in Biological Systems." *Proceedings of the National Academy of Sciences of the USA* 98:13763–68.

Gerhart, J. C., and M. W. Kirschner. (1997). *Cells, Embryos, and Evolution: Toward a Cellular and Developmental Understanding of Phenotypic Variation and Evolutionary Adaptability*. Boston: Blackwell Science.

Gilbert, S. F. (2006). *Developmental Biology*. 8th ed. Sunderland, MA: Sinauer.

Gould, S. J., and E. S. Vrba. (1982). "Exaptation: A Missing Term in the Science of Form." *Paleobiology* 8 (1): 4–15.

Griesemer, J. (2002). "What Is 'Epi' About Epigenetics?" *Annals of the New York Academy of Sciences* 981:97–110.

Hansen, T. F., and D. Houle. (2004). "Evolvability, Stabilizing Selection, and the Problem of Stasis." In *Phenotypic Integration: Studying the Ecology and Evolution of Complex Phenotypes*, ed. M. Pigliucci and K. Preston, 130–50. New York: Oxford University Press.

Jablonka, E., and M. J. Lamb. (2002). "The Changing Concept of Epigenetics." *Annals of the New York Academy of Sciences* 981:82–96.

———. (2005). *Evolution in Four Dimensions: Genetic, Epigenetic, Behavioral, and Symbolic Variation in the History of Life*. Cambridge, MA: MIT Press.

Jurica, M. S., and B. L. Stoddard. (1998). "Mind Your Bs and Rs: Bacterial Chemotaxis, Signal Transduction and Protein Recognition." *Structure* 6:809–13.

Kirschner, M., and J. Gerard. (1998). "Evolvability." *Proceedings of the National Academy of Sciences of the USA* 95:8420–27.

Lagercrantz, H., and J-P. Changeux. (2009). "The Emergence of Human Consciousness: From Fetal to Neonatal Life." *Pediatric Research* 65:255–60.

Laland, K., F. Odling-Smee, and F. Feldman. (1996). "The Evolutionary Consequences of Niche Construction: A Theoretical Investigation Using Two-Locus Theory." *Journal of Evolutionary Biology* 9:293–316.

———. (1999). "Evolutionary Consequences of Niche Construction and Their Implications for Ecology." *Proceeding of the National Academy of Sciences USA* 96:10242–47.

———. (2001). "Niche Construction, Ecological Inheritance, and Cycles of Contingency in Evolution." In *Cycles of Contingency: Developmental Systems and Evolution*, ed. S. Oyama, P. E. Griffiths, and R. D. Gray, 117–26. Cambridge, MA: MIT Press.

Lennox, J. G. (1992). "Teleology." In *Keywords in Evolutionary Biology*, ed. E. F. Keller and E. A. Lloyd, 324–33. Cambridge, MA: Harvard University Press.

Levinthal, F., E. Macagno, and C. Levinthal. (1976). "Anatomy and Development of Identified Cells in Isogenic Organisms." *Cold Spring Harbour Symposia on Quantitative Biology* 40:321–31.

Lewontin, R. C. (2000). *The Triple Helix: Gene, Organism, and Environment*. Cambridge, MA: Harvard University Press.

Macagno, E. R., C. Levinthal, and V. Lopresti. (1973). "Structural Development of Neural Connections in Isogenic Organisms: Variations and Similarities in the Optic System of *Daphnia magna*." *Proceeding of the National Academy of Sciences USA* 70:57–61.

McClintock, B. (1984). "The Significance of Responses of the Genome to Challenge." *Science* 226:792–801.

McGrath, A. (2001–3). *A Scientific Theology*. 3 vols. Grand Rapids, MI: Eerdmans.

Monod, Jacques. (1970). *Le hasard et la nécessité*. Paris: Seuil.

Müller, G. B., and S. Newman. (2003). "Origination of Organismal Form: The Forgotten Cause in Evolutionary Theory." In *Origination of Organismal Form*, ed. G. B. Müller and S. Newman, 3–10. Boston: MIT Press.

Newman, S. A., and G. B. Müller. (2000). "Epigenetic Mechanisms of Character Origination." *Journal of Experimental Zoology* (*Molecular Evolution and Development*) 288:304–17.

Odling-Smee, F. J., K. N. Laland, and M. W. Feldman. (2003). *Niche Construction: The Neglected Process in Evolution*. Princeton: Princeton University Press.

Paley, W. (1802). *Natural Theology: or, Evidences of the Existence and Attributes of the Deity, Collected from the Appearances of Nature*. London: Faulder.

Perfeito, L., L. Fernandos, C. Mota, and I. Gordo. (2007). "Adaptive Mutations in Bacteria: High Rate and Small Effects." *Science* 317:813–15.

Petronis, Arturas. (2010). "Epigenetics as a Unifiying Principle in the Aetiology of Complex Traits and Deseases." *Nature* 465:721–27.

Reik, W., and W. Dean. (2002). "Back to the Beginning." *Nature* 420:127.

Rosa, R., J. K. Grenier, T. Andreeva, C. E. Cook, A. Adoutte, M. Akam, S. B. Carroll, and G. Balavoine. (1999). "Hoxgenes in Brachiopods and Priapulids, and Potostomes Evolution." *Nature* 399:772–76.

Rosenblueth, A., N. Wiener, and J. H. Bigelow. (1943). "Behavior, Purpose, and Teleology." *Philosophy of Science* 10:18–24.

Ruse, M. (2006). *Darwinism and Its Discontents*. New York: Cambridge University Press.

Schlosser, G., and G. P. Wagner. (2004). "Introduction: The Modularity Concept in Developmental and Evolutionary Biology." In *Modularity in*

Development and Evolution, ed. G. Schlosser and G. P. Wagner, 1–16. Chicago: University of Chicago Press.

Tanzella-Nitti, G. (2009). *Faith, Reason and the Natural Sciences: The Challenge of the Natural Sciences in the Work of Theologians*. Aurora, CO: Davies Group.

Waddington, C. H. (1953). "Genetic Assimilation of an Acquired Character." *Evolution* 7:118–26, 386–87.

———. (1957). *The Strategy of the Genes*. London: Allen and Unwin.

———. (1961). "Genetic Assimilation." *Advances in Genetics* 10:257–90.

Wagner, A. (2005). *Robustness and Evolvability in Living Systems*. Princeton: Princeton University Press.

———. (2008). "Robustness and Evolvability: A Paradox Resolved." *Proceedings of Biological Science* 275:91–100.

Wagner, G. P., and L. Altenberg. (1996). "Complex Adaptations and the Evolution of Evolvability." *Evolution* 50:967–76.

Weiner, O. D. (2002). "Regulation of Cell Polarity During Eukaryotic Chemotaxis: The Chemotactic Compass." *Current Opinion in Cell Biology* 14:196–202.

West-Eberhard, M. J. (2003). *Developmental Plasticity and Evolution*. Oxford: Oxford University Press.

PART TWO

Humanity

SEVEN

HUMANITY'S ORIGINS

Bernard Wood

One of Charles Darwin's many achievements is that he began the process of converting the Tree of Life (TOL) from a religious metaphor into a biological reality. All types of living organisms, be they animals, plants, fungi, bacteria, or viruses, are at the end of twigs that reach the surface of the Tree of Life, and all the types of organisms that have ever lived in the past are situated somewhere on the branches and twigs within the tree.[1] Darwin was a forceful proponent of the idea that we, humanity, are just one of the many types of life on the surface of the TOL. The extinct organisms on the branch within the TOL that connects humanity directly to the root of the tree are the ancestors of modern humans. Our close non-ancestral relatives are on extinct branches close to ours.

A "very long" version of humanity's origins would be an evolutionary journey that starts approximately three billion years ago at the base of the TOL with the simplest form of life, then passes into the relatively small section of the tree that contains all animals, and moves on into the even smaller section that contains all the animals with backbones. Around 400 million years ago (mya) we would have entered the section of the tree that contains vertebrates with four limbs, then around 250 million years ago into the branch that contains the mammals, and then into successively smaller branches that contain, respectively, the primates, the monkeys and apes, and then

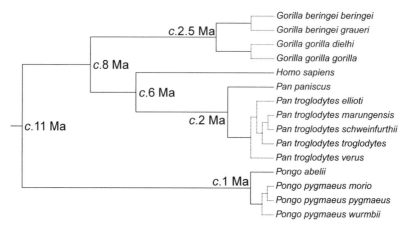

Figure 7.1. Current consensus of the phylogenetic relationships and splitting-times within the great ape clade. The only Asian great ape, the orangutan (*Pongo*), which split off from the African great apes circa 11 million years ago, diverged into the Bornean (*Pongo pygmaeus*) and Sumatran (*Pongo abelii*) orangs circa 1 million years ago. There have been two major and two minor splits in the African ape clade. The first major splitting event, the one leading to gorillas, occurred circa 8 million years ago. The second, leading to modern humans, occurred circa 6 million years ago. The split within gorillas, into mountain (*Gorilla beringei*) and lowland (*Gorilla gorilla*), occurred circa 2.5 million years ago. The split within chimpanzees occurred circa 2 million years ago when the Congo River divided them into bonobos (*Pan paniscus*) to the south and common chimpanzees (*Pan troglodytes*) to the north. The details of the subspecies, along with the timing of any splits, are more conjectural. Figure courtesy of Adam Gordon.

just the great apes. Sometime between 15 and 12 mya we move into the small branch that gave rise to modern humans and to the living African apes. Between 11 and 9 mya the branch for the gorillas split off to leave just a single slender branch consisting of the ancestors and extinct close relatives of chimpanzees and bonobos (chimps/bonobos) and modern humans. Around 8 to 5 mya this very small branch split into two twigs. At the root of the twigs is the most recent common ancestor of chimps/bonobos and modern humans. One of the two twigs ends on the surface of the TOL with the living chimps/bonobos. The other also ends on the surface of the TOL, but with modern humans.

Another, much shorter version of humanity's origins would focus on the base of the twig of the TOL that leads to modern humans, and call that branching point our "origin." Even shorter versions involve focusing on events within the modern human twig of the TOL. From a distance it looks like a simple, straight, 8 to 5 million years–long twig, but when examined in more detail it has its own branching points that lead to smaller twigs that do not reach the surface of the tree. These extinct twigs represent organisms that are our very close relatives, but they are not our ancestors. Researchers familiar with the fossil record suggest that a major branching point in our evolutionary history occurred between 3 and 2 million years ago when our own genus *Homo*, originated. So this version of humanity's origins would focus on the base of the twig that contains all the species included in the genus *Homo*.

The third version of humanity's origins involves looking at the part of the modern human twig of the TOL that lies just beneath the surface of the tree. A few researchers claim that *Homo* has no branches, in which case all the species that have come to be recognized within *Homo*, such as *Homo habilis*, *Homo erectus*, and *Homo neanderthalensis*, are all ancestors of modern humans. But many researchers interpret the fossil evidence differently and suggest that there is at least one, and maybe more, branches within the *Homo* twig of the TOL. This shortest version focuses on the most recent branching event on the *Homo* twig, the one that gave rise to modern humans and to *H. neanderthalensis*, and uses this as the starting point of the journey through the TOL to modern humanity.

This chapter will briefly review the latest scientific evidence for all three versions of the origin of humanity, starting with the one furthest back in time. This means looking at the evidence for when, where, and in what circumstances did the twig, or clade, that leads to modern humans and to all our extinct recent ancestors and close relatives split off from the rest of the TOL?

THE ORIGIN OF THE MODERN HUMAN CLADE

In the nineteenth century, the closeness or otherwise of the relationships between living animals had to be assessed on the basis of how similar they were in terms of their gross morphology. For instance, how much of the skeleton and how many of the soft tissue structures (e.g., muscles, nerves, etc.) resembled each other? The assumption was that the more closely they resembled each other morphologically, the closer the "natural" relationship between the two animals.

One of the first people to undertake a systematic review of the differences between modern humans and the apes, in this case the gorilla and the chimpanzee, was Thomas Henry Huxley. He summarized his views in an essay entitled *On the Relations of Man to the Lower Animals* that formed the central section of a book *Evidence as to Man's Place in Nature* published in 1863 (Huxley 1863). In that essay he concluded that the phenotypic differences between modern humans and the gorilla (and the chimpanzee) were less marked than the differences between the gorilla and the lesser apes (i.e., the orangutan and gibbons). This was the evidence that was later used by Darwin in *The Descent of Man* in 1871 to suggest that, because the African apes were morphologically closer to modern humans than the apes from Asia, the ancestors of modern humans were more likely to be found in Africa than elsewhere. But despite Huxley's prescient observations, until recently it was usual for modern humans to be distinguished from the great apes at the level of the family. The nonhuman great apes (i.e., orangutans, gorillas, and chimpanzees/bonobos) were included in the family Pongidae, and modern humans were placed in a separate family, the Hominidae. That is why modern humans, and all the extinct taxa judged to be more closely related to modern humans than to any other living taxon, are called "hominids."

Developments in immunology and biochemistry during the first half of the twentieth century allowed the focus of the search for better evidence about the nature of the relationships between modern humans and the great apes to be shifted from traditional macroscopic morphology to the morphology of molecules. The earliest attempt to use biochemistry to determine the relationships among primates was made in 1902, but it was in the early 1960s when Morris Goodman and Emil Zuckerkandl began to use molecular methods specifically to address the relationships among the great apes and modern humans (Zuckerkandl, Jones, and Pauling 1960; Goodman 1962, 1963; Zuckerkandl 1963). Morris Goodman used what was then the new technology of immunology, specifically a process called immunodiffusion, to study the affinities of the serum proteins of monkeys, apes, and modern humans. He concluded that because the patterns produced by the albumins of modern humans and the chimpanzee in the immuno-diffusion gels were identical (Goodman 1963, 224), then it is likely that the structures of the albumins were also likely to be, to all intents and purposes, identical. Emil Zuckerkandl used enzymes to break up the hemoglobin protein into its peptide components, and showed that when the peptides were subjected to starch gel electrophoresis, the patterns of the peptides in the gel for modern humans, chimpanzee, and gorilla were indistinguishable (Zuckerkandl 1963, 246).

Proteins like albumin and hemoglobin are made up of strings of amino acids. In many instances one amino acid may be substituted for another without changing the function of the protein. In the 1970s Vince Sarich and Allan Wilson exploited these minor variations in protein structure to determine the evolutionary history of the protein molecules, and therefore, presumably, the evolutionary history of the taxa whose proteins had been sampled. They, too, concluded that modern humans and the African apes, in particular the chimpanzee, were very closely related (Sarich and Wilson 1967). In a later paper Mary-Claire King and Alan Wilson suggested that 99 percent of the amino-acid sequences of chimps and modern humans were identical (King and Wilson 1975).[2]

The discovery by James Watson and Francis Crick of the structure of DNA, and the subsequent discovery by Crick and others of the genetic code, showed that it was the sequence of bases in the DNA molecule that determined the nature of the proteins manufactured within a cell. This meant that the affinities between organisms could be pursued at the level of the genome, thus potentially eliminating the need to rely on morphological proxies, be they traditional morphology or the morphology of proteins, for information about relatedness. The DNA within the cell is located either within the nucleus as nuclear DNA (nDNA) or within the mitochondria as mitochondrial DNA (mtDNA). Comparisons between the DNA of organisms can be made using two methods. In DNA hybridization, all the DNA is compared, but at a relatively crude level. In the early days of DNA sequencing, the base sequences of relatively small amounts of DNA were determined and then compared. In brief, DNA hybridization tells you "a little about a lot" of DNA, whereas the early sequencing methods used to tell you "a lot about a little piece" of DNA. Nowadays technological advances mean that whole genomes can be sequenced at levels of precision that were unthinkable in the 1980s.

Sequencing is favored because the knowledge about the type of differences between the base sequences provides some clues about the steps that are needed to produce the observed differences. This is because the base changes called transitions ("A to G" and "T to C") readily switch back and forth, whereas transversions ("A to C" and "T to G") less readily switch back and forth, and thus they are more reliable indicators of "genetic distance." These methods, hybridization (e.g., Caccone and Powell 1989) and sequencing (e.g., Bailey et al. 1992; Horai et al. 1992), have been applied to the living great apes and modern humans (for reviews see Uddin et al. 2004 and Bradley 2008). Information from both nuclear and mtDNA suggests that modern humans and chimpanzees are more closely related to each other than either is to the gorilla (e.g., Ruvolo 1997; Li and Saunders 2005; Fabre, Rodrigues, and Douzery 2009; Prado-Martinez et al. 2013). When these differences are calibrated using the best paleontological evidence for the split between the apes and the Old World Monkeys, and if the DNA differences are assumed to be neutral, then this predicts that the hypothetical ancestor of modern humans and the chimpanzee lived between about 8 and 5 mya, and probably closer to 5 than to 8 mya. When other, even older, calibrations are used the predicted date for the split is somewhat older (e.g., >10 mya, Arnason and Janke 2002; Venn et al. 2014).

Thus, there is now overwhelming evidence that chimpanzees and bonobos are the closest living relatives of modern humans, and it is very likely that the modern human twig, or clade, separated from the rest of the TOL as late as 5–6 million years ago. Because chimpanzees and bonobos are both only found in Africa, and because the earliest evidence for creatures that *might* belong to the modern human clade comes from Africa (e.g., Brunet et al. 2002: Senut et al. 2001; White et al. 2009), then, as Darwin predicted, Africa is likely to have been the continent where the modern human clade emerged.

Given the abundant evidence for a closer relationship between *Pan* and *Homo* than between *Pan* and *Gorilla* (see above), many researchers have concluded that the human clade should be distinguished in the Linnaean hierarchy beneath the level of the family. The researchers who have come to that conclusion mostly now interpret the family Hominidae more inclusively to include *all* of the great apes (including modern humans), and they use the subfamily Homininae for just *Pan* and *Homo*. The researchers who restrict the subfamily Homininae to just *Pan* and *Homo* mostly recognize the *Homo* clade at the level of the tribe, as the Hominini, with the individuals and taxa within it referred to as "hominins," and chimpanzees and bonobos are referred to as "panins"; we will use this terminology for the rest of this chapter.

THE ORIGIN OF THE GENUS *HOMO*

The most widely used genus concept is the one suggested by Ernst Mayr, who proposed that "a genus consists of one species, or a group of species of common ancestry, which differ in a pronounced manner from other groups of species and are separated from them by a decided morphological gap" (Mayr 1950, 110), and he went on to state that the species united in a such a genus must "occupy an ecological situation which is different from that occupied by the species of another genus, or, to use the terminology of Sewall Wright, they occupy a different adaptive plateau" (110). Thus, according to Mayr, a genus is a group of species of common ancestry that is adaptively both homogeneous and distinctive. Wood and Collard (1999) suggested that Mayr's definition of the genus should be modified so that only clades should qualify, and they saw no reason why the shared adaptive zone had to be unique. Thus, the sensible strategy is to adopt the adage that "all genera should be clades, but not all clades are genera."

There are two options for putting the principles of genus identification (i.e., an adaptively coherent clade) into practice. You can either start in the present or in the past. If one starts in the present, and adopts the "top down" option, one begins with the type species. In the case of the genus *Homo* one takes stock of the derived morphology and behavior of *H. sapiens*, decides on the cardinal features and behaviors one will use to determine the adaptive zone of *H. sapiens*, and then chooses a way of generating hypotheses about which species should be included in the *Homo* clade (the technical term for this is monophyly). Then one works backwards into the past, and by applying the same two tests to each hominin taxon one encounters (i.e., starting from the present, they are *H. neanderthalensis*, *H. heidelbergensis*, *H. erectus*, and *H. habilis*) it is asked whether there is reliable evidence that the taxon is in the same adaptive zone (i.e., reliable qualitative or quantitative proxies of important behaviors) and in the same clade as *H. sapiens*.

If the "bottom up" approach is adopted, one has to make a subjective judgment about whereabouts in the past one should start to pick up the trail leading to *Homo*. One then works toward the present applying the tests set out above to the hominin taxa that are encountered. The difference between this approach and the "top down" option is that in the "bottom up" approach the evidence is sketchier, and thus the likelihood that one can satisfy the "reliability" criterion of the two tests, monophyly and adaptive coherence, is diminished.

Ironically, there have been very few attempts to formally assess the relationships of modern humans with respect to *H. neanderthalensis* and *H. erectus*. Eldredge and Tattersall (1975) included all three taxa in the cladogram (see Eldredge and Tattersall 1975, fig. 4) presented in their seminal paper that pioneered the application of cladistic methods to hominin relationships. However, the authors did not carry out a formal analysis of the relationships among the taxa, nor did they refer to any specific characters when considering the merits of different branching patterns (called cladograms) for expressing the relationships among the premodern *Homo* taxa within the hominin clade. Since most researchers then considered the hypothesis of monophyly of later *Homo* (i.e., *H. sapiens*, *H. neanderthalensis*, *H. heidelbergensis*, *H. erectus*) to be so well supported that the matter did not require formal investigation. Although there are grounds for adding *H. habilis* to the *H. sapiens*, *H. neanderthalensis*, *H. heidelbergensis*, and *H. erectus* clade, I think even the supporters of such an interpretation would accept that the evidence for doing so is not as strong as the evidence for including *H. neanderthalensis* and *H. erectus* within the clade that includes modern humans. Thus, as far as relationships are concerned, there seem to be two options. One either draws the lower boundary of the genus *Homo* so that it includes *H. habilis*, or one draws it beneath early African *H. erectus* so that it excludes *H. habilis*.

As far as adaptive grade is concerned, the problem is more complicated. If the criteria are restricted to what can be deduced about the adaptive grade of a taxon from its morphology, then it could be argued that if the combination of a modern human-sized brain and obligate long range bipedalism are the criteria, then the boundary of *Homo* would be set so that it includes *H. heidelbergensis,* but not *H. erectus* or *H. floresiensis*. If a modern human body shape and obligate bipedalism are deemed to be the criteria, then the boundary would be set so that *Homo* would include early African *H. erectus*, but not *H. habilis* (but see Haeusler and McHenry 2004, 2007, for an alternative interpretation).

But even that solution results in a hominin genus that embraces a substantial range of life histories (Robson and Wood 2008). If *H. habilis* is included in *Homo* for relationship reasons, this poses problems for any genus definition that insists on adaptive coherence, for the same genus would include taxa with a range of cranial, dental, and postcranial morphology and relative size relationships (including very different semicircular canals and limb strength proportions) that imply different dietary and locomotor adaptations. Furthermore, the adaptive strategies of *H. habilis* are probably closer to the adaptive strategy of the type species of the genus *Australopithecus* (i.e., *Au. africanus*) than they are to *H. sapiens*, the type species of *Homo*.

THE ORIGIN OF MODERN HUMANS

Just what are the features of the cranium, jaws, dentition, and the postcranial skeleton that are only found in *H. sapiens*, and what are the limits of living *H. sapiens* variation? How far beyond these limits, if at all, should we be prepared to go and still be prepared to assign the fossil evidence to *H. sapiens*? These are simple enough questions, to which one would have thought there would be ready answers, yet the assembly of a set morphological criteria for "modern human-ness" is a surprisingly difficult task, and little progress has been made since W. W. Howells's seminal study of modern human cranial variation (Howells 1973, 1989). Using a comprehensive sample of modern human cranial measurements, Howells showed that the totality of variation as measured in Mahalanobis D^2 distances among his twenty-eight groups is comparable to the distance that separates all modern human crania from his relatively small sample of Neanderthal crania. Small-bodied modern humans tend to have smaller crania, but overall there is very little among-sample difference in the overall size of the modern human cranium. Howells comments that modern human crania share a "universal loss of robustness," and goes on to write that within modern humans "variation in shape seems to be largely located in the upper face, and particularly the upper nose and the borders of the orbits" (Howells 1989, 83). Others have attempted to specify acceptable ranges of morphometric variation for the cranium of *H. sapiens* (e.g., Stringer, Hublin, and Vandermeersch 1984; Day and Stringer 1991), but the latter authors conceded that a sample need comply with only about 75 percent of the defining characteristics in order to qualify for inclusion in *H. sapiens*. In a review of variation in regional samples of modern human crania, Lahr (1996) emphasized that regional peculiarities should not be incorporated into criteria for inclusion in *H. sapiens*. Lieberman distilled existing cranial definitions of *H. sapiens* and suggested that to be regarded as "anatomically modern human," crania need to have "a globular braincase, a vertical forehead, a diminutive browridge, a canine fossa and a pronounced chin" (Lieberman 1998, 158). Others suggested that all these features may be related in one way or another to a reduction in facial projection (Spoor et al. 1999), and Lieberman, McBratney, and Krovitz (2002) suggested that what modern human crania really have in common is an unusually globular neurocranium.

Dentally, the postcanine teeth of modern humans are notable for the absolutely and relatively small size of their crowns, and for a reduction in the number of cusps and roots (Hillson 1996); presumably this would also be the same for fossil representatives of *H. sapiens*. As for the postcranial skeleton in comparison with Neanderthals and what little is known of the postcranial skeleton of *H. heidelbergensis*, anatomically modern humans have elongated distal limb bones (Trinkaus 1981), limbs that are long relative to the trunk (Holliday 1995), a relatively narrow trunk and pelvis, and low body mass relative to stature (Ruff, Trinkaus, and Holliday 1997). Many of these traits cause the earliest fossil modern humans (e.g., those from Skhul and Qafzeh) to resemble living modern humans from hot, arid climates, and the contrasts in postcranial morphology between modern humans and Neanderthals probably have more to do with the uniqueness and distinctiveness of Neanderthal morphology than with the ability of researchers to define the distinctive characteristics of *H. sapiens* (Pearson 2000). In summary, compared to their more archaic immediate precursors, modern humans are characterized postcranially by their reduced body mass (e.g., Ruff, Trinkaus, and Holliday 1997), their more linear physique, and a distinctive pelvic shape that includes a short, stout, pubic ramus, and a relatively large pelvic inlet (Pearson 2004).

So when and where do we see the earliest evidence of modern human morphology in the fossil record? The simple answer is Africa, where at two sites in Ethiopia, the circa 170 thousand years ago (ka) Herto site (White et al. 2003) and the circa 190 ka site at Omo-Kibish (McDougall, Brown, and Fleagle 2005), there is good evidence of modern human-like crania. There is also molecular evidence that is consistent with a circa 300–200 ka African origin for modern humans (Pearson 2004).

A DIFFERENT SOLUTION

The three choices provided by fossil and molecular evidence for humanity's origin—circa 6–5 mya for the origin of the human clade, circa 2 million years ago for the origin of our own genus, or circa 200,000 years ago for the origin of modern human morphology—are not the only ones available. Modern humans are distinguished from all other living animals by their behavior as well as by their morphology. The extent of our behavioral uniqueness has almost certainly been exaggerated, and the more that is found out about the behavior of other primates, and especially the behavioral repertoire of the great apes, the more researchers realize that our distinctiveness is a matter of degree rather than kind (Haslam et al. 2009). Archeologists try diligently, but not always successfully, to seek for evidence of symbolism and language in the archeological record, but there is accumulating evidence that just as modern human *morphology* seems to be emerging in Africa, modern human behavior may do so as well, but intriguingly the evidence for the latter may antedate the former (McBrearty and Brooks 2000).

NOTES

I thank the John J. Reilly Center for Science, Technology and Values at the University of Notre Dame for support to attend the Notre Dame conference and the George Washington University Signature Program and the GW Provost for research support.

1. See figure 16.1, from the *Origin of Species*, in this volume.
2. The "1 percent difference" conclusion had become so well known that when King took her daughter to Ireland after graduating from high school, they went round a bend in the road to see a large advertisement for Guinness that proclaimed the 1 percent difference!

REFERENCES

Arnason, U., and A. Janke. (2002). "Mitogenomic Analyses of Eutherian Relationships." *Cytogenetic and Genome Research* 96:20–32.
Bailey, W. J., K. Hayasaka, C. G. Skinner, S. Kehoe, L. C. Sieu, J. L. Slightom, and M. Goodman. (1992). "Reexamination of the African Hominoid Trichotomy with Additional Sequences from the Primate B-globin Gene Cluster." *Molecular Phylogenetics and Evolution* 1:97–135.
Bradley, B. (2008). "Reconstructing Phylogenies and Phenotypes: A Molecular View of Human Evolution." *Journal of Anatomy* 212:337–53.
Brunet M., F. Guy, D. Pilbeam, et al. (2002). "A New Hominid from the Upper Miocene of Chad, Central Africa." *Nature* 418:145–51.

Caccone, A., and J. R. Powell. (1989). "DNA Divergence among Hominoids." *Evolution* 43:925–42.

Chimpanzee Sequencing and Analysis Consortium. (2005). "Initial Sequence of the Chimpanzee Genome and Comparison with the Human Genome." *Nature* 437:69–87.

Day, M. H., and C. B. Stringer. (1991). "Les restes crâniens d'Omo-Kibish et leur classification à l'intérieur du genre *Homo*." *L'Anthropologie* 95:573–94.

Eldredge, N., and I. Tattersall. (1975). "Evolutionary Models, Phylogenetic Reconstruction, and Another Look at Hominid Phylogeny." In *Contributions to Primatology 5: Approaches to Primate Paleobiology*, ed. F. S. Szalay, 218–242. Basel: Karger

Fabre, P-H., A. Rodrigues, and E. J. P. Douzery. (2009). "Patterns of Macroevolution among Primates Inferred from a Supermatrix of Mitochondrial and Nuclear DNA." *Molecular Phylogenetics and Evolution* 53:808–25.

Goodman, M. (1962). "Immunochemistry of the Primates and Primate Evolution." *Annals of the New York Academy of Science* 102:219–34.

———. (1963). "Man's Place in the Phylogeny of the Primates as Reflected in Serum Proteins." In Washburn 1963, 204–34.

Haeusler, M., and H. M. McHenry. (2004). "Body Proportions of *Homo habilis* Reviewed." *Journal of Human Evolution* 46:433–65.

———. (2007). "Evolutionary Reversals of Limb Proportions in Early Hominids? Evidence from KNM-ER 3735 (*Homo habilis*)." *Journal of Human Evolution* 53:383–405.

Haslam, M., A. Hernández-Aguilar, V. Ling, S. Carvalho, I. de la Torre, A. DeStefano, A. Du, B. Hardy, J. Harris, T. Matsuzawa, W. McGrew, J. Mercader, R. Mora, M. Petraglia, H. Roche, E. Visalberghi, and R. Warren. (2009). "Primate Archaeology." *Nature* 460:339–44.

Hillson, S. (1996). *Dental Anthropology*. Cambridge: Cambridge University Press.

Holliday, T. W. (1995). "Body Size and Proportions in the Late Pleistocene Western Old World and the Origins of Modern Humans." Unpublished Doctoral Dissertation. University of New Mexico, Albuquerque.

Horai, S., Y. Satta, K. Hayasaka, R. Kondo, T. Inoue, T. Ishida, S. Hayashi, and N. Takahata. (1992). "Man's Place in Hominoidea Revealed by Mitochrondrial DNA Genealogy." *Journal of Molecular Evolution* 35:32–43.

Howells, W. W. (1973). *Cranial Variation in Man: A Study by Multivariate Analysis of Pattern of Differences Among Recent Human Populations*. Cambridge, MA: Harvard University Press.

———. (1989). *Skull Shapes and the Map: Craniometric Analyses in the Dispersion of Modern* Homo. Cambridge, MA: Harvard University Press.

Huxley, T. H. (1863). *Evidence as to Man's Place in Nature*. London: Williams and Norgate.

King, M-C., and A. C. Wilson. (1975). "Evolution at Two Levels in Humans and Chimpanzees." *Science* 188:107–16.

Lahr, M. M. (1996). *The Evolution of Modern Human Diversity: A Study of Cranial Variation.* Cambridge: Cambridge University Press.

Li, W.-H., and M. A. Saunders. (2005). "The Chimpanzee and Us." *Nature* 437:50–51.

Lieberman, D. E. (1998). "Sphenoid Shortening and the Evolution of Modern Human Cranial Shape." *Nature* 393:158–62.

Lieberman, D. E., B. M. McBratney, and G. Krovitz. (2002). "The Evolution and Development of Cranial Form in *Homo sapiens.*" *Proceedings of the National Academy of Sciences* 99:1134–39.

Mayr, E. (1950). "Taxonomic Categories in Fossil Hominids." *Cold Spring Harbor Symposium on Quantitative Biology* 15:109–18.

McBrearty, S., and A. S. Brooks. (2000). "The Revolution That Wasn't: A New Interpretation of the Origin of Modern Human Behavior." *Journal of Human Evolution* 39:453–563.

McDougall, I., F. H. Brown, and J. G. Fleagle. (2005). "Stratigraphic Placement and Age of Modern Humans from Kibish, Ethiopia." *Nature* 433:733–36.

Pearson, O. M. (2000). "Postcranial Remains and the Origins of Modern Humans." *Evolutionary Anthropology* 9:229–47.

———. (2004). "Has the Combination of Genetic and Fossil Evidence Solved the Riddle of Modern Human Origins?" *Evolutionary Anthropology* 13, no. 4:145–59.

Prado-Martinez, J., et al. (2013). "Great Ape Genetic Diversity and Population History." *Nature* 499:471–75.

Robson, S. L., and B. A. Wood. (2008). "Hominin Life History: Reconstruction and Evolution." *Journal of Anatomy* 219:394–425.

Ruff, C. B., E. Trinkaus, and T. W. Holliday. (1997). "Body Mass and Encephalization in Pleistocene *Homo.*" *Nature* 387:173–76.

Ruvolo, Maryellen. (1997). "Molecular Phylogeny of the Hominoids: Inferences from Multiple Independent DNA Sequence Data Sets." *Molecular Biological Evolution* 14, no. 3:248–65.

Sarich, V. M., and A. C. Wilson. (1967). "Immunological Time Scale for Hominid Evolution." *Science* 158:1200–1203.

Senut, B., M. Pickford, D. Gommery, P. Mein, K. Cheboi, and Y. Coppens. (2001). "First Hominid from the Miocene (Lukeino Formation, Kenya)." *Comptes rendus de l'Académie des sciences, Paris* 332:137–44.

Spoor, F., P. O'Higgins, C. Dean, and D. E. Lieberman. (1999). "Anterior Sphenoid in Modern Humans." *Nature* 397:572.

Stringer, C. B., J. J. Hublin, and B. Vandermeersch. (1984). "The Origin of Anatomically Modern Humans in Western Europe." In *The Origins of*

Modern Humans: A World Survey of the Fossil Evidence, ed. F. H. Smith and F. Spencer, 51–135. New York: Alan R. Liss.

Szalay, R. (2009). "Approaches to Primate Paleobiology." *Contributions to Primatology* 5:218–42.

Trinkaus, E. (1981). "Neanderthal Limb Proportions and Cold Adaptation." In *Aspects of Human Evolution*, ed. C. B. Stringer, 187–224. London: Taylor and Francis.

Uddin, M., et al. (2004). "Sister Grouping of Chimpanzees and Humans as Revealed by Genome-Wide Phylogenetic Analysis of Brain Expression Profiles." *Proceedings of the National Academy of Sciences* 101:2957–62.

Venn, O., I. Turner, I. Mathieson, N. de Groot, R. Bontrop, and G. McVean. (2014). "Strong Male Bias Drives Germline Mutation in Chimpanzees." *Science* 344:1272–75.

Washburn, S. L., ed. (1963). *Classification and Human Evolution*. Chicago: Aldine.

White, T. B., D. Asfaw, H. DeGusta, G. D. Gilbert, G. Richards, G. Suwa, and F. C. Howell. (2003). "Pleistocene *Homo sapiens* from Middle Awash, Ethiopia." *Nature* 423:742–47.

White, T. D., et al. (2009). "*Ardipithecus ramidus* and the Paleobiology of Early Hominins." *Science* 326:75–86.

Wood, B. A., and M. C. Collard. (1999). "The Human Genus." *Science* 284:65–71.

Zuckerkandl, E. (1963). "Perspectives in Molecular Anthropology." In Washburn 1963, 243–72.

Zuckerkandl, E., R. T. Jones, and L. Pauling. (1960). "A Comparison of Animal Hemoglobins by Tryptic Peptide Pattern Analysis." *Proceedings of National Academy of Science* 46:1349–60.

EIGHT

DARWIN'S EVOLUTIONARY ETHICS
The Empirical and Normative Justifications

—————

Robert J. Richards

In the increasingly secular atmosphere of the nineteenth century, intellectuals grew wary of the idea that nature had any moral authority. In an earlier age, one might have looked upon the dispositions of nature as divinely sanctioned, and thus one could call upon natural law to ground moral judgment. Certain behaviors, for instance, might have been declared "against nature" or denominated "unnatural acts," and thus morally forbidden. And one could have looked to examples drawn from nature for indications of virtue. For instance, in his little book *Quod animalia bruta saepe ratione utantur melius homine* (1654), Hieronymus Rorarius, papal nuncio of Clement VII to the court of Ferdinand of Hungry, compared the perfidy of human beings to the probity of elephants, "who care for their weak" and who "not only do not know of adultery, but think this act one of turpidity; for after copulating with females they do not return to their herd until they wash themselves in a stream" (Rorarius 1654, 21, 70). With the advent of Darwinian evolutionary theory in the mid-nineteenth century, nature seemed to lose its benign authority. Stripped of divine control, nature began to look like the enemy of humane morality.

Thomas Henry Huxley, Darwin's friend and critical defender, declared in a lecture given just before his death that individuals were obliged to fight against the "cosmic process," since nature was completely indifferent to human welfare. In his lecture "Evolution and Ethics" (1893), Huxley maintained:

> The struggle for existence tends to eliminate those less fitted to adapt themselves to the circumstances of their existence. The strongest, the most self-assertive, tend to tread down the weaker. . . . Social progress means a checking of the cosmic process at every step and the substitution for it of another, which may be called the ethical process; the end of which is not the survival of those who may happen to be the fittest, . . . but of those who are ethically the best. (Huxley 1902, 9:81)[1]

Huxley launched a singular objection to any effort at constructing an evolutionary ethics, an objection that would gather potency during the twentieth century, namely, that any ethics based on evolution would commit the "naturalist fallacy." Huxley didn't call it that; and the philosopher who named the fallacy, G. E. Moore, did not exactly formulate it in the manner of Huxley—though both had evolutionary ethics as their target (Richards 1987, chap. 7 and app. 2).[2] But it's essentially Huxley's formulation that provides the modern definition. The naturalistic fallacy, as it has come to be known, is the supposed derivation of a moral prescription—an "ought" statement—from premises that are descriptive only of empirical facts—"is" statements.[3] Huxley granted that we might have altruistic instincts, as well as selfish impulses. But simply having altruistic or nominally moral impulses does not entail that we "ought" to follow them. That decision requires another judgment, a moral judgment, and cannot flow from the simple fact of having a particular set of evolved instincts, even if those be instincts to act for the benefit of others. The older modes of natural law theory, as well as the Darwinian appeal to our evolved nature, now, in the twenty-first century, seem highly suspect philosophically. The dangers of the easy identification of moral attitudes with natural dispositions appear obvious from the events of the late 1930s and early 1940s in Germany.

The Nazi biology of the period seems to offer the cautionary example of not heeding Huxley's warning. The moral of the tale is encapsulated in the title of a book by Richard Weikart: *From Darwin to*

Hitler (Weikart 2004). A more recent volume trumps the first: *Hitler's Ethic: The Nazi Pursuit of Evolutionary Progress* (Weikart 2009). These volumes lay the blame for Hitler's actions, partly at least, on the doorstep of Down House. The path from that doorstep led to infamy: "No matter how crooked the road was from Darwin to Hitler, clearly Darwinism and eugenics smoothed the path for Nazi ideology, especially for the Nazi stress on expansion, war, racial struggle, and racial extermination." More precisely the source of Hitler's regime of evil was ethics of a particular sort: "Evolutionary ethics drove him [Hitler] to engage in behavior that the rest of us consider abominable" (Weikart 2004, 6; 2009, 2). Though analyses of the kind offered by Weikart pretend to be disinterested, simply a factual account of the historical trajectory of Darwinian ethical theory, it is clear they are intended also to morally indict Darwin's general theory of evolution and his specific conception of ethics (Richards 2013, 192–242).

Even many of those who regard Darwin favorably assume that his ethical considerations were rooted in selfishness. As Michael Ghiselin, a Darwinian scholar, so delicately put it: "Scratch an altruist and watch a hypocrite bleed" (Ghiselin 1974, 247). Michael Ruse and E. O. Wilson, two scholars who couldn't be more committed to Darwinian evolutionary theory, nonetheless argue that our evolved morality has deceived us into thinking we act benevolently and altruistically for good, objective reasons, whereas our decisions flow from inherited epigenetic rules that "(unknown to us) ultimately serve our genetic best interest" (Ruse and Wilson 1986, 179).[4] They maintain, in short, that what we might believe to be an altruistic act directed to the welfare of another is really a selfish act to enhance the propagation of our genes. In order that we guard against this Darwinian original sin of intrinsic selfishness, Richard Dawkins presumes that we have to escape our endowed nature and acquire, through learning, a second nature: "we must *teach* our children altruism," he urges, "for we cannot expect it to be part of their biological nature" (Dawkins 1976, 139).

Thus the usual view of Darwin's accomplishment is that he not only eviscerated nature of moral value, he left man morally naked to the world. I think it will come as something of a surprise to the proponents of Darwinism that Darwin, on the contrary, reconstructed nature with a moral spine in the *Origin of Species* (Richards 2013, 13–54), and that in the *Descent of Man* he conceived his moral theory as removing "the reproach of laying the foundation of the most noble part of our nature in the base principle of selfishness" (Darwin 1871, 1:98). In what follows, I will sketch out Darwin's theory of evolutionary ethics and attempt to show that with some mild philosophical therapy it escapes the usual objections brought against it. First, however, a word about two kinds of justification.

EMPIRICAL AND MORAL JUSTIFICATION OF ETHICAL THEORY

In discussing the viability of a theory of evolutionary ethics, we must distinguish between a descriptive justification and a normative justification. A descriptive justification, say by a historian or anthropologist, would consist of two parts: a description of certain actions that members of a society would call moral or ethical (as well as those they would term immoral or non-ethical)—that is, a description of what society members believe someone in their community ought or ought not to do simply by reason of being a member of their social (as opposed to professional or legal) group; and then, the historian or anthropologist would have to provide empirical evidence that such descriptions are valid (i.e., members of the society do so characterize their behaviors—or might do so if required) and further evidence that the behaviors have causally derived from evolved moral attitudes or instincts.

Such an empirical justification might fail in two ways: the descriptions of the behaviors might be inadequate, that is, might not capture what members of the society actually regard as behaviors governing all members of their group; or the supposed causes of the behaviors are not what the justification claims them to be. Various evolutionary thinkers have advanced theories of moral evolution—Charles Darwin, Herbert Spencer (Spencer 1893), E. O. Wilson, Michael Ruse, Marc Hauser (Hauser 2006), Jonathan Haidt (Haidt 2007), just to name a few—all of whom assume that social groups have acquired, over long periods of time, behavioral attitudes that could nominally be called "moral," and that these attitudes explain the altruistic behaviors of the groups. Darwin and Spencer relied on both natural selection and the inheritance of acquired characters to explain the evolution of moral behavior; modern scholars, of course, have dispensed with the Lamarckian device. Other auxiliary causes are also invoked by advocates, from cultural learning and early training to linguistic codes and behavioral examples.

A thinker could advance an empirically justified moral theory, and yet a critic might still ask for normative justification. T. H. Huxley, in his lecture "Evolution and Ethics," found that evolutionary ethics failed in just this manner. He had originally directed his animus against his old friend Herbert Spencer, but it's obvious that his objections were also relevant to Darwin's proposals in the *Descent of Man*. In the following, I will describe Darwin's theory of morality, which is an empirical theory, and then consider whether a theory of the sort he devised can be empirically justified and then whether it can, despite Huxley's objections, be morally justified.

DARWIN'S MORAL THEORY

Just before he read Malthus in late September 1838, and began to develop his ideas about natural selection, Darwin had considered what his incipient transmutation theory meant for human beings. It would have to explain the distinctive feature of the human animal, which Darwin and his contemporaries took to be moral behavior, not rational behavior. After all, in the British Empiricist tradition, reasoning amounts to the association of ideas, which are only faint sensory images. Animals are quite capable of such associations; many aristocratic Englishmen thought their hunting dogs showed as keen a rational ability as their valets. So it was moral behavior that seemed to distinguish decisively the hounds from the master of the hounds.

Initially, Darwin explained moral behavior following the utilitarian ideas of William Paley, the philosopher whose work he knew best from his Cambridge days. Paley defined the moral as what, in the long run, would be really useful for the individual or the community.

Darwin gave Paley's rule of "expediency" a biological interpretation. He supposed that the social instincts—friendship, cooperation, parental care—which had been honed over long periods of time, were ultimately what was useful, and therefore what we meant by morally good (Darwin *Notebook M*, in Barrett et al. 1987, 552).[5] Darwin read Malthus on 28 September 1838, and three days later further advanced his ideas about moral behavior. He believed he had discovered the roots of conscience in unrequited social instincts. He imagined what would count as the prick of conscience in an animal—being a sporting Englishman, he not surprisingly thought of his dog:

> Octob. 3d. Dog obeying instinct of running hare is stopped by fleas, also by greater temptation as bitch. . . . Now if dogs mind were so framed that he constantly compared his impression, & wished he had done so & so for his interest, & found he disobeyed a wish which was part of his system, & constant, for a wish which was only short & might otherwise have been relieved, he would be sorry or have a troubled conscience.—Therefore I say grant reason to any animal with social & sexual instinct he *must* have conscience—this is capital view. (563–64)

At this stage in his theorizing, Darwin supposed that the social instincts were those impulses we regarded as quintessentially moral; and if a persistent urge to perform some such social act were thwarted by a stronger impulse, then the nagging, unfulfilled social instinct would be what we meant by a feeling of guilt.

Darwin further expanded his considerations of conscience in light of his reading of James Mackintosh's *Dissertation on the Progress of Ethical Philosophy* (1836). Mackintosh, a distant relative of Darwin by marriage, sided with the likes of Shaftesbury and Hutcheson, contending that we had a moral sense for right conduct; he disputed Paley's notion that pleasure or utility guided us in moral action. Mackintosh's analysis of moral sense fit neatly into Darwin's conception of the moral impulse as based in persistent social instincts: "Butler & Mackintosh," he jotted in his notes during the summer of 1839, "characterize the moral sense, but its 'supremacy',—I make its supremacy, solely due to greater duration of impression of social instincts, than other passions, or instincts" (Darwin, "Old and Useless Notes," in Barrett et al. 1987, 628).

Though he had already formulated his principle of natural selection by the time he read Mackintosh, Darwin continued to suppose that inherited habit was the root of the other-regarding instincts, presumably because his new device seemed to establish traits that benefited the actor, not the recipient of the action—he assumed the highest moral behavior aimed principally to aid others, not self. Darwin would only be able to apply the device of natural selection to explain moral instincts when he solved this problem of traits that benefited the recipient and not the actor. And he did solve the problem, but only in the short period before the publication of the *Origin of Species.*

The solution came with the unraveling of another difficulty. In the 1840s, Darwin began reading up on the social insects, especially honey bees and ants. In studying, for example, William Kirby and William Spence's *Introduction to Entomology* (1818), he puzzled over the origin of instincts in neuter insects—especially the instincts characteristic of the various castes of ants within a nest. Since natural selection operates on traits that allow the individual to reach reproductive age and pass on those traits, the instincts of worker ants and bees, which are neuters, present an obvious problem. As Darwin scribbled in the margin of volume 3 of Kirby and Spence's *Introduction*: "Neuters do not breed! How instinct acquired" (Darwin annotation on Kirby and Spence 1818, 2:55). The problem was compounded in the case of soldier bees, who sacrificed their lives in stinging intruders to the hive and thus eviscerated themselves. They instinctively acted, but with extreme prejudice to themselves.

Darwin came to his solution only in the throes of composing his *Big Species Book*, the abridgement and extension of which became the *Origin of Species*. In the chapter on instinct, he hit on the answer to the puzzle. He wrote:

> The principle of selection, namely not of the individual which cannot breed, but of the family which produced such individual, has I believe been followed by nature in regard to the neuters amongst social insects. (Darwin 1975, 510)[6]

With the solution to the problem of the neuter insects, Darwin had a way of applying natural selection to explain traits that were not of an advantage to the individual but to the community. Natural selection worked on the whole hive or nest. Those hives or nests that by chance had individuals that cooperated in the work and protection of the community would have a better chance of surviving than hives or nests lacking such individuals. As the successful hives and nests sent out queens to establish new communities, the continued operations of natural selection would hone instincts and anatomy of workers to finely carve out distinctive castes with distinctive traits. In this way, the properties of the neuter workers would evolve, and other-regarding traits would be explained. Darwin would recruit this kind of community selection to explain those human social instincts that were costly to self but provided advantage to the family and wider community. And in the *Descent of Man*, he quite explicitly drew on the model of the social insects to lay the foundation for his theory of moral conscience.

Since his earliest theorizing about the transmutation of species and the place of human beings in nature, Darwin realized he would have to give an account of the distinctive character of the human animal, which for him was not rational ability but moral capacity. Darwin, as I've suggested above, was schooled in the British Empiricism of his grandfather Erasmus Darwin and was a careful reader of David Hume. Neither he nor others of this school would hesitate to attribute a modicum of reason even to insects, since reason could only be the more or less simple association of ideas, which themselves were but faint images of sensation—certainly characteristics not denied to lower animals. The unique feature of human nature, in Darwin's estimation, was the capacity for moral judgment as realized in the operations of conscience. And in the *Descent*, he devoted considerable space to developing an account of conscience.

Four features of the capacity for moral deliberation seemed to Darwin requisite for an adequate account of conscience: sufficient intelligence and memory to be able to compare past with present deliberations; language ability to codify rules of conduct for a social group; the acquisition of habit to refine behavior in line with community rules; and, most importantly, social instincts that regarded the welfare of the family and the community (Darwin 1871, 1:72). This last note was perhaps the defining characteristic: the unselfish impulse "to act for the good of the community" (1:72). In order to explain the evolution of this kind of instinct, Darwin proposed a scenario based on the model of the social insects.

He imagined a community of social animals advancing toward a stage in which the social instincts had become well developed; then, with the gradual acquisition of higher reasoning ability and finally language, human beings, as we now recognize them, would slowly emerge. But how to explain the binding instincts that formed those proto-men into a real human community? Darwin thought the principal explanatory force would be natural selection operating on small clans. Those groups that by chance had individuals that cooperated with one another and sought the welfare of their kin and that of their fellows in the group—they would have the advantage over other clans that lacked altruistic individuals. Here is how Darwin put it in the *Descent*:

> It must not be forgotten that although a high standard of morality gives but a slight or no advantage to each individual man and his children over the other men of the same tribe, yet an increase in the number of well-endowed men will certainly give an immense advantage to one tribe over another. There can be no doubt that a tribe including many members who, from possessing in a high degree the spirit of patriotism, fidelity, obedience, courage, and sympathy, were always ready to give aid to each other and to sacrifice themselves for the common good, would be victorious over most other tribes; and this would be natural selection. At all times throughout the world tribes have supplanted other tribes; and as morality is one element in their success, the standard of morality and the number of well-endowed men will thus everywhere tend to rise and increase. (Darwin 1871, 1:166)

Darwin added that the fundamental altruistic impulse would be augmented by two processes: individuals becoming responsive to "praise and blame" and individuals recognizing that other-regarding action might well be reciprocated. While both of these principles would undoubtedly be operative in a community, he thought that such principles, in themselves, produced a "low motive" for moral behavior (1:163). Authentic altruism was rooted in natural selection. The natural selection of those communities exhibiting members who were morally sensitive would ensure that moral behavior was not adventitious, not something that was only contingently related to human beings; rather, selection would instill such attitudes into the very fabric of human nature. In Darwin's estimation, morality was bred in the bone.

In his theory, Darwin had a place for learning and what we might consider cultural evolution. He envisioned proto-human groups, small tribal clans undergoing natural selection for altruistic behavior. But then the gradual development of intellect and the establishment of cultural habits would play a role: individuals in these groups would come to learn just who their brothers and sisters were and what measures were truly beneficial. They would eventually discover that superficial differences of skin color and head shape did not distinguish their group from the other group across the river. Thus as societies developed, members of tribal clans would expand their other-regarding behavior to neighbors, then to larger community groups, finally to members of the same nation and other nations—to mankind at large (1:100–101).

Darwin was persuaded that the histories of peoples, explored within the framework of evolution and natural selection, supported his theory of moral development. Such histories supplied an empirical justification of his moral theory. Today, a good many studies in behavioral biology and behavioral economics seem to lend general support to the empirical justification of Darwin's general view (Katz 2002). But what about moral justification? That is, though we may have evolved to have the impulse to help our fellows on this or that occasion, the question remains: "Ought we to help them?" And more generally, though human beings may have a set of altruistic impulses, is there a general argument that suggests those impulses "should" be followed? Put another way, can the Darwinian approach escape committing the naturalistic fallacy of moving from what evolution as a matter of fact has instilled as motives for behavior to justifying those motives as norms of right action?

Darwin himself seems to have anticipated the issue of moral justification, at least in an inchoate way. Recognizing the special character of the imperious "ought," he asked in the *Descent of Man*: "Why should a man feel that he ought to obey one instinctive desire rather than another? Why does he bitterly regret if he has yielded to the strong sense of self-preservation, and has not risked his life to save that of a fellow-creature?" (Darwin 1871, 1:87). These would also be Huxley's questions. Darwin answered by suggesting that the deeply instilled social instincts would be more persistent than the momentary though stronger desires for pleasure or other kinds of gratification. When these latter faded and the underlying social instinct remained unfulfilled,

> man will then feel dissatisfied with himself, and will resolve with more or less force to act differently for the future. This is conscience; for conscience looks backwards and judges past actions, inducing that kind of dissatisfaction, which if weak we call regret, and if severe remorse. (Darwin 1871, 1:91)

The attentive philosopher will not, of course, be content with this answer, since it remains at the level of empirical fact. Yet, as I believe, it offers the opening wedge for a normative justification.

NORMATIVE JUSTIFICATION OF EVOLUTIONARY ETHICS

The crux of the assumed problem of normative justification is the gap between empirical fact and normative command. It seems as if no justification of norms based on facts can obtain. But part of the difficulty, I believe, has to do with the concept of justification itself. What does it mean to justify a proposition? Presumably it means providing compelling evidence for it. Of course, we want, in a regressive way, to make sure the standards for compelling evidence are appropriate, that they themselves are justified. As propaedeutic to tackling the problem of the nature of justification in the moral realm, we might first consider standards for evidence in scientific justifications of hypotheses.

Usually some relationship, principle, or law is justified through experiment or controlled observation. Popperians are wont to point out that such efforts at justification commit the logical fallacy of affirming the consequent (i.e., the problem of induction: no limited empirical observations can allow, according to the norms of modern logic, to conclude to a universal proposition). Yet the whole edifice of science is based exactly on justifying the laws of science by experiment and observation. The practice and success of science sets the standards for accepting laws and other norms. So even if one points out that according to the rules of modern logic, one cannot go from factual premises to normative conclusions, this does not settle the question. It certainly didn't settle it in the general case of the establishment of scientific laws.

Quite typically we advance the moral proposition "You should not cheat" by asking: "What if everyone cheated?" The implicit answer that society would be in chaos is, of course, a factual reason for accepting the norm. That's the justification, and, of course, it's a typical strategy for justifying norms. When immigrants to the U.S. swear allegiance to the Constitution, they presumably do so for the reason that, all things considered, the society the U.S. Constitution mandates is of a certain empirical kind—that is to say, an orderly, Constitution-governed society supplies the justification for accepting the norms embodied in the Constitution. The ubiquity and success of the practice, as in the case of science, sets the standards for justification in the moral sphere.

Of course, one may be called on to justify the desire to live in a particular kind of society, but that's another, independent justification. It's comparable to asking why one should use a T-test as opposed to a Chi-square test in accepting a hypothesis in science, after the T-test has justified the hypothesis. In the case of the new immigrants assenting to the Constitution, they might simply say that, as a matter of fact, they prefer the kind of society that the U.S. Constitution mandates. So here is a typical two-stage justification, appealing only to matters of fact. So the first step in overcoming the naturalistic fallacy is to recognize there is no general fallacy of this kind (though there may be fallacious arguments in particular instances of going from facts to values).

Looking more precisely at Darwin's argument that moral instincts are persistent and that this supplies their unique character, a Kantian argument comes to the fore. Kant justified the a priori status of the categories of cause, substance, unity, and so on, by arguing that it made sense of or explained the character of experience, at least in the theoretical realm. But if we have evolved in the manner that Darwin has maintained, then we have come equipped with a set of persistent social instincts, which we are ready to identify as moral impulses because they have that persistent, special character. Now in a cool hour, if we reflect on these instincts and ask, should I follow these impulses as opposed to others, then the reflective answer must be "Yes, you ought"—since we have evolved so as to regard them as special, privileged, and of ultimate consequence. Thus by reason of our nature—that is, by reason of who we are—we recognize the morally imperative power of these instincts; we must recognize that power because that's the kind of creature we are. But what more is there to justification? One could keep asking the regressive question: "Yes, but what of the reflective judgment that these instincts have normative force—what justifies that judgment?" The regressive question can always be asked in any effort of justification; so either the justificatory buck must stop at some point or no justification is possible in any domain. But the latter is quite unacceptable; so the appeal to further justification must cease. Here is a case in which, from a meta-ethical standpoint, a normative framework that allows the justification of lower ordered norms is itself dependent on an empirical justification, namely, that of the evolution of human beings as altruists.

A related objection might yet be voiced, of the sort played out in Ronald Dworkin's book *Justice for Hedgehogs* (2011). The objector might say that an individual who binds himself or herself under the norms of the Constitution does so because of yet more remote values, say, the value of living a life that protects individual liberties. Dworkin remains convinced that the is-ought divide cannot be bridged. Only the whole system of values can justify a particular value that fits coherently into the system. I believe this tactic, however, is liable to the problem of infinite regress: attempting to justify each lower level value by a higher level value, that itself stands in need of justification. Unless one breaks the circle with an appeal to principles outside the circle, the problem of justification would fall into Zeno's trap. The circle is broken by recognizing that typically we justify complexes of norms by appeal to empirical situations, and often justify single norms in the same fashion.

When Aristotle was formulating the norms of syllogistic logic, he sought to examine those kinds of arguments that people generally regarded as sound. His normative principles were justified by demonstrating that they rendered valid those arguments that individuals intuitively regarded as valid. That is, he justified norms by appeal to empirical facts. The individual who binds himself or herself under the rule of the Constitution does the same.

As I've already mentioned, Michael Ruse and E. O. Wilson have an objection to the normative justification of evolutionary ethics. This is based on the supposed non-objective character of the values instilled by evolution. In their article "Moral Philosophy as Applied Science," the authors grant that the empirical evidence is mounting that human beings have evolved so as to have altruistic impulses as part of their genetic endowment (Ruse and Wilson 1986). They maintain that, as part of this endowment, humans have been led to believe that moral prescriptions are objective, that they are given independently of human beliefs and desires:

> Human beings function better if they are deceived by their genes into thinking that there is a disinterested objective morality binding upon them, which all should obey. We help others because it is "right" to help them and because we know that they are inwardly compelled to reciprocate in equal measure. (179)

Ruse and Wilson thus maintain that because moral impulses are part of our hereditary legacy, they are contingent: "No abstract moral principles exist outside the particular nature of individual species." They believe that this "is obviously quite inconsistent with the notion of morality as a set of objective, eternal verities" (186). The authors agree with the tired philosophical idea that values cannot be justified by facts.

The first response to the position of Ruse and Wilson is that they have a strange idea about what "objective" means. It's rather surprising that they harken back to the religious idea that if God promulgated the moral code, then it would be objective—that is, eternal and binding on all human beings. Yet, one can immediately inquire about their notion of "objective" when it is applied to other kinds of propositions. "Objective" in mathematics, logic, and science means "not a matter of individual whim," "recognized as valid independently of individual preferences," "publically testable and confirmable." Mathematical demonstrations and logical proofs can be perfectly objective without any external propositions against which they need be measured. In arithmetic, for instance, it is an objective fact that changing the order of the operands in addition or multiplication does not change the results, since the commutative law sanctions such moves—it justifies them. One can ask for the justification of the commutative and other such rules, but that is another matter, though one which can be adjudicated objectively as well—one could bring evidence to show using such rules produces a useful formal instrument. Just so, an act conferring benefit on another can be objectively justified if done intentionally and with an altruistic motive—such is the justification of moral behavior. To ask for a justification of the motive—why act altruistically?—is another request for justification, which then moves to the empirical realm: objective evidence indicates that is the way we have evolved. And so, when we reflectively consider our altruistic instincts, we naturally conclude they ought to be followed. But could we have evolved otherwise?

Ruse and Wilson seem to think that we could have evolved into some other kind of creature than an altruist; and it is also this possibility, they suggest, that makes our morality non-objective: we are contingent creatures. But if we evolved into another kind of creature, we wouldn't be talking about human beings: the very meaning of "human" includes responsiveness to the needs of others. It could also be that humans, as we now known them, might in future evolve into a different kind of beast; but again, that would have no bearing on the objective assessment of humans as they now exist. If all copper suddenly disappeared from the universe, it would still be true to say that copper conducts electricity; just so, if we evolved into a different kind of creature, it would still be objectively true that intentionally acting for motives of altruism is morally sanctioned. If we could see into the future and observe individuals who had our bodily form but were pitiless in the presence of others' pain, who neglected the welfare of their children, who failed completely to cooperate with others—if such creatures were observed we would not likely call them human beings, but rather refer to them as some weird zombielike creatures, pod people who only look like us.

But this is science fiction. As far as we know, a race that lived in groups, had our mating characteristics, and produced offspring requiring long-term care—such a race that failed to cooperate with one another and neglected their offspring would, under the terms of natural selection, quickly become extinct. We can imagine a future universe in which fire did not consume human flesh, but in a real universe that would require the abrogation of the laws of physics and biology. The conclusion to be drawn is, I believe, simply this: the fact that humans have evolved—even contingently evolved—does not preclude there being perfectly objective truths about their nature and behavior, especially their moral behavior.

DARWINIAN MORALITY JUSTIFIED

The empirical justification of Darwin's theory in most areas of biology has been established conclusively over the last 150 years since the publication of the *Origin of Species*. The *Descent of Man* extended his considerations to human beings and offered a compelling account of human moral behavior. In the last twenty years or so the evidence of the general empirical validity of Darwin's conception as applied to the explanation of human moral behavior has gained apace, without concern for the philosopher's problem of fact-norm justification.[7]

On the assumption that we have, as the evidence indicates, evolved basically in the way Darwin has argued, then we come possessed with a set of dispositions that constitute a necessary feature of what it is to be human. So the simple argument for normative justification runs:

if you want to be human—and you cannot help but want that—you ought to act altruistically in the appropriate circumstances. The judgment has the form of what is usually called an instrumental ought— for example, if you want to cut wood in the most efficient way, you ought to use a saw, since that's the best way to do it. It is sometimes thought that the *ought* of the moral injunction should be absolute: you simply ought to act altruistically. But we would not say that of an individual who lacks the requisite mental functioning, an individual so deficient as not to be able to discern an altruistic act or its circumstances. We, of necessity, make the argument in relation to the empirical subject before us—ourselves or others—and presume that individual to be a well-functioning human being. Then we put it to the person: if you want to be true to your human nature—and you do (since that is the way you are evolutionarily constituted)—then you ought to act altruistically.

One could yet ask the regressive question: But why should I as a human believe either that I do have such desires or that I have evolved in the way suggested? But at this stage, the moral justification has become the empirical justification, and the regressive questioning has been staunched. It should be remembered that most all practical efforts at justification—from the promptings of one's mother to the coaxing of one's parole officer—are attempts to get the individual to recognize his or her own humanity and its deepest features. And this typical practice of human beings is the final justification.

NOTES

1. Huxley's animus was initially directed at Herbert Spencer's system of evolutionary ethics; but it was clear his objections would also have told against Darwin's conception. I have discussed Huxley's objections and Darwin's position in Richards 1987, chap. 7 and app. 2.

2. I have discussed Moore's position in Richards 1987, 323–25.

3. Sometimes this form of the so-called fallacy is attributed to David Hume. See Hume 1888, 469: "In every system of morality, which I have hitherto met with, I have always remarked, that the author proceeds for some time in the ordinary ways of reasoning, and establishes the being of a God, or

makes observations concerning human affairs; when all of a sudden I am surprised to find, that instead of the usual copulations of propositions, is, and is not, I meet with no proposition that is not connected with an ought, or an ought not. This change is imperceptible; but is however, of the last consequence. For as this ought, or ought not, expresses some new relation or affirmation, 'tis necessary that it should be observed and explained; and at the same time that a reason should be given; for what seems altogether inconceivable, how this new relation can be a deduction from others, which are entirely different from it." It should be noted that Hume does not reject this kind of inference; he only asks that it be explained. And he has his own explanation, which is not far from Darwin's—namely, it is based on a kind of moral instinct.

4. Ruse and Wilson argue that our genes deceive us into thinking we are making objective judgments in making moral decisions: "human beings function better if they are deceived by their genes into thinking that there is a disinterested objective morality binding upon them, which all should obey" (Ruse and Wilson 1986, 179).

5. Darwin in Barrett et al., 1987, 552: "Sept. 8th. I am tempted to say that those actions which have been found necessary for long generation, (as friendship to fellow animals in social animals) are those which are good & consequently give pleasure, & not as Paley's rule is those that on long run *will* do good—alter *will* in all cases to *have* & *origin* as well as *rule* will be given."

6. A comparable passage can be found in Darwin 1859, 242.

7. For example, the collection *Evolutionary Origins of Morality* has only one, passing mention of the naturalistic fallacy (Katz 2002).

REFERENCES

Barrett, P., et al., eds. (1987). *Charles Darwin's Notebooks, 1836–1844.* Ithaca, NY: Cornell University Press.

Darwin, C. (1859). *On the Origin of Species.* London: Murray.

———. (1871). *Descent of Man and Selection in Relation to Sex.* 2 vols. London: Murray.

———. (1975). *Charles Darwin's Natural Selection, being the Second Part of his Big Species Book Written from 1856–1858.* Edited by R. Stauffer. Cambridge: Cambridge University Press.

Dawkins, R. (1976). *The Selfish Gene.* New York: Oxford University Press.

Dworkin, R. (2011). *Justice for Hedgehogs.* Cambridge, MA: Harvard University Press.

Ghiselin, M. (1974). *The Economy of Nature and the Evolution of Sex.* Berkeley: University of California Press.

Haidt, J. (2007). "The New Synthesis in Moral Psychology." *Science* 316:998–1002.

Hauser, M. (2006). *Moral Minds: How Nature Designed Our Universal Sense of Right and Wrong.* New York: HarperCollins.

Hume, D. (1888). *A Treatise of Human Nature.* Edited by L. A. Selby-Bigge. Oxford: Clarendon Press.

Huxley, T. H. (1902). "Evolution and Ethics." In *Collected Essays*, 9 vols., 9:46–116. New York: D. Appleton, 1896–1902.

Katz, L., ed. (2002). *Evolutionary Origins of Morality.* Thorverton: Imprint Academic.

Kirby, W., and W. Spence. (1818). *Introduction to Entomology.* 2nd ed. 4 vols. London: Longman, Hurst, Rees, Orme, and Brown.

Mackintosh, J. (1836). *Dissertation on the Progress of Ethical Philosophy*, with a preface by William Whewell. Edinburgh: Adam and Charles Black.

Richards, R. J. (1987). *Darwin and the Emergence of Evolutionary Theories of Mind and Behavior.* Chicago: University of Chicago Press.

———. (2013). *Was Hitler a Darwinian? Disputed Questions in the History of Evolutionary Theory.* Chicago: University of Chicago Press.

Rorarius, H. (1654). *Quod animalia bruta saepe ratione utantur melius homine.* Edited by Gabriel Naudé. Amsterdam: Ravesteinium.

Ruse, M., and E. O. Wilson. (1986). "Moral Philosophy as Applied Science." *Philosophy* 61:173–92.

Spencer, H. (1893). *Principles of Ethics.* 2 vols. Indianapolis: Liberty Classics, 1978.

Weikart, R. (2004). *From Darwin to Hitler: Evolutionary Ethics, Eugenics, and Racism in Germany.* New York: Palgrave Macmillan.

———. (2009). *Hitler's Ethic: The Nazi Pursuit of Evolutionary Progress.* New York: Palgrave Macmillan.

NINE

CROSSING THE MILVIAN BRIDGE
When Do Evolutionary Explanations of Belief Debunk Belief?

Paul E. Griffiths and John S. Wilkins

But then with me the horrid doubt always arises whether the
convictions of man's mind, which has been developed from
the mind of the lower animals, are of any value or at all trustworthy.
Would any one trust in the convictions of a monkey's mind,
if there are any convictions in such a mind?

—Darwin to Graham, 3 July 1881

EVOLUTIONARY DEBUNKING ARGUMENTS IN THREE DOMAINS

Two traditional targets for evolutionary skepticism are religion and morality. Evolutionary skeptical arguments against religious belief are continuous with earlier genetic arguments against religion, such as that implicit in David Hume's *Natural History of Religion* (Hume [1757] 1956; Kahane 2011, 121n10). Evolutionary arguments are also commonly used to support moral skepticism. For example, Richard Joyce's influential *The Evolution of Morality* argues "that descriptive knowledge of the genealogy of morals (in combination with some philosophizing) should undermine our confidence in moral judgments" (Joyce 2006, 223). In contemporary philosophy, however, the most widely discussed form of evolutionary skepticism is probably that proposed by Alvin Plantinga (1993). He argues that if the mind has evolved by natural selection and if there is no creator God, then we have no reason to suppose that any of our beliefs are true. Plantinga, of course, does not actually advocate evolutionary skepticism. He uses it as a stick with which to beat the view that there is no creator God.[1] The relevance of his argument to the present chapter is that it shows evolutionary skepticism can be directed against science and commonsense as well as its traditional targets.

Guy Kahane (2011) has outlined the general form of what he terms "evolutionary debunking arguments":

Causal premise. S's belief that p is explained by X
Epistemic premise. X is an off-track process
Therefore
S's belief that p is unjustified.

An "off-track" process is one that does not track truth, that is to say, it produces beliefs in a manner that is not sensitive to whether those beliefs are true or not.

In the next section we will present the most straightforward reply to evolutionary debunking arguments. This is to deny the epistemic premise in Kahane's schema. Evolution is not an off-track process with respect to truth in the domain of beliefs under attack. Evolution will favor organisms that form true beliefs in that domain. In the following section we show that the standard arguments that evolution will not track truth rest on misunderstandings of natural selection, and we define in what sense evolution does, indeed, track truth. We conclude that with this definition of truth-tracking, there is a plausible defense of commonsense beliefs. Next we examine just how far such a defense will take us, and tentatively suggest that it can be extended to beliefs derived from the sciences. In later sections we consider a subtler way to deny the epistemic premise in Kahane's schema, namely, by giving a deflationary account of the meaning of truth-claims in the relevant domain. Kahane has explored this response to evolutionary skepticism about morality; we explore it as a response to evolutionary skepticism about religion.

IN HOC SIGNO VINCES: TRUTH AND PRAGMATIC SUCCESS

> Constantine . . . is reported to have seen with his own eyes the luminous trophy of the cross, placed above the meridian sun, and inscribed with the following words: BY THIS CONQUER. . . . Christ . . . directed Constantine to frame a similar standard, and to march, with the assurance of victory, against Maxentius and all his enemies. (Gibbon [1782] 1850, 2. XX.iii)

When Constantine fought the battle at the Milvian bridge in 312, he adopted a new battle standard: the cross. Constantine won and went on to found the Byzantine Roman Empire. Did he win because of the power of the sign and the truth it denoted, or because his largely Christian soldiers were inspired to fight more effectively? Traditionally, many Christians have assumed the former. Constantine was successful because his beliefs were true: God was on his side.

We call an argument that links true belief to pragmatic success a "Milvian bridge" argument. The specific kind of pragmatic success with which we will be concerned is evolutionary success. To defeat evolutionary skepticism, true belief must be linked to evolutionary success in such a way that evolution can be expected to produce organisms that have true beliefs. However, it would obviously be too much to require that evolution produce organisms all of whose beliefs are true. Evolutionary theory must explain the world as it actually is, and we know that people and animals often form false beliefs. It would also be too much to require that evolution produce organisms whose beliefs are formulated in an ideal conceptual scheme—it ought to be possible for someone other than God or an ideal epistemic agent speaking at the "end of inquiry" to have true beliefs.[2] We suggest that a reasonable formulation of the Milvian bridge principle would be something like this:

Milvian Bridge: The X facts are related to the evolutionary success of X beliefs in such a way that it is reasonable to accept and act on X beliefs produced by our evolved cognitive faculties.

We do not believe that a Milvian bridge can be constructed linking true religious beliefs to evolutionary success. Even a cursory examination of the leading contemporary accounts of the evolution of religious belief makes it clear that none of them make any reference to the truth or falsity of those beliefs when explaining their effects on reproductive fitness (Atran 2002; Boyer 2001; Wilson 2002). Conversely, although a Milvian bridge argument has been endorsed by some religious thinkers in the past, and may persist in the vulgar theology of some religious traditions, few if any contemporary theologians accept the claim that the relative truth of two religions can be decided in battle, or by counting their adherents. Believers may be guaranteed success in the afterlife, but they are not guaranteed the kind of success that is relevant to reproductive fitness.

While the Milvian bridge has no serious standing in theology, however, it continues to be taken seriously as argument for the truth of scientific beliefs. Richard Dawkins has repeatedly made use of this argument to contrast science and religion: "If all the achievements of scientists were wiped out tomorrow, there would be no doctors but witch doctors, no transport faster than horses, no computers, no printed books, no agriculture beyond subsistence peasant farming. If all the achievements of theologians were wiped out tomorrow, would anyone notice the smallest difference?" (Dawkins 1998, 6). In the philosophy of science the so-called "ultimate argument" or "miracle" argument for scientific realism is closely related to the Milvian bridge. According to this argument, unless something very like the entities referred to by scientific theories actually exists, and unless those theories are at least approximately true, then the pragmatic success of the technologies derived from those theories would be miraculous (Smart 1963; Putnam 1975). However, while this argument is still taken seriously, it is notoriously hard to formulate a version of the argument that does not prove either too much or too little (Musgrave 1988; Psillos 1999). Too much if it suggests that technological success establishes the truth of the science from which it is derived. The Industrial Revolution, after all, was founded on Newtonian theories that we now know to be fundamentally mistaken. So we have good reason to suspect that even the most successful scientific theories are only stepping stones to new and different theories. Too little because once we water down the notion of truth to avoid the problem just raised, we risk defining truth in terms of pragmatic effectiveness and rendering the argument circular.

Fortunately, it is not necessary for us to settle the realism debate in the philosophy of science. In the next section we will construct a Milvian bridge linking commonsense beliefs about the world around us to evolutionary success. Having done so, we will sketch how the bridge might be extended to scientific beliefs. We will be content to show that it is reasonable to accept and act on scientific beliefs, and will not attempt to establish any particular form of scientific realism.

BUILDING THE MILVIAN BRIDGE: WHY EVOLUTION TRACKS TRUTH

Many authors have argued that evolution will not produce cognitive systems that are truth-tracking. Evolution will favor cognitive adaptations that produce beliefs that maximize an organism's fitness irrespective of whether those beliefs are true. Hence, we should expect cognitive adaptations to be fitness-tracking rather than truth-tracking. We know that selection will often favor unreliable cognitive systems, which produce many false beliefs, over more reliable cognitive systems that would eliminate those false beliefs. Hence we should not expect our evolved cognitive adaptations to track truth.[3]

The extensive psychological literature on heuristics and biases in human cognition is a rich source of examples to underpin this argument. Human beings perform very badly on apparently simple reasoning tasks, committing a range of well-known fallacies. These effects are so widespread and so systematic that they are overwhelmingly likely to be intrinsic to the design of the human mind. People also exhibit a broad range of self-serving cognitive biases, giving them unrealistically positive views of themselves and their prospects. These traits correlate with mental well-being, giving rise to the phenomenon of depressive realism, in which mildly depressed people have more accurate self-perceptions. Hence these traits are also likely to be part of the design of the mind.

But despite these facts, the fundamental selection pressure driving the evolution of cognition is truth-tracking. The very idea that fitness-tracking is an alternative to truth-tracking is confused. When the relation between the two is properly formulated it becomes clear that the various circumstances in which selection favors unreliable cognitive mechanisms all involve obtaining as much truth as possible given the constraints. All selection processes are constrained, or else organisms "would live for ever, would be impregnable to predators, would lay eggs at an infinite rate, and so on" (Maynard Smith 1978, 32). An unconstrainedly optimal cognitive system would have every true belief relevant to its activities and no false beliefs, but this is not possible. Evolution selects for truth-tracking in the same sense that it optimizes any other trait under selection—it does the best it can given the constraints.

Why Truth-Tracking and Fitness-Tracking Are Not Alternatives

It is an error to contrast truth-tracking and fitness-tracking because, as one of us has discussed at length elsewhere (Goode and Griffiths 1995), this is to treat complementary explanations at different levels of analysis as if they were potential rivals at the same level of analysis. It is perfectly sensible to ask which of various properties of a trait under selection is the "target of selection": does blood contain hemoglobin because it binds oxygen, or because it makes it appear red, or both? Such questions have answers, at least in principle. They ask whether either or both of these properties enter into some lawlike generalization about selection, so that they can figure in a selective explanation of the trait. But such questions presuppose that the two properties and the corresponding selection explanations are potential alternatives to one another. It makes no sense to ask if hemoglobins were selected for binding oxygen or for enhancing fitness. To regard these as alternative hypotheses about the evolution of hemoglobins is to confuse two, separate levels of explanation.

The classic way to determine which of two properties is the target of selection (or if both are) is to pose the counterfactual question whether, if either property had occurred without the other, the trait would have followed the same trajectory in the population (Sober 1984). The presence of an atom of iron in a hemoglobin molecule causes it to bind oxygen, but also modifies the absorption spectrum of the molecule so that it appears red. If hemoglobins bound oxygen equally efficiently but did not appear red, this would have no effect on selection. After all, many organisms—octopus, for example—use copper-based respiratory proteins (hemocyanins) that do not appear red. If hemoglobins bound oxygen less efficiently, but were just as red, however, then this would have an immediate effect on selection. Such questions are meaningful when asked about properties that are potentially alternative targets of selection: it might be one or the other or both that has some nomic connection with fitness, and we want to know which. But if we ask "would this trait have followed the same evolutionary trajectory if it lacked such-and-such a physical property but this made no difference to its fitness," we will always get the same answer: yes, and trivially so.

An explanation of why one trait was selected rather than another in terms of their relative fitness is the most abstract level of evolutionary explanation. Treating traits as mere bearers of fitness values makes sense if one wishes to access the generalizations of population genetics while abstracting away from details that, in conventional evolutionary theory at least, make no difference to the evolutionary trajectories of populations.[4] Given some number of alternative traits, their heritabilities, initial frequencies, fitness functions, the effective population size, and so forth, we can compute their likely frequencies at some future time. But this explanation is in no sense a rival to an explanation that includes the underlying reasons why in some particular case the alternative traits have those relative fitnesses. The second explanation is an instance of the first, more general explanation.[5]

In summary, it is senseless to set up "fitness-tracking" as an alternative to "truth-tracking" because truth-tracking is a property at a lower level of explanation. It is admittedly still quite an abstract property. It is best regarded as a general measure of a certain kind of ecological interaction with the environment, akin to "foraging efficiency" or "respiratory efficiency," as we will see below. The claim that an organism succeeded because it was better than its rivals at tracking truth has just as much empirical content as the claim that it succeeded because it was a more efficient forager or had a more efficient respiratory system. Abstract as they may be, these claims nevertheless stand to the claim that the organism succeeded because it was *fitter* than its rivals as potential instances of this still more abstract explanation.

Why Evolution Selects for Truth-Tracking

So the proposal that our evolved cognitive adaptations do not track truth cannot mean that they track fitness instead. It must mean that they track some other property that is a genuine alternative to truth at the same level of explanation. But, we will argue, the cognitive adaptations that give rise to the commonsense beliefs with the help of which we and other animals act on an everyday basis are not tracking any such alternative to truth. If they fail to track truth as effectively as they might, it is because they are tracking truth subject to constraints. The currency of evolutionary success in the domain of cognition remains truth-tracking.

The most fundamental constraint is cost. Cognition is very costly. The human brain makes up about 2 percent of body mass but accounts for about 20 percent of oxygen consumption. Having beliefs, whether true or false, comes at a high price and the currency is glucose. Because cognition is so costly we can immediately rule out some evolutionary scenarios that have been proposed by evolutionary skeptics (Plantinga 2002). The hypotheses that belief has no effect on behavior, or that having beliefs reduces fitness, are completely absurd. If this were the case then there would be strong selection for not having beliefs, as beliefs are very costly to acquire and maintain.[6] The proposal that beliefs evolved by genetic linkage also has no plausibility: the relevant neural traits are complex or quantitative, and their genetic basis is widely distributed across the genome. A surprising amount of attention has been given to Plantinga's suggestion that most beliefs could be false but that organisms might have wacky desires that, when added to the false beliefs, give rise to adaptive behavior: "Perhaps Paul very much *likes* the idea of being eaten, but when he sees a tiger, always runs off looking for a better prospect, because he thinks it unlikely the tiger he sees will eat him. This will get his body parts in the right place so far as survival is concerned, without involving much by way of true belief" (Plantinga 1993, 225).

But the issue is not whether there is some combination of false beliefs and matching desires that could generate adaptive behavior. The issue is whether evolution could design cognitive adaptations that consistently produce adaptive behavior by producing beliefs that are false and adjusting desires to fit. We submit that this is obviously not possible. The only way to do it would be to have some *other* cognitive mechanisms that tracked truth and that adjusted the desires in the light of the actual goals of the organism so as to ensure that the false beliefs nevertheless produced adaptive behavior. In that case, it would be the states of this second mechanism that would be the effective beliefs and desires, and the false beliefs and wacky desires envisaged by Plantinga would be a bizarre and expensive detour between the effective beliefs and desires and the organism's motor systems.

We can safely conclude that beliefs are the output of a set of cognitive adaptations. Those adaptations are not designed to produce only true beliefs, or to produce all of the relevant true beliefs on every occasion. But this is not because they are tracking some property other than truth. It is because they are tracking truth in a constrained manner.

Truth-tracking is strongly constrained by cost because organisms have limited resources and truth-tracking is not the only thing they need to do to survive. Resources allocated to forming true beliefs are resources unavailable for making sperm or eggs, or fighting off the effects of ageing by repairing damaged tissues. Modern humans in first-world countries lead a sheltered life, and it is hard for us to appreciate just how direct these trade-offs can be. A dramatic example comes from a small Australian mammal, the Brown Antechinus (*Antechinus Stuartii*). In this and several related species a short, frenzied mating season is followed by a period during which the male's sexual organs regress and their immune system collapses. Then all the males in the population die. The Antechinus has little chance of surviving to the next breeding season, and so it allocates all of its resources to the reproductive effort and none to tissue maintenance. There can be little doubt that if, like us, the Antechinus had a massively hypertrophied cortex and engaged in a lot of costly abstract reasoning, it would allow that neural tissue to decay in the mating season so as to allocate more resources to sperm production and sexual competition.

As Gigerenzer and collaborators have long argued, many of the best-known human failures of rationality can be understood as heuristics that sacrifice being right all the time for being right most of the time at a greatly reduced cost (Gigerenzer, Todd, and the ABC Research Group 1999; Gigerenzer and Selten 2001). A heuristic is a method for obtaining truth that does not guarantee a correct answer every time, but which gets it right often enough that there is no point in adopting a more reliable but more costly method. A heuristic is not a method for obtaining something other than truth. So, while our use of simple heuristics does, indeed, show that truth is being traded off against fitness, what this means is only that truth-tracking, a component of fitness, is being traded off against other components of fitness, such as sperm production. The same is true of our adaptations for locomotion—we could have stronger legs, but the cost would not be justified. This does nothing to question the view that the adaptive purpose of legs is locomotion. Just so, the existence of "bounded rationality" does nothing to question the view that the adaptive purpose of the cognitive traits that give rise to beliefs is truth-tracking.

Another constraint arises from the intrinsic logical structure of many cognitive tasks. Type one errors involve accepting something that is not true. Type two errors involve rejecting something that is true. When a task requires an organism to act, or to form a potentially action-guiding belief, on less than perfect information, it is logically impossible to reduce the probability of committing a type one error without *increasing* the probability of committing a type two error (and vice-versa). Since organisms often need to act before absolutely conclusive information is available, they have to accept some risk of error. The unavoidable evolutionary problem they face is that of trading off type one errors against type two errors.

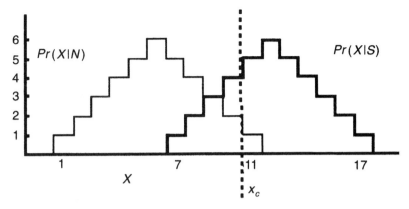

Figure 9.1. A signal detection problem. Probability (vertical axis) of receiving sensory input of type X at varying strength (horizontal axis), either as noise (leftmost curve), or as a veridical signal (rightmost curve). From Godfrey-Smith (1991, 714), used with permission.

The application of this observation to evolutionary psychology has been termed "error management theory" and is the subject of a small scientific literature (Haselton 2007). Here, however, we want to emphasize the importance of this observation for actually defining what it means to "track truth." This point has been clearly made by Godfrey-Smith using a class of evolutionary models called "signal detection theory" (Godfrey-Smith 1991). Organisms have to act on the basis of signals that are less than perfectly reliable indicators of relevant states of affairs in the environment. From an information theoretic point of view, the problem facing an organism that receives a signal is whether it is a veridical signal or whether it is noise (fig. 9.1).

Evolution will select organisms that decide to act, or that form an action-guiding belief, at whatever point along the horizontal axis is evolutionarily optimal (X). This point can be determined by combining the probabilities in figure 9.1 with the values of the four possible outcomes: acting when the signal is veridical, acting when the signal is noise (type one error), not acting when the signal is veridical (type two error), and not acting when the signal is noise. The costs and benefits associated with these four outcomes are summarized in table 9.1. The net values of the four outcomes, and the probabilities

Table 9.1. Costs and benefits of acting or not acting in response to either a veridical signal or noise.

	Veridical Signal	*Noise*
Act	Benefit of successful action minus cost of action	No benefit, but still incur the cost of action
Not Act	No cost of action, but opportunity cost—loss of fitness relative to an organism which *does* act	No benefit, no cost

of their obtaining for any value of X, determine the average payoff to the organism if that value is chosen as the threshold for action (or action-guiding belief). The value of X with the highest average payoff will be favored by natural selection.

This simple signal detection model embodies the lesson that tracking fitness and tracking truth are not alternatives. Clearly, the model assumes that organisms are tracking fitness—the selected value of X is the one with the highest average payoff. But if we ask why a given value of X has a given average payoff, the explanation will be in terms of truth-tracking. The explanation of the value of each cell in the payoff matrix is the fit, or lack of fit, between what the organism does and how the world is. Increases in fitness are explained by successful truth-tracking and reductions in fitness by failures in truth-tracking. Why does acting when the signal is veridical yield a gain in fitness? Because the animal acts as if the world is a certain way when the world is that way, and thus achieves its goals. Why does acting when the signal is noise yield a reduction in fitness? Because such an action would be frustrated by the way the world is, and thus constitute a waste of precious resources (see table 9.1). Godfrey-Smith puts the point like this: "correspondence truth has a definite place in the signal detection approach. For, if C *[the mental state formed on the basis of X]* is regarded as representing S, hits and correct rejections are truths, while misses and false alarms are errors. So truth remains a definite goal of cognition, though the truth-linked virtue of 'reliability' is taken apart and overhauled" (Godfrey-Smith 1991, 722).

In the light of the two classes of unavoidable constraints that we have identified, we can conclude that "truth-tracking" should be understood to mean obtaining as much truth and as little error as possible, given the intrinsic trade-off between these two, with the balance between them determined by the value of the truths and the cost of the errors, and with solutions constrained by the cost of cognitive resources. Somewhat more succinctly we might say that organisms track truth because they obtain as much *relevant* truth as they can afford and tolerate no more error than is needed to obtain it.[7]

We propose that, with this definition of truth-tracking, it is overwhelmingly likely that commonsense beliefs are produced by cognitive adaptations that track truth. By "commonsense," we mean those everyday beliefs that guide mundane action and whose subjective certainty was famously appealed to by G. E Moore (Moore 1925). Moore's examples included the existence of his body, and of other human bodies and inanimate bodies, all arranged in space and time, as well as the fact that those other human bodies knew similar things. Our argument is simply that any plausible account of the evolution of these kinds of beliefs in humans and other animals will be of the kind described in the last section. At the heart of the explanation will be the fact that animals can increase their fitness by detecting states of affairs in the world and matching their actions to those states of affairs.

A BRIDGE TOO FAR? COMMONSENSE AND SCIENCE

Commonsense and the Human *Umwelt*

Commonsense beliefs are couched in commonsense concepts, not the concepts of our best current science. Moore was certain that he had two hands, not that he had two instances of the distal portion of the pentadactyl limb. Our commonsense concepts are themselves an evolutionary inheritance, and we know that they differ systematically from those of other animals. So it is plausible that, had our evolution followed a different course, we would have a different conceptual scheme. Our first point in this section is that the evolution of cognition tracks truth in a sense that does not discriminate between the conceptual schemes of different species. To take a trivial example, the description of a bird's plumage in my field guide is not false because it omits the patterns in the ultraviolet spectrum that the bird itself perceives. Nor are botanical illustrations fanciful because they omit the markings that are visible to insect pollinators! A more interesting example comes from the work of the pioneering ethologist Konrad Lorenz on the "world of the bird" (Lorenz 1935). Lorenz argued that his birds (primarily jackdaws, *Corvus monedula*) do not perceive other members of their species as instances of a single kind of animal with which they might, on different occasions, flock, fight, mate, or engage in parenting behavior. Instead, they occupy a world consisting of various "companions." A "companion" is an array of stimulus features that identifies something as the appropriate object of a suite of behaviors. So in Lorenz's account of the bird's world, a jackdaw is not surrounded by other jackdaws, but by flocking companions, mating companions, parenting companions, and so forth.[8] From a human perspective, the jackdaws are missing something. But Lorenz's jackdaws are not *mistaken* about their social world, any more than we are mistaken about the colors of birds and flowers.

Lorenz's work was an explicit application of the Estonian biologist Jacob von Uexküll's concept of the *Umwelt* (Uexküll 1957).[9] The *Umwelt* of a species is the world described using the perceptual and conceptual categories available to that species. While the term itself may not be widely used in modern behavioral biology, the concept remains essential: "The lesson from the past decade of studies on bird behavior is: if you want to get inside an animal's mind, it helps to see the world through its eyes" (Dalton 2004, 597). This is because an animal responds to the world as the animal sees and conceptualizes it, and not as we do.

From an evolutionary perspective, the world of commonsense can only be seen as the human *Umwelt*. Whatever ontological authority may attach to the concepts and categories of science, the commonsense way in which we see the world has no more or less ontological authority than the way in which birds see the world. A related point was famously made by the physicist Arthur Eddington when he contrasted the commonsense understanding of his writing table as a substantial object with the scientific understanding of it as an area of mostly empty space of which the best that could be said was that the probability of his elbow sinking through it was small enough to be neglected for the purpose of writing his lecture (Eddington 1928). There is a naïve response to facts of this kind according to which the idea that grass is green or that tables are solid are illusions foisted on us by our selfish genes, and there are no green or solid objects, only electromagnetic radiation and quantum interactions (Johnston 1999). But there is no reason to reject the world of commonsense, as long as we are prepared to accept that we are not the only animal whose evolved perceptual and conceptual schemes can stand alongside the measurement and conceptual schemes of science. One of the triumphs of science is that it allows us to appreciate the *Umwelt* of another species. In the human *Umwelt*, male and female bluetits (*Cyanistes caeruleus*) are visually indistinguishable. In the bluetit *Umwelt*, the two sexes have different markings. This does not mean that truth is relative to a conceptual scheme, only that there are more truths to be had than can be captured in the evolved conceptual scheme of either species. There are many ways of classifying the world that are not purely arbitrary, and the fact that these classifications are constrained by reality explains why they have some degree of pragmatic utility.

The obvious philosophical framework in which to make sense of the fact that organisms can have true beliefs in many different evolved conceptual schemes would be some form of structural realism. That is to say, we should have the same attitude toward beliefs formulated in an evolved conceptual scheme that we have to the content of a successful but now superseded scientific theory like Newtonian physics. The fact that it is possible to have true beliefs using this conceptual apparatus is to be understood in terms of some degree of structural resemblance between that apparatus and the structure of the world (or of our current best theory, which we can treat as a surrogate for how things really are). However, there are many competing versions of structural realism and substantial ongoing philosophical debates about the position, so we will not attempt to develop this connection in more detail in this essay (see Ladyman 2009).

The fact that humans track truth about human colors and bluetits track truth about bluetit colors can both be explained by selection for truth-tracking. Will our evolved faculties allow us to move beyond the world of commonsense, however? Darwin worried about this, as did Alfred Russel Wallace: our faculties will be adequate for practical life, but why suppose they are adequate for higher mathematics or metaphysics? Wallace famously thought evolution insufficient to account for these capacities, and so concluded that some higher process—Spirit—caused them. It is to this issue we now turn. Can selection for truth-tracking explain how science tracks truth: can the Milvian bridge that connects commonsense beliefs to evolutionary success be extended to reach scientific beliefs?

Darwin's Real Doubt: How Far Will Our Evolved Faculties Take Us?

We began this chapter with the quotation that has come to be known as "Darwin's doubt." It is supposed to show that Darwin himself entertained evolutionary skepticism. The context of the quotation, however, makes it clear that Darwin was skeptical about how far beyond the world of commonsense our evolved faculties would take us. Darwin was responding to the author of a philosophical work. He writes:

> You would not probably expect anyone fully to agree with you on so many abstruse subjects; and there are some points in your book which I cannot digest. The chief one is that the existence of so-called natural laws implies purpose. I cannot see this. . . . But I have had no practice in abstract reasoning and I may be all astray. Nevertheless you have expressed my inward conviction, though far more vividly and clearly than I could have done, that the Universe is not the result of chance. But then with me the horrid doubt always arises whether the convictions of man's mind, which has been developed from the mind of the lower animals, are of any value or at all trustworthy. Would any one trust

in the convictions of a monkey's mind, if there are any convictions in such a mind? (Darwin to Graham, 3 July 1881, in F. Darwin 1887, 1:315)[10]

Another famous letter gives some insight into Darwin's attitude to the limitations of the human intellect:

> I am inclined to look at everything as resulting from designed laws, with the details, whether good or bad, left to the working out of what we may call chance. Not that this notion *at all* satisfies me. I feel most deeply that the whole subject is too profound for the human intellect. A dog might as well speculate on the mind of Newton.—Let each man hope & believe what he can. (Darwin to Gray, 22 May 1860, in Burkhardt and Smith 1984–, 8:224)

Darwin is suggesting an epistemic boundedness thesis. We know there are limitations on the conceptual abilities of other animals. Dogs will never be able to master calculus. This leads to the suspicion that there may be truths that we are similarly constitutionally unable to entertain. But Darwin was quite confident that our faculties were adequate to unravel the workings of nature. The historian Jon Hodge has described how the young Darwin argued on theological grounds for both the impossibility of knowledge of the mind of the creator and the power of the human mind to understand creation itself, a position Hodge describes as "cognitive optimism about nature" (Hodge 2009, 62).

In this section we will argue that whether or not there are some ultimate limits to our ability to improve our conceptual scheme, we are not simply confined to our evolved conceptual scheme and we have not yet reached the limits of our ability to bootstrap ourselves into more adequate conceptual schemes. We will then argue that, given that our cognitive adaptations track truth in the realm of commonsense, we have reasons to believe that we can derive reliable knowledge in these more adequate conceptual schemes.

Our argument that humans are not confined to their evolved conceptual scheme is an appeal to the history of science. No human being had the concepts of differentiation and integration before Leibniz and Newton's introduction of the calculus. After two thousand years of speculation about physical theory, the space of possible dynamical theories was completely altered. As a result, for the past four hundred years educated people have regularly had thoughts that no hominoid had in the previous four hundred thousand. Other dramatic examples include the "probability revolution" of the nineteenth century (Hacking 1990), and the spatialization of time. As John Bigelow has pointed out, the idea of travelling to a distant time as one travels to a distant country is now commonplace, but has no precedent before the nineteenth century (Bigelow 2001). Something in the concept of time changed. One plausible explanation of this involves what historians of geology have dubbed "the discovery of deep time" by the new science of geology (Rudwick 1992).

In this respect, human beings are very different from bluetits, and even from other primates. There is no precedent in other animals for radical changes in the structure of the *Umwelt* caused by individual cognitive innovations that spread by cultural diffusion. The argument that this process has not come to an end is a simple appeal to the current state of science and to informed opinion about its likely future. It is obvious that more conceptual innovations like those just described are likely to occur, and we will not labor the point.

If human beings are able to supplement their evolved conceptual scheme with new concepts of their own invention, should we have confidence that our cognitive faculties can track truth in this new, enriched conceptual framework? We have argued above that our cognitive faculties were selected because they tracked truth in the human *Umwelt*. But they were not selected for their ability to do calculus, or to reason using the probability calculus (as we see from our systematic neglect of base rates and similar phenomena), or for their ability to use very indirect evidence to reconstruct the distant past.

So there is no direct Milvian bridge linking these particular cognitive processes to pragmatic success. Instead, there is an indirect Milvian bridge. There is a Milvian bridge connecting our commonsense beliefs to pragmatic success, and we can use commonsense to justify the methods by which we arrive at our scientific beliefs. The reasons we have to think that our scientific conclusions are correct, and that the methods we use to reach them are reliable, are simply the data and arguments that scientists give for their conclusions, and for their methodological innovations. Ultimately, these have to be able to stand up to the same kind of commonsense scrutiny as any other addition to our beliefs. The conviction that the base rate fallacy is a fallacy, and that we should guard against our tendency to commit it when reasoning informally, does not rest on a decision to follow probabilistic reasoning wherever it leads, but on evaluating the argument. When a piece of very similar reasoning leads to a conclusion that does not stand up to commonsense evaluation—such as the argument for buying a ticket in the St. Petersburg game[11]—we conclude that there is an error somewhere in that reasoning.

Thus, if evolution does not undermine our trust in our cognitive faculties, neither should it undermine our trust in our ability to use those faculties to debug themselves—to identify their own limitations, as in perceptual illusions or common errors in intuitive reasoning. Nor should it undermine our confidence in adopting new concepts and methods that have not themselves been shaped by the evolution of the mind but whose introduction can be justified using our evolved cognitive faculties.

Darwin's suspicions about the limitations of the human mind seem just about right. There are very powerful commonsense arguments for supposing that the natural sciences reveal facts about the structure of the natural world. So Darwin's "cognitive optimism" about our capacity to use our evolved minds to understand nature is a reasonable attitude. But there are no similar commonsense arguments for the view that the inability of centuries of theology and metaphysics to establish any firm conclusions is merely a temporary glitch. Darwin's pessimism about this seems eminently reasonable and is shared by many philosophers. The mere fact that the human mind has evolved from that of a lower animal does not add any weight to the commonsense grounds for this pessimism, but it had some relevance in Darwin's day because he would have viewed his theory as discrediting the traditional idea that the mind has been stamped by its creator with innate knowledge of that creator.

EVOLUTIONARY SKEPTICISM AND ETHICS

In the previous two sections we have argued that evolutionary skepticism about commonsense and science fails. Evolutionary skepticism about commonsense is defeated by the existence of a Milvian bridge connecting the reliability of our cognitive faculties in the world of commonsense—the human *Umwelt*—to evolutionary success. Evolutionary skepticism about science is defeated by the fact that our trust in science can be justified using commonsense arguments. In this section we turn to evolutionary skepticism about ethics.

Evolutionary accounts of ethics precede Darwin, being the basis of Herbert Spencer's ethical system (Spencer 1851). After Darwin, Spencer championed a slightly revised version of his evolutionary ethics based on natural selection (Spencer 1879). But both evolutionists, such as Thomas Huxley (Huxley 1893), and moral philosophers, most notably G. E. Moore (Moore 1903), soon reached the conclusion that moral value was something that evolution, and indeed nature in general, could not deliver. G. E. Moore's well-known Naturalistic Fallacy, which purports to show that Good is not a Natural property, was targeted directly at Spencer's evolutionary progressivism (Moore 1903; see also Cunningham 1996). Evolutionary ethics has had some lukewarm revivals since that time, most recently in the mid-1980s following work by Robert Richards and Michael Ruse (Richards 1986, 1987; Ruse 1986).[12]

In his recent paper, Kahane considers the possibility that evolution not only fails to provide a foundation for morality but actually debunks moral truth (Kahane 2011). Evolutionary debunking arguments (EDAs) in support of a moral "error theory" have been produced before (Joyce 2000, 2001, 2006), but the argument Kahane considers goes further: evolution makes moral realism very unlikely. His argument is one that will by now be familiar: evolution by natural selection tracks fitness rather than truth. The evolution of the moral sense is an "off-track" process because it has no intrinsic tendency to produce a moral sense that tracks moral truths. Kahane gives the argument as follows:

1. *Causal premise*. Our evolutionary history explains why we have the evaluative beliefs we have.
2. *Epistemic premise*. Evolution is not a truth-tracking process with respect to evaluative truth.
C. *Evaluative skepticism*. None of our evaluative beliefs is justified.

The idea that evolution is an off-track process with respect to evaluative truth can be seen at work in Darwin's classic discussion of the evolution of morality:

> In the same manner as various animals have some sense of beauty, though they admire widely different objects, so they might have a sense of right and wrong, though led by it to follow widely different lines of conduct. If, for instance, to take an extreme case, men were reared under precisely the same conditions as hive-bees, there can hardly be a doubt that our unmarried females would, like the worker-bees, think it a sacred duty to kill their brothers, and mothers would strive to kill their fertile daughters; and no one would think of interfering. (C. Darwin [1871] 1981, 1:83)

Darwin argues that if our ecology had been different, then we would judge different things to be right and wrong, just as different species of animals judge different things to be beautiful. Animals are aesthetically attracted to things to which it is fitness-enhancing for them to be attracted. Just so, Darwin argues, they will morally approve of actions that it is fitness-enhancing for them to approve. This would seem to imply either that evolution is an off-track process with respect to evaluative truth, or that evaluative truths are truths about what maximizes reproductive fitness.[13]

There is no Milvian bridge available to connect moral truth to pragmatic success and thus defend it from evolutionary skepticism. This is because contemporary evolutionary explanations of morality, just like Darwin's, do not involve any adaptive advantages produced by detecting and acting in accordance with objective moral facts (Sober and Wilson 1998; Ridley 1997). But Kahane notes that the assumption that moral truths correspond to objective moral facts is one that is questioned by many moral philosophers for independent reasons. The evolutionary debunking argument against ethics would therefore be better stated as follows:

1. *Causal premise.* Our evolutionary history explains why we have the evaluative beliefs we have.
2. *Epistemic premise.* Evolution is not a truth-tracking process with respect to evaluative truth.
3. *Metaethical assumption.* Objectivism (moral realism) is the correct account of evaluative discourse.
C. *Evaluative skepticism.* None of our evaluative beliefs is justified.

If we deny the assumption that evaluative beliefs denote moral realities, then the conclusion fails to follow. Noncognitivist ethical theories, according to which the function of ethical judgments is not to express facts but to do something like express allegiance to a norm, remain a significant force in contemporary moral philosophy (van Roojen 2009). So the evolutionary debunking argument is best conceived as an argument against moral realism rather than simply against moral truth.

The case of ethics makes clear that there are actually two ways to respond to an evolutionary debunking argument. The first is to build a Milvian bridge and argue that, for some cognitive domain X, our beliefs are related to the facts in such a way that evolution will select cognitive faculties that track truth in that domain. The second is to argue that what is meant by "truth" in a certain cognitive domain is not a matter of tracking some external state of affairs, so that the question of whether evolution is an off-track process in that domain does not arise. In the next section we ask if either of these responses is available when evolutionary skepticism is applied to religious beliefs.

RELIGION—A HARD PLACE BETWEEN SCIENCE AND ETHICS?

As we stated in the second section, we do not believe that a Milvian bridge is available for religious beliefs, because none of the leading contemporary accounts of the evolution of religious belief makes any reference to the truth or falsity of those beliefs when explaining their effects on reproductive fitness. This is true both of evolutionary theories that explain religion as a side-effect of other adaptations and those that explain it as an adaptation in its own right. For example, in one approach to the evolution of religion, belief in God is an adaptation in its own right. However, this is because it is fitness-enhancing to advertise in a clear and hard-to-fake manner your membership of your community, so that you can access its support and protection when you need it (Sosis and Alcorta 2003; Bulbulia 2004). On this theory, Constantine's new battle standard was a signal to his Christian troops that members of their community would receive the protection of the state, as indeed they did in an imperial edict of the following year. His victory was due to social solidarity rather than divine intervention.

The prominent theory of David Sloan Wilson shows the same lack of concern with the specific content of religious beliefs. On this account, religion evolved through a process of multilevel selection. Its evolution was driven by the benefits that social cohesion and prosocial behavior provide at the level of the group (Wilson 2002). Nothing in this explanation discriminates between true and false religious beliefs. According to this theory, Constantine won because his predominantly Christian army approximated a superorganism like an ant colony slightly more closely that did the predominantly pagan army of Maxentius, not because Christianity was true and paganism false.

Contemporary theories that explain the evolution of religion as a side-effect are equally undiscriminating. The idea that religious belief is to a large extent the result of mental adaptations for agency detection has been endorsed by several leading evolutionary theorists of religion (Atran 2002; Guthrie 1993; Barrett 2005). Broadly, these theorists suggest that there are specialized mental mechanisms for the detection of agency behind significant events. These have evolved because the detection of agency—"who did that and why?"—has been a critical task facing human beings throughout their evolution. Religious belief has been jokingly described as "taking the universe personally," and on this account, that is precisely correct.

None of the contemporary evolutionary explanations of religious beliefs hypothesizes that those beliefs are produced by a mechanism that tracks truth. This may seem puzzling, given that we have argued above that the evolution of cognition is driven by truth-tracking. But the contradiction is only an apparent one. The side-effect explanation of religious belief is completely in line with the models we have sketched above. If the agency detection account is correct, then people believe in supernatural agents that do not exist for the same reason that birds sometimes mistake objects passing overhead for raptors. These beliefs are type one errors, and they are the price of avoiding more costly type two errors. The adaptive explanations of religion work somewhat differently. They identify a way in which a sophisticated cognitive system could evolve a positively selected departure from truth-tracking. But this explanation presumes that the underlying cognitive system has evolved in the way we sketched above, with truth-tracking as its fundamental aim. This pattern of dependency between the two explanations cannot be reversed, any more than the explanation of the pseudo-penis and pseudo-scrotum of female hyenas (Kruuk 1972) could be turned into an explanation of the evolution of the penis and scrotum, with male, intromissive penises developing as a secondary adaptation! Like the pseudo-penis, religious beliefs are a specialized secondary adaptation found in a small number of species and based on a more basic adaptation that can be found in very many species and that evolved for the same fundamental reason in all those species. It is entirely consistent to argue that penises evolved for intromission, but that pseudo-penises evolved through mimicry. In the same manner, a single, integrated account of the evolution of cognition can argue that the basic evolutionary dynamic that produced cognition is truth-tracking, but that certain, specialized classes of beliefs evolved as secondary adaptations for promoting social solidarity.

If a Milvian bridge cannot be constructed linking the truth of religious beliefs to evolutionary success, then the alternative is to argue that the truth of religious beliefs is not a matter of their tracking some state of affairs in the world. There have been several attempts by academic theologians to purge religion of its claims about the supernatural, one of the best known being the "Sea of Faith" movement headed by Don Cupitt (2008). The evolutionary theorist of religion David Sloan Wilson has also suggested that the theological beliefs associated with a religious tradition may be more or less epiphenomenal with respect to its functioning as a social institution and that they may be perceived by the adherents of the faith as mere "preacher talk" (Wilson 2007, 265). However, liberal theologians like Cupitt have usually been perceived as heretical by ordinary believers. We are skeptical that the vast majority of religious believers could be persuaded to accept a noncognitivist or a fictionalist theology.

Religious beliefs thus emerge as peculiarly vulnerable to evolutionary debunking arguments. The truth of religious beliefs does seem to be a matter of tracking some external state of affairs, so that the question of whether evolution is an off-track process with respect to religious beliefs is one that cannot be sidestepped. But the leading evolutionary explanations of these beliefs all suggest that they are produced by cognitive adaptations which are not designed to track truth.

CROSSING THE MILVIAN BRIDGE, FINALLY

Evolutionary debunking arguments suggest that the evolutionary origin of our cognitive faculties should undermine our confidence in the beliefs those faculties produce. Perhaps unsurprisingly, the force of this skeptical argument depends on the specific class of beliefs. We have argued that it has no force against commonsense, factual beliefs. In this cognitive domain natural selection will design cognitive faculties that track truth in the sense that they obtain as much relevant truth as the organism can afford, and tolerate no more error that is needed to obtain it. This is enough to build what we have called a "Milvian Bridge": the commonsense facts are related to the evolutionary success of commonsense beliefs in such a way that it is reasonable to accept and act on commonsense beliefs produced by our evolved cognitive faculties.

We have further argued that evolutionary skepticism about scientific beliefs is unsuccessful because there are commonsense justifications of the processes by which we arrive at our scientific beliefs.

Evolutionary debunking arguments have more force when applied to ethical beliefs. Drawing on Kahane's work we have argued that evolutionary accounts of the origins of moral intuitions may undermine confidence in those intuitions if moral beliefs are given a strongly realist interpretation. But noncognitivist moral philosophers, and perhaps some less ambitious moral realists, are unaffected by the evolutionary debunking argument, since they reject the idea that moral beliefs are in the business of tracking moral facts.

Finally, we have argued that religious beliefs emerge as particularly vulnerable to evolutionary debunking arguments, since neither class of counterargument seems to be viable in that domain. Current evolutionary theory really does support the view that human beings would have religious beliefs even if all religious beliefs were uniformly false.

But debunking is not disproving. If there are independent reasons for religious belief, their cogency is not removed by the fact that religious beliefs have evolutionary explanations. As Darwin wrote to his Presbyterian friend Asa Gray, "Let each man hope & believe what he can" (Darwin to Gray, 22 May 1860, in Burkhardt and Smith 1985–, 8:223).

NOTES

This research was supported under Australian Research Council's Discovery Projects funding scheme (DP0984826). JSW acknowledges the invitation to deliver a keynote address to the 3rd National SoFiA Conference in Melbourne, "Science and Faith: An Open Dialogue?" 19–21 September 2008, in which these ideas were first mooted.

1. For an introduction to the extensive philosophical literature on Plantinga's argument see Beilby (2002).

2. It has been argued that Plantinga's evolutionary skepticism rests on just such an overblown conception of "true belief" (Ruse 2004).

3. For a good introduction to these ideas, see Downes (2000). Serious criticisms of many suggested scenarios in which evolution might favor false beliefs can be found in McKay and Dennett (2009).

4. We are ignoring all the issues raised by evolutionary developmental biology and other attempts to enrich traditional neo-Darwinian evolution-

ary theory. Introducing these would massively complicate the exposition and we are confident that the differences are simply not germane to the topic of this essay.

5. These are "robust process" and "actual sequence" explanations respectively. See Sterelny and Griffiths (1999, 84).

6. We are not addressing the strand of Plantinga's thought that envisages beliefs stripped of their semantic properties but otherwise precisely functionally equivalent, so that this loss makes no difference to evolution. If semantics cannot be naturalized, then naturalism is in trouble for many reasons besides Plantinga's skeptical argument.

7. McKay and Dennett (2009) conclude that the most plausible cases of actual selection for false belief are the biases that give people unrealistically positive views of themselves and their prospects. They note that there is no direct evolutionary benefit to getting things wrong: "it's not clear that there is anything adaptive about trying and failing" (506). Instead, typical explanations of positive illusions rely on an asymmetry between the costs and benefits of overestimating your chances of success in various projects (and acting accordingly) and the those of underestimating those chances (and acting accordingly). This would seem broadly in line with our framework.

8. Whilst this is in no sense "current science," it was a key founding document of the modern, Darwinian study of behavior. It was the first of Lorenz's works to be translated into English, and the theory it embodied proved highly productive in the "classical" ethology of the mid-twentieth century. We use this theory as our example because Lorenz is exploring the "conceptual scheme" of another species, something a contemporary behavioral ecologist, with their more functional, ecological focus, would be unlikely to do.

9. Note that this translation was published in a conventional scientific volume alongside the early papers of the Nobel Prize winners Lorenz and Niko Tinbergen. The original practical, scientific use of the *Umwelt* concept is often forgotten because of its later use in philosophy and literary theory.

10. Also available in a newly edited version on the Darwin Correspondence database at http://www.darwinproject.ac.uk/entry-13230.

11. In this game a coin is tossed until it comes up heads. You receive $1 if the first toss is heads, $2 if the second toss is heads, $4 if the third toss is heads, and so forth. An (overly) simple application of decision theory suggests that the expected outcome of this game is an infinite amount of money, and so it would be rational to pay any amount of money to be allowed to play.

12. A good summary can be found in Farber (1994), as well as van der Steen (1999). Arguments in both directions can be found in Nitecki and Nitecki (1993), and see Richards, this volume.

13. For a modern version of Darwin's argument, see Ruse and Wilson (1986, 186).

REFERENCES

Atran, S. (2002). *In Gods We Trust: The Evolutionary Landscape of Religion.* New York: Oxford University Press.

Barrett, J. L. (2005). "Counterfactuality in Counterintuitive Religious Concepts." *Behavioral and Brain Sciences* 27:731–32.

Beilby, J. K., ed. (2002). *Naturalism Defeated? Essays on Plantinga's Evolutionary Argument against Naturalism.* Ithaca, NY: Cornell University Press.

Bigelow, J. (2001). "Time Travel Fiction." In *Reality and Humean Supervenience: Essays on the Philosophy of David Lewis,* ed. G. Preyer and F. Siebelt, 57–91. Lanham, MD: Rowman and Littlefield.

Boyer, P. (2001). *And Man Creates God: Religion Explained.* New York: Basic Books.

Bulbulia, J. (2004). "Religious Costs as Adaptations That Signal Altruistic Intention." *Evolution and Cognition* 10:19–42.

Burkhardt, F., and S. Smith, eds. (1985–). *The Correspondence of Charles Darwin.* Cambridge: Cambridge University Press.

Cunningham, S. (1996). *Philosophy and the Darwinian Legacy.* Rochester: University of Rochester Press.

Cupitt, D. (2008). *Above Us Only Sky: The Religion of Ordinary Life.* Santa Rosa, CA: Polebridge.

Dalton, R. (2004). "True Colours." *Nature* 428:596–97.

Darwin, C. R. ([1871] 1981). *The Descent of Man, and Selection in Relation to Sex.* Facsimile of the 1st ed. Princeton: Princeton University Press.

Darwin, F., ed. (1887). *The Life and Letters of Charles Darwin: Including an Autobiographical Chapter.* 2 vols. New York: Appleton.

Dawkins, R. (1998). "The Emptiness of Theology." *Free Inquiry* 18:6.

Downes, S. M. (2000). "Truth, Selection and Scientific Inquiry." *Biology and Philosophy* 15:425–42.

Eddington, A. S. (1928). *The Nature of the Physical World.* New York: Macmillan.

Farber, P. L. (1994). *The Temptations of Evolutionary Ethics.* Berkeley: University of California Press.

Gibbon, E. ([1782] 1850). *History of the Decline and Fall of the Roman Empire.* Vol. 2. New York: Harper Brothers.

Gigerenzer, G., and R. Selten. (2001). *Bounded Rationality: The Adaptive Toolbox.* Cambridge, MA: MIT Press.

Gigerenzer, G., P. M. Todd, and the ABC Research Group. (1999). *Simple Heuristics That Make Us Smart.* New York: Oxford University Press.

Godfrey-Smith, P. (1991). "Signal, Decision, Action." *Journal of Philosophy* 88:709–22.

Goode, R., and P. E. Griffiths. (1995). "The Misuse of Sober's Selection of/ Selection for Distinction." *Biology and Philosophy* 10:99–108.

Guthrie, S. (1993). *Faces in the Clouds: A New Theory of Religion.* New York: Oxford University Press.

Hacking, I. (1990). *The Taming of Chance.* Cambridge: Cambridge University Press.

Haselton, M. G. (2007). "Error Management Theory." In *Encyclopedia of Social Psychology,* ed. R. F. Baumeister and K. D. Vohs, 311–12. Thousand Oaks, CA: Sage Publications.

Hodge, M. J. S. (2009). "The Notebook Programmes and Projects of Darwin's London Years." In *The Cambridge Companion to Darwin,* 2nd ed., ed. M. J. S. Hodge and G. Radick, 44–72. Cambridge: Cambridge University Press.

Hume, D. ([1757] 1956). *The Natural History of Religion.* London: A. & C. Black.

Huxley, T. H. (1893). *Romanes Lecture 1893: Evolution and Ethics.* London: Macmillan.

Johnston, V. S. (1999). *Why We Feel: The Science of Human Emotions.* Cambridge, MA: Perseus Books.

Joyce, R. (2000). "Darwinian Ethics and Error." *Biology and Philosophy* 15:713–32.

———. (2001). *The Myth of Morality.* New York: Cambridge University Press.

———. (2006). *The Evolution of Morality.* Cambridge, MA: MIT Press.

Kahane, G. (2011). "Evolutionary Debunking Arguments." *Noûs* 45:103–25.

Kruuk, H. (1972). *The Spotted Hyena: A Study of Predation and Social Behaviour.* Chicago: University of Chicago Press.

Ladyman, J. (2009). "Structural Realism." In *The Stanford Encyclopedia of Philosophy,* ed. E. N. Zalta. Available at http://plato.stanford.edu/archives/sum2009/entries/structural-realism/.

Lorenz, K. Z. (1935). "Der Kumpan in Der Umwelt Des Vogels." *Journal für Ornithologie* 83:137–213.

Maynard Smith, J. (1978). "Optimization Theory in Evolution." *Annual Review of Ecology and Systematics* 9:31–56.

McKay, R. T., and D. C. Dennett. (2009). "The Evolution of Misbelief." *Behavioral and Brain Sciences* 32:493–510.

Moore, G. E. (1903). *Principia Ethica.* Cambridge: Cambridge University Press.

———. (1925). "A Defence of Common Sense." In *Contemporary British Philosophy (Second Series),* ed. J. H. Muirhead, 192–233. London: George Allen & Unwin.

Musgrave, A. (1988). "The Ultimate Argument for Scientific Realism." In *Relativism and Realism in Science,* ed. R. Nola, 229–53. Boston: Kluwer.

Nitecki, M. H., and D. V. Nitecki, eds. (1993). *Evolutionary Ethics.* Albany: State University of New York Press.

Plantinga, A. (1993). *Warrant and Proper Function.* New York: Oxford University Press.

———. (2002). "The Evolutionary Argument against Naturalism." In Beilby 2002, 1–13.

Psillos, S. S. (1999). *Scientific Realism: How Science Tracks Truth.* New York: Routledge.

Putnam, Hilary H. (1975). "What Is Mathematical Truth?" In *Mathematics, Matter and Method,* 60–78. Cambridge: Cambridge University Press.

Richards, R. J. (1986). "A Defense of Evolutionary Ethics." *Biology and Philosophy* 1:265–93.

———. (1987). *Darwin and the Emergence of Evolutionary Theories of Mind and Behavior.* Chicago: University of Chicago Press.

Ridley, M. (1997). *The Origins of Virtue: Human Instincts and the Evolution of Cooperation.* New York: Viking.

Rudwick, M. J. S. (1992). *Scenes from Deep Time: Early Pictorial Representations of the Prehistoric World.* Chicago: University of Chicago Press.

Ruse, M. (1986). "Evolutionary Ethics: A Phoenix Arisen." *Zygon* 21:95–112.

———. (2004). "The New Creationism: Its Philosophical Dimension." In *The Cultures of Creationism: Anti-Evolutionism in English-Speaking Countries,* ed. S. Coleman and L. Carlin, 175–92. Ashgate: Aldershot.

Ruse, M., and E. O. Wilson. (1986). "Moral Philosophy as Applied Science." *Philosophy* 61:173–92.

Smart, J. J. C. (1963). *Philosophy and Scientific Realism.* London: Routledge and Kegan Paul.

Sober, E. (1984). *The Nature of Selection: Evolutionary Theory in Philosophical Focus.* Cambridge, MA: MIT Press.

Sober, E., and D. S. Wilson. (1998). *Unto Others: The Evolution and Psychology of Unselfish Behavior.* Cambridge, MA: Harvard University Press.

Sosis, R., and C. Alcorta. (2003). "Signaling, Solidarity and the Sacred: The Evolution of Religious Behavior." *Evolutionary Anthropology* 12:264–74.

Spencer, H. (1851). *Social Statics, or, the Conditions Essential to Human Happiness Specified, and the First of Them Developed.* London: John Chapman.

———. (1879). *The Data of Ethics.* London: Williams and Norgate.

Sterelny, K., and P. E. Griffiths. (1999). *Sex and Death: An Introduction to Philosophy of Biology.* Chicago: University of Chicago Press.

Uexküll, J. von. (1957). "A Stroll through the Worlds of Animals and Men: A Picture Book of Invisible Worlds." In *Instinctive Behavior: The Development of a Modern Concept,* ed. C. H. Schiller, 5–80. New York: International Universities Press.

van der Steen, W. J. (1999). "Methodological Problems in Evolutionary Biology. XII. Against Evolutionary Ethics." *Acta Biotheoretica* 47:41–57.

van Roojen, M. (2009). "Moral Cognitivism vs. Non-Cognitivism." In *The Stanford Encyclopedia of Philosophy*, ed. Edward N. Zalta. Available at http://plato.stanford.edu/archives/fall2009/entries/moral-cognitivism/.
Wilson, D. S. (2002). *Darwin's Cathedral: Evolution, Religion, and the Nature of Society*. Chicago: University of Chicago Press.
———. (2007). *Evolution for Everyone: How Darwin's Theory Can Change the Way We Think About Our Lives*. New York: Delacorte Press.

QUESTIONING THE ZOOLOGICAL GAZE
Darwinian Epistemology and Anthropology

Phillip R. Sloan

> *This great question [of the moral sense] has been discussed by many*
> *writers of consummate ability; and my sole excuse for touching on*
> *it is the impossibility of here passing it over, and because, as far as I*
> *know, no one has approached it exclusively from the side of natural*
> *history. The investigation possesses, also, some independent interest,*
> *as an attempt to see how far the study of the lower animals can throw*
> *light on one of the highest psychical faculties of man.*
> —Darwin, *Descent of Man* ([1871] 1981, 1:71)

This quotation from Darwin's *Descent of Man* illuminates an under-
explored issue in Darwin's work—not the issue of evolutionary ethics
itself, but the epistemology of experience assumed in his work, and the
consequences of his application of this "zoological gaze" to human beings. I will term this epistemological stance in this chapter "natural historical realism" (*NHR* subsequently). It is the claim of this chapter
that although this epistemological stance may be unproblematic when
involved in the study of objects within the scope of Darwin's general

natural historical interests, it encounters several significant problems when turned upon human beings. In a wider view of my arguments, I will also suggest that these same problems confront several dimensions of contemporary sociobiology and its successors.

As we envision the developments of evolutionary science in its application to human beings in the coming decades, a growing body of empirical work along the lines of sociobiology, evolutionary psychology, and the "molecular biology" of consciousness is to be expected, typically with reductivist conclusions. The argument of this essay is that a meaningful dialogue between the humanities and the natural sciences is dependent not on the subsumption of the humanities under the sciences, but on reexamining critically the epistemological stance toward the human being in relation to the underlying assumptions of evolutionary biology. In this territory also lies, it is argued, a framework for affirming human transcendence while accepting the evolutionary view of humanity bequeathed us by the Darwinian revolution. Although the arguments made here are compatible with the concerns of Christian anthropology to assert the special status of human beings in nature and its defense of human dignity against a growing tendency of biological reductionism, the analysis here does not presuppose the acceptance of a theistic interpretation of human origins, and it is compatible with naturalistic assumptions. Ideally it offers a direction for a fruitful dialogue between science, theology, and philosophy around these questions into the future.

The need for historical reference if we are to acquire some purchase on contemporary and future discussions of the place of human beings in an evolutionary universe is also highlighted here. Even though Darwin's works are historical documents, the return of major architects of sociobiology to the works of Darwin for inspiration (e.g., E. Wilson 2006) indicates the need to reexamine these sources for some reorientation of perspective. It is also evident that many of the methodological moves made by Darwin in such works as the *Descent of Man* in applying his theory to human beings set up the project that is now being explored by those pushing Darwinian perspectives into domains traditionally occupied by the humanistic disciplines.

To approach this issue, I will give a definition of the epistemic stance of *NHR* and illustrate its role in the work of Charles Darwin, first as it is manifest in his work as a natural historian and then as it is applied to human beings. The second section of the paper will then develop an argument for a more adequate philosophical anthropology that returns to insights developed within the phenomenological tradition, and draws upon select developments within a tradition of Continental philosophical anthropology. It is claimed that these perspectives offer a more satisfactory epistemology for dealing effectively with the human phenomenon in the wake of Darwin's revolutionary work than that of *NHR*.

NATURAL HISTORICAL REALISM

My initial focus is on the epistemological stance widely involved in the "observational" sciences such as natural history, field ecology, paleontology, comparative anatomy, animal behavior studies, and taxonomy. That this is a distinct epistemic stance is an issue little discussed in the literature, and seems more presumed as unproblematic in these scientific inquiries, rather than examined in any depth.[1]

I will take as an example of what I mean the views of E. O. Wilson in a semipopular essay that summarizes claims he makes elsewhere in greater depth (E. Wilson 1998, 1978). In this essay, Wilson wishes to distinguish his evolutionary approach to ethics, grounded upon "observation" and "experience," from the "transcendental" arguments made by theologians and philosophers who allegedly appeal to principles outside of experience. As a trained entomologist, his meaning of "experience" is closely tied to that scientific background, rather than derived from the highly interpreted and instrument-mediated concept of "experience" as this might be conceived by a theoretical physicist or molecular biologist. Rather than being simply a naïve or accidental point of view, however, I argue that the "natural historical realist" tradition has a long history behind it that reaches back to the split of the natural-historical and experimental sciences in the early modern period. It is also not simply to be equated with "empiricism" as this has been generally understood since Locke.

Although space will not allow elaboration of my historical claim (see Sloan 2006a, 1995; Larson 1979, 1971), it is my argument that the "natural historical" sciences did not participate, or at least did not participate deeply, in the major epistemological and methodological changes that emerged from the transformations in natural philosophy in the seventeenth century. As a result, they do not share in the philosophical developments that accompanied these changes associated with the names of Descartes, Galileo, Gassendi, Hobbes, and Locke. In fact, to have absorbed the claims of these better-known philosophical traditions would have done much to undermine the fundamental scientific aims of the natural historical sciences (Sloan 1972). These sciences drew instead upon select developments of the moderate realism of Aristotle as it was refracted through some of the subsequent scholastic tradition and then through Renaissance neo-Aristotelians such as Andreas Cesalpino, forming an alternative epistemology to that developed in the physical sciences.[2]

I characterize this form of knowing by the following properties. First, it assumes a theory of unproblematic access to the underlying structure of the world through trained observation and description. One can observe and discover natural truths about the world without significant worries about pyrrhonnian skepticism, doubts about causality, problems raised by the mediation of instruments like the microscope, or antirealist and conventionalist views of science. Reality is "perceived" and not "constructed." In the words of one of the great eighteenth-century architects of the natural history tradition, Linnaeus, "The first advance in Wisdom is to know things in themselves. This notion consists in a true idea of objects" (Linnaeus 1758, A5).[3]

Second, unlike traditional "empiricism," as this term has been typically understood since the seventeenth century, the mediation of experience by "ideas" does not entail the mitigated skepticism that developed in the historical tradition extending from Gassendi and Locke to Berkeley, Condillac, Hume, Mill, Mach, and logical empiricism. Although the tradition of "Natural Historical Realism" is strongly "empiricist" in the sense that it is based on *experientia* in the classical sense, and even if, as we see in the aphorism of Linnaeus, there may be reference to the mediation of such experience by "ideas," this does not imply the mitigated skepticism of the modern empiricist tradition. Behind it is an optimistic conception of knowledge in which the ideal is to discover the true "natural system" by reason and observation. Nature is knowable in its inner self through such observation. The "natural" system is thereby distinguishable from one based on human convention or arbitrary construction.[4] For this reason, it is incorrect to understand this tradition through the categories of modern empiricism. One of Darwin's fundamental claims was that his evolutionary theory had finally solved this problem of the true "natural system," a claim that could not be sustained on the foundations of Lockean and Humean empiricism (Darwin [1859] 1964, 413).

Third, *NHR* makes no commitment to an underlying mathematical ontology of nature in the Platonic-Galilean sense that was so important in the developments of the seventeenth-century physical sciences. For the latter sciences, mathematical idealizations are the privileged means of understanding the natural world, revealing a hidden world behind that relying on the secondary qualities of sensation. This point must be understood precisely. Mathematics has indeed come to play a large role, via statistics, population dynamics, and mathematical model-building, in evolution, ecology, and taxonomy. But mathematics in these sciences provides methods for the analysis of data obtained by empirical observation. This is not the same as the claim of the tradition of mathematical physics that the underlying ontology of the world is fundamentally mathematical in character and is therefore best grasped by mathematical analysis. *NHR* even obtains its force in the Enlightenment period, especially in the natural history of Linnaeus's contemporary Georges de Buffon (1707–1788), through the claim that this epistemological position is superior to, and more certain than, the understanding of nature revealed by mathematical physics (Hoquet 2005, chap. 8; Sloan 2006a).

Fourth, the framework of scientific understanding and explanation within *NHR* is primarily classificatory, qualitative, and descriptive rather than "experimental" in one important sense. Experiments, such as, for example, those that Darwin may have performed on seeds, orchids, and earthworms, those that a primatologist may conduct in studying the behavior of a chimpanzee colony, or those of a field ecologist who may test the impact of a chemical agent on an ecosystem, are indeed experiments, but these are not efforts to create new phenomena by manipulation of a hidden world behind appearances.

Finally, *NHR* is "precritical" in the sense that it does not engage the Kantian problem of the structuring of experience by a transcendental subject, paradigms, *Denkstilen*, social structures, or other a prioris, with the potential gap between things as they may be *an sich* and things that can be known with some certainty by human cognition. It ignores or brackets the problem of human subjectivity in the scientific process itself and the importance of intentionality in human knowing. To utilize Edmund Husserl's language, it profoundly exemplifies the "natural attitude" as a starting point for empirical science (Husserl 1931, I. Sects. 27–32).

These are large claims that I can defend only in a limited way in this chapter. To illustrate my points with some greater refinement, I will turn to Darwin's work directly.

DARWIN AND NATURAL HISTORICAL REALISM

Darwin absorbed the basic epistemic stance of *NHR* from the preceding natural history tradition into which he was initiated first at Edinburgh as a student from 1825 to 1827, where he attended the lectures of natural historian Robert Jameson, and there entered into close association with the invertebrate specialist Robert Edmond Grant. It was in this period that we first see the emergence of Darwin's expertise as an invertebrate zoologist (Sloan 1985). An example of the kind of epistemological framework into which Darwin first developed his scientific vision can be drawn from his careful descriptive analysis of reproduction in deep-water invertebrates of the genus *Flustra* that formed his first scientific contribution in a paper read to the student Plinian Society in Edinburgh in 1825. As he writes:

Having procured some specimens of the Flustra Cabacea (Lam:) from the dredge boats at Newhaven; I soon perceived without the aid of a microscope small yellow bodies studded in different

directions on it.—They were of an oval shape & of the colour of
the yolk of an egg, each occupying one cell. Whilst in their cells I
could perceive no motion; but when left at rest in a watch glass,
or shaken they glided to & fro with so rapid a motion, as at some
distance to be distinctly visible to the naked eye [. . . .] That such
ova had organs of motion does not appear to have been hitherto
observed either by Lamarck [,] Cuvier [,] Lamouroux or any
other author. (Darwin 1825, DAR 118: fols. 5–6; quoted by permission of the Syndics of Cambridge University Library.)

There are two points to which I draw attention here. First, there
is no uncertainty manifest in what Darwin claims he is seeing. Second, there are some low-level theoretical conclusions drawn from
these observations that display the way in which theory is being built
upon this solid observational base.

These observational skills that we find first manifest in the Edinburgh period were then refined during Darwin's years at Cambridge
from 1829 to 1831 where his scientific abilities were developed under
the tutelage of John Stevens Henslow, Adam Sedgwick, and the London and Cambridge entomologists with whom he was closely associated. Darwin's work in this period, captured particularly in his
extensive correspondence in these years with his cousin W. D. Fox
(Burkhardt et al. 1985–, vol. 1), shows the development of the acute
gaze and skilled vision of an attentive field observer who could discern the smallest differences between species of coleopterans and the
minute declinations of geological strata and formations, skills that
made him a remarkable field naturalist in his years on the HMS *Beagle*.
These skills mark his full entry into what he later would characterize
as the expertise of a "competent" naturalist, skills that were even more
deeply refined after the *Beagle* by his extensive taxonomic revision
of the subclass Cirrhipedia (barnacles) from 1846 to 1854.

I emphasize these points alongside the philosophical and conceptual formation of Darwin as a "philosophical" naturalist who had
been first exposed to some of the currents of Scottish empiricism and
common-sense realism at Edinburgh (Manier 1978; Pajewski 2012).
At Cambridge he had also imbibed the writings of Alexander von

Humboldt, John Herschel, and then, on the voyage, those of Charles Lyell, all of which fashioned him into an especially gifted naturalist (Sloan 2009; Hodge 2009; Hodge and Radick 2009, introduction). This expanding philosophical layer in his thought indeed extends the sophistication of his analyses and the range of his generalizations, but it does not lead him to question or challenge the foundations of *NHR* itself. It only reinforces his view that through the optimistic epistemological assumptions of *NHR* he can penetrate to the underlying *vera causae* of natural phenomena. In other words, he does not seem to have been affected either by the problems within empiricism articulated by Hume, or by the critiques of the foundations of empiricism articulated by Kant, which do seem to have some impact on authors like Humboldt, Cuvier, and Herschel, on whom Darwin often drew for inspiration (Sloan 2006b).

We see this *NHR* epistemic stance in place in the post-Cambridge writings as Darwin begins noting down observations of marine organisms soon after leaving England on the HMS *Beagle* in December of 1831. His *Zoology Diary* is filled with minute descriptive details on animals collected and studied intently both with the naked eye and with the aid of microscopes. As one example, we can take an entry entitled "Luminous Sea" dated August 22:

> the sea was very luminous: light, pale, sparkling, but not as in Tropics either milky or in flashes.—The Luminous particles passed through fine gauze.—In the water were some minute Crustaceae of the genus Cyclops. I should not be surprised if these added to the effect.—During the day the sea has abounded with Dianoea—& I find these when kept in water till they are dead render it luminous.—can this be the cause of the appearance in the ocean.—
> (Darwin DAR 30.1, fol. 73, in Keynes 2000, 67)

Another entry of the same date under "Polype undescribed" gives us another illustration:

> This specimen agreed with those found at the Abrolhos.—Pl. 1. Fig.1. I have drawn the posterior half of animal.—The Tail, or that

part which the central intestinal tube does not penetrate is filled with a fine granular pulpy matter. With .3 focal distance lens, a longitudinal division & one on each side of this might be seen. (68)

Similar examples can be multiplied without limit from the zoological and geological writings of this period. They display repeated examples of the increasing refinement of his observational skills and his care with details that marked him as an outstanding naturalist of the period to which was added an increasingly sophisticated theoretical component of reflection.

TURNING THE ZOOLOGICAL GAZE ON HUMANS

When Darwin first turned his form of inquiry onto human beings during the *Beagle* period, we see this same observational stance employed. Humans can also be dealt with as zoological specimens, observed and described in empirical qualitative detail, and valid conclusions can then be drawn about them from these observations. In his first encounters with the aboriginals of Tierra del Fuego, he records in his *Diary of Observations*:

> I would not have believed how entire the difference between savage & civilized man is.—It is greater than between a wild & domesticated animal, in as much as in man there is greater power of improvement.—The chief spokesman was old & appeared to be head of the family. . . . The skin is dirty copper colour. Reaching from ear to ear & including the upper lip, there was a broad red coloured band of paint.—& parallel & above this, there was a white one; so that the eyebrows & eyelids were even thus coloured; the only garment was a large guanaco skin, with the hair outside. . . . Their very attitudes were abject, & the expression distrustful, surprised & startled. (Darwin, entry for 18 December 1832, in Keynes 1988, 122)

As we see in this passage, there is no difference in descriptive language from that employed when viewing any zoological object.

The Fuegian is essentially a zoological specimen, separated by a wide gap from civilized man, and therefore open to the same kind of natural historical gaze as employed elsewhere in the descriptions from the *Beagle* period. Although recognizing these aboriginals as members of the same zoological species as himself, there is a fundamental gap between observer and observed that constitutes the grounds for this *NHR* stance.

This all might seem unproblematic to read—just Darwin describing carefully as a skilled natural historian the world revealed to his senses. But these examples highlight the way in which Darwin's observation of the world is assumed to be epistemically unproblematic, forming a solid base upon which further reasoning is built. Darwin's expanding theoretical vision during these years, developing theories of geological formation, continental elevation and subsidence, reflections on the relationships between the animal and plant kingdom, and theories of biogeography, all mark remarkable events in Darwin's development during the *Beagle* years. What they do not indicate is any fundamental change in his adherence to the assumptions of *NHR*.

When Darwin opened his transformist reflections in the spring and summer of 1837, his descriptive language involving plants, animals, and human beings undergoes no discernible change, even though the broad generalizations he draws from such observations, of course, become extensive and far-reaching. The breakdown of the ontological boundary between humans and animals necessarily involved him in reflection on a wide range of questions, and as his conclusions on the permeable boundaries of organic species developed, we also see him engage the anthropological meaning of these inquires. What does not occur to him until writings late in his life is any reflection on the possible revisions this might require in *NHR* itself when the observer's own consciousness is put into some kind of continuity with the things being observed.

When he first addressed the issues raised by transformism in a famous passage from the later section of Notebook B opened in July of 1837, we see first emerge the view that inner human properties must also be continuous with those of animals:

Animals—whom we have made our slaves we do not like to consider our equals.—<<Do not slave holders wish to make the black man other kind?>>Animals with affections imitation, fear <of death>. pain. sorrow for the dead.—respect

....

<<the soul by consent of all is superadded, animals not got it, not look forward>>if we choose to let conjecture run wild then <our> animals our fellow brethren in pain, disease death & suffering <<& famine>>; our slaves in the most laborious work, our companion in our amusements, they may partake, from our origin in<there> one common ancestor we may be all netted together.— (Darwin Notebook B, fols. 231–32, in Barrett et al. 1987, 228–29)

The M and N Notebooks of the late 1830s, and the collection of notes known as "Old and Useless Notes" from this same period contain the first statements of the main body of Darwin's views on human beings later synthesized in the *Descent of Man* and in the *Expression of the Emotions* of the 1870s. From this early period, Darwin's repeated claim is that his zoological perspective, when combined with the thesis of continuity, becomes the solution to a host of philosophical questions: "He who understands baboon <will> would do more towards metaphysics than Locke" (Darwin Notebook M, fol. 84c, in Barrett et al. 1987, 539). But as the reflections in Notebook M of 1838 bring out most clearly, adopting the standpoint of the "metaphysical baboon" requires Darwin to take on an enormous number of issues—ethical reasoning, abstract reflection, and religious and aesthetic sentiment among the most pressing. By abolishing the traditional distinctions of reason and instinct, of free will and deterministic behavior, accepted by much of prior naturalistic and secular, as well as theological, tradition, differences that traditionally defined the distinction between humans and animals, Darwin had to confront numerous questions. In his words, it required the "study of instincts, heredetary [*sic*], & mind heredetary, whole [of] metaphysics" (Notebook B, fol. 228, in Barrett et al. 1987, 227).

The first issue encountered was that of the relation of body and mind within a historical transformist perspective. This issue first comes to the surface in Notebook N:

> To study Metaphysic, as they [*sic*] have always been studied appears
> to me to be like puzzling at Astronomy without Mechanics.—
> Experience shows the problem of the mind cannot be solved by
> attacking the citadel itself.—the mind is function of body.—we
> must bring some *stable* foundation to argue from. (Notebook N,
> fol. 5, in Barrett et al. 1987, 564)

These passages are followed by reflections inspired by phrenologist
Johann Casper Lavater's *Essays on Physiognomy* in the second edition of 1804, in which Lavater argued for a close identity of somatic
and mental properties. Darwin also begins to speak of mental properties like "will" as properties of lower organisms, involved in such
actions as the phototropic properties of hydra, and "memory" in the
repetitive actions of plants (Barrett et al. 1987, 567, 577). Furthermore, and I underline this point, the degree to which such properties
are manifest is *directly related* to the degree of organizational complexity of the organism:

> We must believe, that it require[s] a far higher & far more com-
> plicated organization to *learn* Greek, that [*sic*] to have it handed
> down as an instinct.—Instinct is a modification of bodily struc-
> ture <<(connected with locomotion.).>.<<no, for plants have
> instincts>><<either>>to obtain a certain [end]: & intellect is a
> modification of <intellect> <<instinct>>—an unfolding & gen-
> eralizing of the means by which an instinct is transmitted.—
> (Barrett et al. 1987, 567)

It is also in the Notebook period that we can follow Darwin's attraction to forms of metaphysical monism as a solution to this problem.[5]
He seems to have found this best expressed by the German *Natur-
philosoph* Carl Gustav Carus, whose "Kingdoms of Nature: Their
Life and Affinity" he read in May or June of 1838. As he comments
in an entry in Notebook C:

> There is one living spirit, prevalent over this wor[l]d . . .which
> assumes a multitude of forms <<each having acting principle>>

according to subordinate laws. There is one thinking . . . principle (intimately allied to one kind of organic matter.—brain. & which . . . thinking principle, seems to be given or assumed according to a more extended relations [*sic*] of the individuals, whereby choice with memory. or reason? is necessary).—which is modified into endless forms, bearing a close relation in degree & kind to the endless forms of the living beings.—We see thus Unity in thinking and acting principle in the various shades of <dif>separation between those individuals thus endowed, & the community of mind, even in the tendency to delicate emotions between races, & recurrent habits in animals. (Darwin, Notebook C, fols. 210e–211, in Barrett et al. 1987, 305)

This monism seems to have furnished Darwin at a very early point in his theoretical development with a way of directly linking stages of physical organization and stages of mental development with a universal "thinking principle" as the glue holding these together. It extends the inner, mental properties of humans in some way to all of life. Hence we read in the "Old and Useless Notes":

A Planaria [i.e., flatworm] must be looked at as an animal, with consciousness,, it choosing food—crawling from light.—Yet we can split Planaria into three animals, & this consciousness become multiplied with the organisms structure, it looks as if consciousness an effects [*sic*] of sufficient perfection of organization & if consciousness, individuality. (Darwin, OUN, fol. 16, in Barrett et al. 1987, 604)

This psycho-physical identity becomes a fundamental premise of Darwin's reflections on the body-mind question, laid down very early, and never rejected. It is also closely allied to one meaning Darwin gives to the concept of "nature" (Sloan 2005). This enables us to see the foundation for the otherwise puzzling moves that Darwin makes in the discussions of the *Descent of Man* of 1871. How Darwin eventually links up his relation of the physical and mental constitutes one of the main conceptual problems facing his anthropology.

NHR IN THE *DESCENT OF MAN*

The discussion to this point has been in many respects preliminary to an analysis of the ways in which *NHR* comes to function in the published *Descent* itself. Here the many strands of argument first sketched out in the Notebooks were combined with the theoretical apparatus of natural selection theory and its companion, sexual selection theory, now supported by a massive body of observations. In the interim between the Notebook period and the composition of the *Descent* also intervened Darwin's numerous publications between 1842 and 1871, including the first five editions of the *Origin* with its numerous changes, and also the broad and omnivorous reading Darwin conducted on human as well as general biological subjects that commenced particularly in the 1840s. These readings included the writings of many authors who dealt in some way with human questions. His concern with the polygenetic challenge put forth by pro-slavery authors was also of some importance in these readings (Desmond and Moore 2009).[6]

Hidden from view in the original *Origin*, Darwin's zoological perspective on human beings formed an issue to which he returned in February of 1867. At that time Darwin decided to pull out material relating to human beings from his overly long manuscript of the *Variation of Plants and Animals Under Domestication* to form a "very small volume, 'an essay on the origin of mankind'" (CD to Hooker 8 February 1867, and CD to Turner 11 February 1867, in Burkhardt et al. 1985–, 15:74, 80). In 1867 he also sent out a questionnaire to a wide network of correspondents asking for information on various aspects of human emotions that he incorporated into the *Expressions of the Emotions*. His inquiry into human questions blossomed into a much more ambitious project that taxed his abilities. Ethics particularly agitated him: "the difficulties of the Moral sense has [*sic*] caused me much labour" (CD to Gray 15 March 1870, in Burkhardt et al. 1985–, 18:68). The two-volume work that resulted from this, made even shorter than the original manuscript after he removed for separate publication his work on the expression of the emotions, was sent to the printer in June of 1870 (CD to Haeckel 23 June 1870, in Burkhardt et al. 1985–, 176).

In sheer bulk, the *Descent* is as much a zoological treatise as one devoted to anthropological questions, with 66 percent of the total of 828 pages of text of the first edition devoted to expanding on the general concept of sexual selection and its effects on molluscs, crustaceans, insects, amphibians, birds, and mammals. The original "Essay on Man," as it might be termed, formed primarily the material occupying the central five chapters (2–7) of volume one. These dealt serially with the manner of development from some lower form (chap. 2); mental powers (3); moral sense (4); the development of intellectual and moral faculties during primeval and civilized times (5); affinities and genealogy of man (6); and a chapter on races (7), with a final concluding chapter to the second volume on the more general application of sexual selection to humans and its bearing on aesthetics, musical powers, and secondary sexual characteristics.

Darwin's lengthy treatment of human questions reads generally as an expansion of remarks and proposals we can trace back to M and N Notebooks and the "Old and Useless Notes About the Moral Sense" of the late 1830s, and embodies the same *NHR*. But there are also major developments beyond these foundations encountered in the *Descent*, including the important additions of natural selection theory and the new emphasis on sexual selection that had played only an ancillary role in the *Origin*. The main ingredients of his general argument are drawn together in his final summary discussion of volume two. The general question he thinks he has addressed is "how far the principle of evolution will throw light on some of the more complex problems in the natural history of man," with the evolutionary origin of humans considered so well founded in the judgment of "competent" naturalists as to "never be shaken" (Darwin [1871] 1981, 2:385).

THE QUESTION OF HOMOLOGIES

As we pursue Darwin's general application of his theory to human beings, we see that this involves an inferential leap in his work that is neither immediately evident nor argued for, yet it reaches to some of the deeper levels of the argument taking place in both the *Descent* and *Expression of the Emotions*. This leap has been so deeply ingrained in the tradition that has built upon his arguments that it is rarely questioned. This concerns the application of the concepts of "homology" and "analogy" to human beings. At the deepest level it concerns the meaning of "sameness" in biology (Ramsey and Peterson 2012).

The critical conceptual distinction between homology and analogy entered Darwin's work in a prominent way after his encounter with Richard Owen's mature formulation of this distinction in 1845, which Owen published in a long monograph in the *Reports of the British Association* (Owen 1846).[7] The use of these new technical terms has been widespread in biology since Owen's work, often employed, however, in inconsistent ways (Laubichler 2000). Through this distinction, Owen claimed to give a theoretical solution to the problem of "sameness" in biology, seeking to resolve in this way the famous Cuvier-Geoffroy dispute of the 1820s and '30s (Appel 1987). To accomplish this, Owen limited the term "analogy" to *functional* similarities of parts. In a common example, one could speak of the wing of a bird and the pectoral fin of a flying fish as "analogous," since they accomplish the same function, but do not have any common parts and emerge from different structures embryologically. This meaning of "analogy" was then differentiated by Owen from genuine anatomical identity, his meaning of "homology." For Owen, this latter form of identity held through "every variety of form and function." The flipper of a whale, the forelimb of a human being, and the wing of a bat would be examples of homology, "special" homology in Owen's definitions.[8] The long-standing conceptual problem of mid-nineteenth-century comparative anatomy had been to explain how this kind of similarity could be distinguished from functional resemblance. In offering a solution to this question, Owen grounded it upon the identity of parts with those of a theoretical transcendental archetype that served both as an ideal type and also as a dynamic immanent law of historical development (Rupke 1993; Sloan 2003). Darwin subsequently historicized and materialized the archetype concept to designate this as a literal historical ancestor (Darwin [1859] 1964, 434–39). This then enabled him to explain the relationships of homology as due to historical derivations from a historical common source. The designation of the "same" parts by common names across groups no longer needed to be considered "metaphorically," as he termed it, but could have a plain literal meaning. One can speak univocally of a "femur" being present in a whale and a bat, and this presence implies, for Darwin, one of the primary proofs of common historical ancestry.

Since 1859, the exploration of this issue has, to be sure, deepened considerably. These analyses have also transformed the concept from its original definition at the level of gross anatomy to that of intimate cellular identities and homologies of genetic structures and developmental genetic modules such as the *Pac6* gene complex that governs the formation of eyes across a wide spectrum of groups (Gilbert, chap. 2, this volume; Lynch and Wagner 2010). This has meant that structures, such as the camera eye of the cephalopod mollusc and of the fish, which might have been considered "analogous" in an older comparative anatomical account, have been interpreted as homologies from a developmental genetics perspective.

But the warrant for extending this same distinction to mental and behavioral properties is not so transparently obvious, yet it is crucial for Darwin's anthropological project. It cannot be dismissed as simply folk metaphor or loose anthropomorphic language when Darwin attributes "play" to ants, "love" to dogs, "aesthetic sense" to bower birds, and "bravery" to monkeys. These must be, in the technical sense developed above, homologous, and not simply analogous relationships, if the argument he proceeds to build in the *Descent* is to be convincing. This point is today considered justifiable by some recent animal behaviorists and primatologists on the scientific principle of parsimony. By this they defend the extension of animal properties to humans and the use of anthropomorphic language in scientific contexts as fully warranted by an evolutionary perspective (e.g., Waal 1997, 2006a).

Surprisingly, however, there is almost a complete absence of the technical language of "homology" in Darwin's writings when speaking of these inner properties of human beings. Nor does one find an explicit argument for the presumed parallelism of anatomical and inner resemblances elsewhere in his writings, even though the reasoning must hold if Darwin's anthropological project is to push through.[9] Darwin's arguments on this point are much looser than expected. In keeping with his earlier language of the M and N Notebooks, and "Old and Useless Notes" on which he was also drawing heavily in composing the *Descent*,[10] he commonly employs instead a broader and nontechnical concept of "analogy" to discuss these resemblances, as when he speaks of the songs of birds offering "in several respects the nearest analogy to language" (Darwin [1871] 1981, 1:55).

This is more than a minor textual detail or literary oversight on Darwin's part. If there is to be some parallel transference of the *anatomical* evidence for homological identity to *mental* and other nonmaterial properties of human beings through reasoning upon observations made about the behavioral properties of other organisms, the homology-analogy distinction must hold in this mental and inner domain as well. In other words, there cannot be some fundamental difference in kind involved that renders human properties only similarities of *function* rather than resemblances of genuine *identity* in the sense of evolutionary homology.

Darwin seems to accomplish this linkage not by a substantive argument but rather, as we have developed above, by the reliance upon a deeper unstated metaphysical monism—an underlying identity of mind with the degree of material organization that harkens back to the claims we find in Notebook C quoted above and elsewhere in Notebooks M, N, and OUN.[11] But the crucial question is whether this is adequate warrant for dealing with the more complex human phenomena in zoological terms in keeping with the framework of *NHR*.

RECONCEPTUALIZING HUMAN TRANSCENDENCE

The foregoing excursion into historical analysis and the foundations of Darwin's zoological perspective on human beings has been intended, by going back to the sources, to open up space for reconceptualizing the relationship of humans to nature consistent with an evolutionary perspective on human beings. This is, to be sure, an enormous problem for future development that requires extended analysis and a critique of the adequacy of *NHR* in the human sciences. I offer here only some preliminary suggestions of ways this alternative might be developed, which I will apply to the limited range of issues I have developed concerning the "analogy-homology" distinction made previously.

The long-standing impasse over the issue of human transcendence in relation to evolutionary biology, typically pitting "creationist" and "naturalistic" accounts against one another, is familiar to all. In addition to the familiar disputes, this opposition also has often been couched as a claim that evolutionary history requires that we must "decenter" human importance within a generally naturalistic perspective if we are to acquire a deeper respect for nature and for other forms of life consistent with an ethic of global and ecological responsibility (Baxter 2007; D. Wilson 2007). I suggest that there are ways to reconceptualize the issue of human transcendence in a form that can both support a robust view of human dignity and moral and ethical freedom that is also consistent with a deep respect for the world in which we live without these Darwinian reductionisms.[12]

To focus this discussion, I will draw selectively upon certain themes that have been pursued in greater philosophical depth within Continental philosophical anthropology as these have developed from foundations laid down in a German context by Edmund Husserl, Adolph Reinbach, Max Scheler, and from the French side particularly by Maurice Merleau-Ponty.[13] This is to argue that we must break with a line of philosophical reflection common in Anglo-American philosophy of the life sciences that assumes that any such reflections on human beings necessarily must begin from the data of the natural sciences with the ideal of a reductive naturalistic closure. Following the foundational standpoint developed by Edmund Husserl and others in the phenomenological tradition, I claim here that it is necessary to return to a more fundamental starting point that begins from the experience and description of ourselves as existentially existent, conscious, and self-reflective beings. This starting point must be taken into account in any adequate understanding of what science has to tell us about ourselves. In other words, it is to grant full reality to our prephilosophical experience as human beings in the world, a reality that stands *prior to* our scientific rationalizations about our origins, or to causal explanations of human experience that then might be supplied by the natural sciences. To seek this alternative starting point is not to deny or question the information or insights of the natural sciences, but only to reposition ourselves in relation to these inquiries.

To bring this into conversation with what I have termed Darwin's "natural historical realism," I will utilize a short essay from this Continental philosophical tradition, "On the Upright Posture," initially published in 1952 by the German phenomenological psychiatrist Erwin Straus (1891–1975), a political refugee from National Socialism who came to the United States in 1938. Straus had studied at Göttingen with Husserl and with his lesser known colleague, Adolph Reinach, who developed phenomenological methodology in a more realist and applied direction (Mulligan 1987). He then became for a period a major leader in the development of phenomenological approaches to psychiatry before his forced departure from Germany.[14] Straus's reflections also display debts to arguments reaching back to Johann Herder, Husserl, Jakob von Uexküll, and others not mentioned in the essay, but my concern here is not to work out these filiations. Although Straus engages issues that are developed in greater depth by more major figures in the phenomenological movement,[15] there are specific dimensions of the presentation by Straus that are more concrete and more directly relevant to the issues I am pursuing here than I have found in others.

Straus was concerned, like others developing on the foundations laid by Husserl, to draw attention to the need for philosophy to return to a starting point in the fundamental prereflective and prescientific experience of ourselves as a framework from which all philosophizing and science must ultimately begin. But in the tradition of Reinach, his approach remained more in the tradition of developing a "descriptive psychology" of action rather than moving to the "transcendental turn" of Husserl's later works. In this sense Straus was mainly concerned to develop some consequences of the concrete positioning of the human being in the world, focusing on, among other issues, the unusual anatomical and physical positioning of human beings in the world as upright beings defined by obligate bipedalism.[16]

In his essay Straus argues that it is this unusual way of experiencing the world that most strikingly defines the differences between the human life-world and that of other animals. The physical aspects of human posture, which he analyzes in terms of its gradual acquisition by the child, the distancing and unusual forms of encounter it imposes on human sociality, and its implications for walking, the use of the forelimbs, and developments made possible by the hand, are all considered to define for us a novel world of experience that is not that of the animal, not even that of the higher primates. Obviously developing on some themes set forth by Jakob von Uexküll in his notion of the *Umwelt*, Straus argues:

> Men and mice do not have the same environment, even if they share the same room. Environment is not a stage with the scenery set as one and the same for all actors who make their entrance. Each species has its own environment. There is a mutual interdependence between species and environment. The surrounding world is determined by the organization of the species in a process of selecting what is relevant to the function-circle of action and reaction. Upright posture pre-establishes a definite attitude toward the world; it is a specific mode of being-in-the-world. (Straus 1952, 532)[17]

This may seem an obvious point, but it is not developed by other phenomenologists who may have discussed more generally the position of the embodied self in the world, such as Maurice Merleau-Ponty. One is struck with how often this elementary fact, and its deeper implications, is simply ignored in the efforts to extend Darwin's analysis and the *NHR* perspective to human beings. As philosopher of biology Marjorie Grene summarizes the importance of Straus's phenomenology of the living world for human life:

> Straus tries to lead us . . . to critical and constructive reflection on the nature and presuppositions of action, and, more generally, of what it means to be, both in perception and motility, in sensing and performing, "an experiencing being." (Grene 1974, 295)

Straus is not questioning an evolutionary derivation of human beings, and his argument, and that which I am building upon Straus's suggestive remarks, are compatible with any reconstruction of human phylogenetic history. However we have come historically to our present condition, our way of being-in-the-world defines our orientation to it and to others. The human being existentially and immediately encounters the world from this vertical and gravity-defying perspective of obligate bipedalism. It would over-simplify matters to argue that all human properties are at some point to be attributed simply to this development of the human mode of being sometime in human phylogenetic history. Nonetheless, I argue that recognition of the implications of this form of being has important ramifications for the way we conceptualize the meaning of the continuities of humans and animals, and for scientific analyses of humans more generally.

The failure to attend sufficiently to the experiencing being behind our scientific enquiries reveals the weakness of the "natural attitude" of *NHR* in the Darwinian anthropological project and its contemporary heirs. For what may be considered valid heuristic reasons, *NHR* simply assumes this primordial experience of ourselves as a background given behind and prior to scientific inquiry into nature. For large areas of observational and field science, this works well, and I am not questioning its general adequacy in these domains. Science need not be dissolved into subjectivity, transcendentalism, and strong constructivism if this unexamined primordial experience of the subject is acknowledged. But when *NHR* is directed unreflectively on the one who "knows that one knows," the "zoological gaze" assumed by *NHR* runs into epistemological limits that Darwin himself eventually recognized in his famous "horrid doubt" about the status of his own thinking and the conclusions validly drawn from it (see Griffiths and Wilkins, chap. 9, this volume).

For one thing, *NHR* simply brackets the issue of intentionality and self-reflection in our scientific analysis of nature. Oxford evolutionary anthropologist Robin Dunbar has recently discussed some of the issues I am raising here in terms of orders of intentionality involved in human culture, and the way in which this differentiates human knowing from that of the primates. This builds upon experimental work that has revealed that humans are capable of up to five orders of nested intentionalities, whereas animals may be capable of only one or at most three (Dunbar 2011, 2008. See also Barnard 2012, chap. 1). Through the difference in these orders of intentionality, we engage the complexities of language, the reflexive problem raised by the awareness of our own consciousness, and the different intentional levels of meaning involved in complex culturally situated human exchange. To use Dunbar's example of a human audience watching a Shakespearean play: "The audience watching this play has to *believe* that Iago *intends* that Othello *imagines* that Desdemona *is in love* with Casio and that Casio reciprocates this state of *being in love*" (Dunbar 2011, 280). The italicized words indicate a series of five nested levels of mental states— "believe," "intend," "imagines," "is in love with," and "being in love back"—which are all necessary to understand the play. To miss any of them is to lose the larger intentional meaning. The perspective of *NHR* remains, however, operative only at the first three of these levels of intentionality. These we might summarize as: "I *know* that X property I *observe* in this creature Y *means* Z." Or less abstractly: "Grooming behavior in colonial baboons indicates the evolutionary origins of human altruism." What is not taken into account is the sequence: "*I* as scientific observer for *purpose* Z, *understand* X property to *mean* Y." Reflection on the intentionality of conscious acts in the process of scientific knowing necessarily leads us back to consideration of how this intentionality relates to the way in which we do in fact know the world.

TRANSFORMING HOMOLOGIES INTO ANALOGIES

To factor the full range of human intentionality and its attendant self-reflection into the picture of the human being, and to specify the role these factors play in the world defined by human positioning in the world, is a topic with numerous complexities and ramifications that cannot be explored here. Analyses of the origins of language, sociality, and mental powers are all involved, and are the subjects of a wide literature (e.g., Barnard 2012; Deane-Drummond 2014a; Marks 2002;

Tomasello and Call 1997; Donald 1991). The point of relevance I single out in this wide area of inquiry concerns one key issue: the possibility that properties that may have *originated* as genuine *homologies* in the Darwinian sense can nonetheless be transformed *by entry into this human life-world* effectively into *analogies*, that is, as similarities of function only. The emergence of the human world and the positioning of the human being in relation to the biological world do not, of course, imply a change of underlying biology. The similarities of human and animal anatomy and physiology that can be pursued down to the genetic and biochemical levels remain true homologies. The manifestation of changes in human inner life historically would not be observed in some marked anatomical evidence beyond the fact that they have accompanied in some way, and not necessarily all at once, the development of obligate bipedalism. Where this new world is made manifest is in the remarkable expressions of human self-reflection as found in the manufacture of cultural artifacts, art, religious customs, symbolic language, and depictions of the external world, for example, in the great paleolithic cave art of Lascaux and Chauvet-pont-d'arc in France. If anatomical structures and biochemical and physiological processes remain true homologues through this process, it does not simply follow that behavioral and mental properties also remain homological identities.

My claim is that the transformation of the "inner" properties of human existence revealed by the history of such cultural and technological developments is so profound as to make the assumption of homological identity at this level deeply misleading. For example, ethical behavior, a wide topic of current exploration by evolutionary psychology and sociobiology (Maienschein and Ruse 1999; Katz 2000; Clayton and Schloss 2004; Griffiths and Wilkins, chap. 9, this volume), seems at least difficult to resolve in evolutionary terms. On one hand, evolutionary accounts seem unable to develop a sufficient distinction demanded by human ethical reflection between properties that are of evolutionary advantage from those that might be considered ethically good—the classic issue of the "naturalistic fallacy" as defined by G. E. Moore and others (Farber 1994). On the other hand, the assumption of a homological identity of human ethical properties with benevolent and peaceful behavior of primates (Waal 2006a, Singer 2006) has difficulty developing a sufficiently robust ethical theory that does justice to the ethical reflections that have in fact emerged in human history from the interpenetration of language, culture, and symbolism. Hence even if the primatologists have it correct in seeing interesting continuities between the behavior of higher primates and humans, it is questionable whether the development of these properties within the human life-world, with its complex and dynamic interaction of thought, language, culture, reflection, and physical being, is in any sufficient way "explained" by evolutionary accounts based on the assumption of some kind of homological identity.

Consider, for example, musical expression. The evolutionist can indeed indicate many ways in which music can be appreciated and even developed in many animals and virtually all humans. Neurological studies even suggest that musical appreciation and even musical ability may have neurophysiological and genetic dimensions, suggesting some warrant for a language of homology (Sacks 2007).[18] As Darwin noted extensively in the *Descent*, musical expression is found widely among humans, and there are elements of this evident in animals. The evolutionary story he offers traces musical expression to the probable role of musical sounds in courtship and mating in animals (Darwin [1871] 1981, 2:330–37). But with all the novelty introduced by the human upright positioning in nature, these evolutionary continuities are remarkably transformed, something that led even Darwin to find musical performance "amongst the most mysterious" properties with which humans are endowed (Darwin [1871] 1981, 2:333). The freeing of the hands and the face-to-face encounter of human beings has made it possible to develop all the elaborations of musical performance, both vocal and instrumental. The composition and performance of a piece of modern music requires self-reflection and conscious awareness at all the five levels of intentionality outlined by Dunbar. What this all means is that the performance of a musical piece need no longer serve a mating purpose, and in this sense is not even "analogous" in the sense of common function, but it may have acquired new functions—religious, social, and aesthetic—that exist only in the human life-world. To consider musical expression as "homologous" to animal mating calls is to confuse historical origins with an explanation of what music in fact is in human culture.

It is at this point that *NHR* reveals its inadequacy as an epistemological stance appropriate for analyzing human beings. The concept of "experience" or "empiricism" appealed to by scientists like E. O. Wilson to make the case for evolutionary reductions of human properties to those of animals simply involves a two- or at best three-level intentionality. It takes no account of the fact that it is we, as conscious and self-reflecting beings, who are encountering the world, often, as in science, refracting our experience through complex theories and written products describing scientific observations. For certain intentional goals, I may bracket, or render subsidiary, this fundamental lived experience as I study natural objects, or even human beings themselves. Within the *NHR* perspective the primatologist might teach us many things about the origins of certain behavioral properties. But this is a methodological limiting of vision, not an abolition of the lived human world within which our scientific inquiries are taking place.

To develop a more adequate epistemic stance for analyzing human beings is a project for inquiry in the future rather than something than can be accomplished by this chapter. As we look toward a future where we will assuredly see increasing efforts to subsume the human sciences and indeed all aspects of human experience under an evolutionary perspective, as foreseen in the projections by Jean Gayon in this volume, it is critically important to get certain things correct in this inquiry, and one of these is the need to examine more critically the application of *NHR* to human beings. My suggestion is that there are indeed resources both in the philosophical tradition and in contemporary reflection that open up new ways to think about the way human beings are approached within an evolutionary framework. None of this is to deny the importance of evolutionary science or question its relevance to many inquiries. It is instead an appeal for a reexamination of the presuppositions of scientific inquiry. By redirecting our attention to the starting point of all of our inquiry in human lived experience, we can on one hand avoid the paradoxes either of genetic and neurophysiological determinism or those of reductive sociobiology.

On the other hand, this repositioning of the human being in the natural world allows us to acknowledge the claims of evolutionary science and the remarkable body of work it has accumulated since 1859. By arguing in this chapter for the need for a more "critical" epistemology in dealing scientifically with the human being in the ensuing decades, my argument is that this requires an anthropological vision that fully recognizes the intentional, conscious human being as a free existential agent behind all our scientific rationalizations. In this we also see the opening for a rich philosophical anthropology of the human being wherein the interests of Christian theological anthropology and natural science may indeed find some new common ground for conversation.[19]

NOTES

I am indebted to Lenny Moss, Celia Deane-Drummond, Gerald McKenny, Kathleen Eggleson, Agustin Fuentes, Jonathan Marks, Vittorio Hösle, Michael Ruse, Alessandro Pajewski, and Robert Richards for incisive comments on this chapter. The views here are, of course, my own.

1. This is particularly apparent in Anglophone discussions. Reflecting more the heritage of Gaston Bachelard and Georges Canguilhem, and the Continental tradition of "historical " epistemology, Hans-Jörg Rheinberger has displayed some of the alternative ways of conceptualizing the epistemology of the life sciences within another philosophical framework, although his focus has been principally on the epistemology of genetics and molecular biology. His concern with the constructive nature of scientific epistemology, and his development of this against the background of Continental philosophy has some relevance to my own concerns, although I am drawing on different strands in this tradition. See Rheinberger (2010, 1997).

2. This also has connections with some of the impact of Roman Stoic epistemology on the analysis of experience of some early modern thinkers. On drawing my attention to this point I am indebted to the dissertation by Alessandro Pajewski of the University of Chicago (Pajewski 2012) and his comments on an earlier version of this paper.

3. "Primus Sapientiae gradus est res ipsas nosse. Notitia consistit in vera *idea* objectorum"; italics in original.

4. The differentiation of a "natural," as distinct from an "artificial," system in classical taxonomy was not a simple matter, and this search formed

much of the research program of pre-Darwinian natural history, requiring repeated iterations and revisions of taxonomic and natural history treatises, and the employment of extensive correspondence networks and networks of exchange of specimens. On this see Müller-Wille (2007) and Müller-Wille and Charmantier (2012). I also emphasize the importance of skill and training in this tradition, underlined by Darwin's frequent appeals to "competent" naturalists. Exacting discrimination of natural species, for example, still requires intensive experience and detailed comparative observation, with open-ended provisions for constant correction and revisions. *NHR* is therefore not to be equated with untutored naïve empiricism. Darwin's complex theory of science relies on the ideal of discovering "the real, true, known causes" behind such a natural system (Hodge 2009, 62).

5. A distinction is necessary between the loosely defined monism I see Darwin embracing and "materialism," which is sometimes attributed to Darwin. There is no effort in Darwin's works to reduce mental properties to force, matter, or material substrate in the fashion of the true materialists of his era, such as Ludwig Büchner, or more recent biophysical reductionists. See on this Gregory (1977).

6. I maintain reservations on the degree to which these concerns played a crucial role in the genesis of the *Origin* and the *Descent*.

7. These concepts are first employed by Darwin in the *Monograph on the Subclass Cirrhipedia* of 1851 to work out the homologies that presumably revealed the ancestral source of the cirrhipedes and crustaceans. Some contemporary biologists have even termed the homology-analogy distinction crucial for all subsequent biological theory, with homology functioning as "the central concept for *all* of biology" (Wake 1994, 265). I am indebted to my colleague Grant Ramsey for several references on the current discussion of this topic. See also his examination of this in contemporary philosophy of biology in Ramsey and Peterson (2012).

8. Owen distinguished between "special" homology, such as the comparison made here between different forms of forelimbs, from "serial" homologies, such as the fore and aft similarities of the front and hind limbs of mammals. These are both distinguished from "general" homologies, which referred to the relationship of both special and serial homology to a transcendental archetype of all the vertebrates. Darwin's transformation of Owen's archetype into a material, historical ancestor rendered general homology, and its subordinate special and serial homologies, evidence of historical common descent.

9. All but one of the usages of the term "homology" or its congeners (twelve usages) in the *Descent* are exclusively references to anatomical features. The one explicit use of the homology-analogy distinction beyond this is the application to the homologies between languages, warranting inferences

to common linguistic descent (Darwin [1871] 1981, 1:59). The use of variants of "analogy" (ninety-one examples) is much more common in the text, with only a subset of these usages (twenty) bearing some sense of contrast to homology in Owen's technical meaning. The two uses of "homologous" in the *Expressions* are also exclusively anatomical. For Franz de Waal's views, see Waal (2006a, 59–67; 1996, 17–20).

10. Owen's technical distinction of homology and analogy post-dated Darwin's transformist Notebooks. Darwin does, however, make several uses of a technical meaning of "analogy" in the Notebooks as this had been defined by the entomologist William Sharp MacLeay, where it is contrasted with a technical meaning of "affinity" (MacLeay 1825). As Darwin defined this in the early B Notebook: "Relations in analogy consist in a correspondence between certain insulated parts of the organization of two animals which differ in their general structure" (Darwin B Notebook, fol. 129 and editor's note in Barrett et al. 1987, 201). The most relevant use of the concept of "analogy" in Notebooks M and N is in the comparison of the wrinkling of the brow in a monkey to that of man, which Darwin sees as an "important analogy with man" (Notebook M, fol. 106 in Barrett et al. 1987, 545). There are numerous comparisons of animal and human inner states made by Darwin thoughout these early notebooks that are not stated in the language of analogy, but simply are treated as identities.

11. This underlies the reliance throughout his writings on one meaning of "nature" as a foundation for ethical normativity and even consciousness (Sloan 2005). See also Richards (2002, 2005).

12. I am indebted to my colleague, Celia Deane-Drummond (2014b), for assisting me in seeing the importance of this point. See also her chapter in this volume, and Deane-Drummond (2009).

13. I am careful to note here my selective appeal to the broad enterprise termed "philosophical anthropology" as employed in Continental philosophical literature. I am not lumping this diverse collection of traditions together as presenting a single vision. Joachim Fischer's recent characterization of the project of German philosophical anthropology is useful (Fischer 2009). There he wishes to distinguish this from neo-Kantianism, phenomenology, structuralism, and post-structuralism. As Fischer defines it, the early sources for this tradition are Max Scheler ([1928] 1961); Helmuth Plessner ([1928] 1975); and Arnold Gehlen ([1950] 1988). The line of argument I am developing primarily involves the "realist" interpretations of phenomenology as developed through Adolph Reinach and Scheler. It does not necessarily intersect with the views of Arnold Gehlen and Helmuth Plessner.

14. For brief details on Straus's intellectual biography, see the introduction by David Moss to Straus (1982). For philosophical analysis of his writings see Grene (1968, 1974) and chapters in Griffith (1966). On Reinach, see

Schumann and Smith (1987). I thank Prof. John Crosby for valuable references on Reinach.

15. This is particularly reflected in his ability to connect issues of phenomenology with concrete matters of human anatomy and mode of locomotion. Some of the specific dimensions of his reflections obviously relate to Straus's many years of work on the staff of a veteran's hospital.

16. I should emphasize that I am not interpreting upright posture as mono-causally related historically to the development of human uniqueness. At the same time, I would claim that there are intimate interconnections between these physical-anatomical developments and the eventual development of the phenomena that are associated by workers like Dunbar and Barnard with the human life-world.

17. He cites von Uexküll's *Theoretical Biology* (1926) in his bibliography. For some remarks on the *Umwelt* concept, see Griffiths and Wilkins, chap. 9 in this volume. I am not here attempting to work out the similarities and differences between von Uexküll's more biologically oriented notion of *Umwelt* and Husserl's concept of the *Lebenswelt* that has played a more prominent role in the phenomenological analysis of the human sciences. Straus's arguments seem more closely tied to von Uexküll here. A fuller analysis of the issue of human transcendance as I am developing it would need to explore this similarity and difference more fully. Some beginning of this is to be found in Konopka (2009).

18. I am avoiding entering the complex discussions of "innate" and "acquired" properties that can surround something like musical ability. See on this Bateson and Mameli (2007).

19. For some development of my arguments in this direction, see Sloan (2012). For a different perspective see Deane-Drummond (2014b).

REFERENCES

Appel, T. (1987). *The Cuvier-Geoffroy Debate*. Oxford: Oxford University Press.

Barnard, A. (2012). *The Genesis of Symbolic Thought*. Cambridge: Cambridge University Press.

Barrett, P., P. J. Gautrey, S. Herbert, D. Kohn, and S. Smith, eds. (1987). *Charles Darwin's Notebooks, 1836–1844*. Cambridge: Cambridge University Press.

Bateson, P., and M. Mameli. (2007). "The Innate and the Acquired: Useful Clusters or a Residual Distinction from Folk Biology." *Developmental Psychobiology* 49:818–31.

Baxter, B. (2007). *A Darwinian Worldview: Sociobiology, Environmental Ethics and the World of E.O. Wilson*. Burlington, VT: Ashgate.

Burkhardt, F., et al., eds. (1985–). *The Correspondence of Charles Darwin*, 18 vols. Cambridge: Cambridge University Press.

Clayton, P., and J. Schloss, eds. (2004). *Evolution and Ethics: Human Morality in Biological and Religious Perspective*. Grand Rapids, MI: Eerdmans.

Darwin, C. (1825). "Charles Darwin's Collected Manuscripts." Cambridge University Archives.

———. ([1859] 1964). *On the Origin of Species*. 1st ed., reprint edition. Cambridge, MA: Harvard University Press.

———. ([1871] 1981). *Descent of Man, and Selection in Relation to Sex*. 1st ed., reprint edition. Princeton: Princeton University Press.

Deane-Drummond, C. (2009). *Christ and Evolution: Wonder and Wisdom*. Minneapolis: Fortress.

———. (2014a). *The Wisdom of the Liminal: Evolution and Other Animals in Human Becoming*. Grand Rapids, MI: Eerdmaans.

———. (2014b). "In God's Image and Likeness: From Reason to Revelation in Humans and Other Animals." In *Questioning the Human: Perspectives on Theological Anthropology for the Twenty-First Century*, ed. L. Boeve, Y. De Maeseneer, and E. Van Stichel, 60–78. New York: Fordham University Press.

Desmond, A., and J.R. Moore. (2009). *Darwin's Sacred Cause: How a Hatred of Slavery Shaped Darwin's Views on Human Evolution*. Boston: Houghton Mifflin Harcourt.

Donald, M. (1991). *Origins of the Modern Mind*. Cambridge, MA: Harvard University Press.

Dunbar, R.I. (2008). "Mind the Gap: Or Why Humans Are Not Just Great Apes." *Proceedings of the British Academy* 154:403–23.

———. (2011). "How Humans Came to Be So Different to Other Monkeys and Apes." In *Biological Evolution: Facts and Theories; A Critical Appraisal 150 Years after "The Origin of Species,"* ed. G. Auletta, M. LeClerc, and R.A. Martinez, 275–89. Rome: Gregorian and Biblical Press.

Eldredge, N., and S.J. Gould. (1972). "Punctuated Equilbria: An Alternative to Phyletic Gradualism." In Schopf 1972, 82–115.

Farber, P. (1994). *The Temptations of Evolutionary Ethics*. Berkeley: University of California Press.

Fischer, J. (2009). "Exploring the Core Identity of Philosophical Anthropology through the Works of Max Scheler, Helmuth Plessner, and Arnold Gehlen," *IRIS: European Journal of Philosophy and Public Debate* 1:153–70, available at http://www.fupress.net/index.php/iris/article/view/2860/2992.

Gehlen, A. ([1950] 1988). *Man, His Nature and Place in the World*. Translated by C. McMillan and K. Pillamer. New York: Columbia University Press.

Gregory, F. (1977). *Scientific Materialism in Nineteenth-Century Germany*. Boston: Reidel.

Grene, M. (1968). *Approaches to a Philosophical Biology*. New York: Basic Books.

———. (1974). "The Characters of Living Things II: The Phenomenology of Erwin Straus." In *Marjorie Grene: The Understanding of Nature; Essays in the Philosophy of Biology*, ed. R.S. Cohen, 294–319. New York: Reidel.

Griffith, R. M., ed. (1966). *Conditio Humana: Erwin W. Straus on His 75th Birthday*. New York: Springer.

Gutting, G. (2001). *French Philosophy in the Twentieth Century*. Cambridge: Cambridge University Press.

Hodge, M. J. S. (2009). "The Notebook Programmes and Projects of Darwin's London Years." In Hodge and Radick 2009, 44–72.

Hodge, M. J. S., and G. Radick, eds. (2009). *The Cambridge Companion to Darwin*. 2nd ed. Cambridge: Cambridge University Press.

Hoquet, T. (2005). *Buffon: Histoire naturelle et philosophie*. Paris: Honoré Champion.

Hösle, V., and C. Illies, eds. (2005). *Darwinism and Philosophy*. Notre Dame, IN: University of Notre Dame Press.

Husserl, E. ([1931] 1972). *Ideas: General Introduction to Pure Phenomenology*. Tranlsated by W. R. B. Gibson. Reprint ed. New York: Collier.

Katz, L. D., ed. (2000). *Evolutionary Origins of Morality: Cross-Disciplinary Perspectives*. Bowling Green, OH: Imprint Academic.

Keynes, R. D., ed. (1988). *Charles Darwin's* Beagle *Diary*. Cambridge: Cambridge University Press.

———. (2000). *Charles Darwin's Zoology Notes and Specimen Lists from H.M.S. Beagle*. Cambridge: Cambridge University Press.

Konopka, A. (2009). "The Role of *Umwelt* in Husserl's *Aufbau* and *Abbau* of the *Natur/Geist* Distinction." *Human Studies* 32:313–33.

Larson, James L. (1971). *Reason and Experience: The Representation of Natural Order in the Work of Carl von Linné*. Berkeley: University of California Press.

———. (1979). "An Alternative Science: Linnaean Natural History in Germany, 1770–1790." *Janus* 66:267–83.

Laubichler, M. (2000). "Homology and Development and the Development of the Homology Concept." *American Zoologist* 40, no. 5:777–78.

Linnaeus, C. (1758). *Systema naturae per regna tria naturae*. 10th ed. Stockholm: Laurentii Salvii. Available at http://www.biodiversitylibrary.org/bibliography/542#/summary.

Lynch, V., and G. P. Wagner. (2010). "Revisiting a Classic Example of Transcription Factor Functional Equivalence: Are *Eyeless* and *Pax6* Functionally Equivalent or Divergent?" *Journal of Experimental Zoology. Part B: Molecular and Developmental Evolution* 316B:93–98.

MacLeay, W. S. (1825). "Remarks on the Identity of Certain General Laws which Have Been Lately Observed to Regulate the Natural Distribution of Insects and Fungi." *Transactions of the Linnean Society of London* 14:46–68.

Maienschein, J., and M. Ruse, eds. (1999). *Biology and the Foundation of Ethics*. Cambridge: Cambridge University Press.

Manier, E. (1978). *The Young Darwin and His Cultural Circle*. Boston: Reidel.

Marks, J. (2002). *What It Means to Be 98% Chimpanzee*. Berkeley: University of California Press.

Merleau-Ponty, M. (1962). *The Phenomenology of Perception*. Translated by C. Smith. New York: Humanities Press.

Müller-Wille, S. (2007). "Collection and Collation: Theory and Practice of Linnaean Botany." *Studies in History and Philosophy of the Biological and Biomedical Sciences* 38:541–62.

Müller-Wille, S., and I. Charmantier. (2012). "Natural History and Informational Overload: The Case of Linnaeus." *Studies in History and Philosophy of the Biological and Biomedical Sciences* 43:4–15.

Mulligan, K., ed. (1987). *Speech Act and Sachverhalt: Reinach and the Foundations of Realist Phenomenology*. Boston: Reidel.

Owen, R. (1846). "Report on the Archetype and Homologies of the Vertebrate Skeleton." *Reports of the British Association for the Advancement of Science 1846*: 169–340.

Pajewski, A. (2012). "Analogy and the Face of Nature." Ph.D. dissertation, Program in Conceptual Foundations of Science, University of Chicago.

Plessner, H. ([1928] 1975). *Die Stufen des Organischen und Mensch: Einleitung in der philosophischen Anthropologie*. Reprint of 3rd edition. Berlin: De Gruyter.

Ramsey, G., and A. S. Peterson. (2012). "Sameness in Biology." *Philosophy of Science* 79:255–75.

Rheinberger, H.-J. (1997). *Toward a History of Epistemic Things*. Stanford: Stanford University Press.

———. (2010). *The Epistemology of the Concrete: Twentieth-Century Histories of Life*. Durham, NC: Duke University Press.

Richards, R. J. (1987). *Darwin and the Emergence of Evolutionary Theories of Mind and Behavior*. Chicago: University of Chicago Press.

———. (2002). *The Romantic Conception of Life: Science and Philosophy in the Age of Goethe*. Chicago: University of Chicago Press.

———. (2005). "Darwin's Metaphysics of Mind." In Hösle and Illies 2005, 166–80.

Rupke, N. (1993). "Richard Owen's Vertebrate Archetype." *Isis* 84:231–51.

Sacks, O. (2007). *Musicophilia*. New York: Knopf.

Scheler, M. ([1928] 1961). *Man's Place in Nature*. Translated by H. Meyerhoff. Boston: Beacon.

Schopf, T., ed. (1972). *Models in Paleobiology*. San Francisco: Freeman.

Schumann, K., and B. Smith. (1987). "Adolph Reinach: An Intellectual Biography." In Mulligan 1987, 3–27.

Singer, P. (2006). "Morality, Reason, and the Rights of Animals." In Waal 2006b, 140–58.

Sloan, P. R. (1972). "John Locke, John Ray, and the Problem of the Natural System." *Journal of the History of Biology* 5:1–53.

———. (1985). "Darwin's Invertebrate Program, 1826–36: Preconditions for Transformism." In *The Darwinian Heritage. A Centennial Retrospect*, ed. D. Kohn, 71–120. Princeton: Princeton University Press.

———. (1995). "The Gaze of Natural History." In *Inventing Human Science*, ed. C. Fox, R. Porter, and R. Wokler, 112–51. Berkeley: University of California Press.

———. (2003). "Whewell's Philosophy of Discovery and Owen's Archetype of the Vertebrate Skeleton." *Annals of Science* 60:39–61.

———. (2005). "It Might Be Called Reverence." In Hösle and Illies 2005, 143–65.

———. (2006a). "Natural History." In *The Cambridge History of Eighteenth-Century Philosophy*, vol. 11, ed. K. Haakonssen, 903–38. Cambridge: Cambridge University Press.

———. (2006b). "Kant on the History of Nature: The Ambiguous Heritage of the Critical Philosophy for Natural History." *Studies in History and Philosophy of the Biological and Biomedical Sciences* 37:627–48.

———. (2009). "The Making of a Philosophical Naturalist." In Hodge and Radick 2009, 21–43.

———. (2012). "Being Human and Christian in a Darwinian World." *Logos* 15:150–77.

Smith, D. W. (2008). "Phenomenology." In *Stanford Encyclopedia of Philosophy*. Available at http://plato.stanford.edu/entries/phenomenology/.

Straus, E. (1952). "The Upright Posture." *Psychiatric Quarterly* 26:529–61. Reprinted in *Phenomenological Psychology: Selected Papers of Erwin W. Straus*, New York: Basic Books, 1966.

———. (1982). *Man, Time, and World*. Translated by D. Moss. Pittsburgh, PA: Dusquesne University Press.

Tomasello, M., and J. Call. (1997). *Primate Cognition*. New York: Oxford University Press.

Waal, F. B. M. (1996). *Good Natured: The Origins of Right and Wrong in Humans and Other Animals*. Cambridge, MA: Harvard University Press.

———. (1997). "Are We in Anthropodenial?" *Discover* 18:50–53.

———. (2006a). "Primate Social Instincts, Human Morality, and the Rise and Fall of 'Veneer Theory.'" In Waal 2006b, 1–58.

———, ed. (2006b). *Primates and Philosophers*. Princeton: Princeton University Press.

Wake, D. B. (1994). "Comparative Terminology." *Science* 265:268–69.

Wilson, D. S. (2007). *Evolution for Everyone: How Darwin's Theory Can Change the Way We Think about Our Lives*. New York: Delacorte Press.

Wilson, D. S., and E. O. Wilson. (2007). "Rethinking the Theoretical Foundations of Sociobiology." *Quarterly Review of Biology* 82:327–48.

Wilson, E. O. (1978). *On Human Nature*. Cambridge, MA: Harvard University Press.

———. (1998). "The Biological Basis of Morality." *Atlantic Monthly* 281, no. 4:53–70.

———, ed. (2006). *From So Simple a Beginning: The Four Great Books of Charles Darwin*. New York: Norton.

PART THREE

God

ELEVEN

EVOLUTION AND CATHOLIC FAITH

John O'Callaghan

To begin to examine the relation of orthodox Catholic Christian faith to evolutionary theory and the question of human origins, consider words of the fourth pope, St. Clement:

> Let us fix our gaze on the Father and Creator of the whole world, and let us hold on to his peace and blessings, his splendid and surpassing gifts. Let us contemplate him in our thoughts and with our mind's eye reflect upon the peaceful and restrained unfolding of his plan; let us consider the care with which he provides for the whole of his creation. (Clement 1975, 4:439).

Pope St. Clement presents, however briefly, a portrait of a provident God, who, in the words of scripture, "orders all things mightily," and whose "plan" for creation is amenable to human thought and contemplation. Against the background of neo-Darwinian evolutionary theory ("NET" from here on out),[1] it is sometimes claimed that something like this aspiration of St. Clement's, to discern God's purposes and providential plan in creation, is mistaken when it comes to NET and human origins, for there is an insuperable conflict between religious claims about human origins and evolutionary claims. This essay will argue two theses. The first thesis is that from a Roman

Catholic perspective it is not possible for there to be any conflict between the claims of NET concerning human origins and orthodox Roman Catholic faith (RC from here on out). The second thesis is that there can be a conflict between RC and certain claims made about NET and human origins.

In order to argue for these theses, I need to fill out what often amounts to a cliché in Roman Catholic discussions of science, and that is that God is the author of both sacred revelation and the world studied by the natural sciences. Insofar as He is, He cannot say one thing in revelation and another in the world. Thus there is no conflict between RC and the truth of NET in particular, since the latter is about the world. But, going beyond the cliché, one wants to know what it is about God such that this claim is true. Otherwise it is too often introduced just to end all inquiry, with the result that the Roman Catholic theological and philosophical tradition is often absent from contemporary discussions of NET and religious belief, leaving a vacuum to be filled by religious views on divine action and the world as known by natural science, views held even by some Roman Catholics, that at times are in conflict with RC.

THE FIRST THESIS

Concerning the first thesis, presumably the suggestion that there might be a conflict between RC and NET is driven by concerns about God's creative activity in bringing about the various species of life studied within NET, and, in particular, the existence of the human species—human origins. It is thought, no doubt due to the claims of some cultural popularizers of NET, that the success of NET implies the exclusion of divine activity in the bringing about of the species of life. Religious believers understandably react against such popular claims about NET from the perspective of biblical revelation, a revelation that is quite clear about God's intimate relation to all creatures, living and non-living, in their genesis and history. However, those religious believers can themselves fall prey to conceiving of God's creative activity in ways that share the presuppositions of the cultural popularizers more than it does RC. It is part of my thesis that particularly for advocates of RC it is a proper understanding of God as creator that eliminates these confusions.

It is a tenet of RC that God is the cause of the existence of all beings and all being other than Himself (Vatican 2012, 4. II: #290, 4. IV: #296–97; Sokolowski 1982; Ott 1960, chap. 1). This is the traditional doctrine of creation *ex nihilo* that is developed within the theological and philosophical tradition of reflection upon the Jewish and Christian scriptures, particularly in the first four centuries of the history of the church amidst the battle with gnosticism (May 1994). While Thomas Aquinas argues that in principle this is a doctrine that can be grasped by anyone upon sufficient philosophical reflection about the world and without the aid of religious faith, as a matter of history the idea of creation *ex nihilo* is unknown to pre-Christian Greek thought. Creation accounts as one sees in such pre-Christian settings are always a creating out of a preexisting stuff that is acted upon, not a causing to be "from nothing" (Sedley 2007). This doctrine must be clearly distinguished from so called "creationism" or "creation science," which really amounts to a thesis about the hermeneutics of biblical interpretation rather than science. Broadly, creationism is the thesis that the account of origins in Genesis needs to be read and understood in a more or less "according to the letter" sense—if it says that the world was created in six days, then that means in something like six solar days of twenty-four hours each, or various attenuations of that into ages, but as described in Genesis according to the letter.[2] And of course with that hermeneutic in place, creationism might then look for and interpret natural phenomena to support that "according to the letter" reading of scripture, as well as to discredit phenomena taken to undermine it. But the RC doctrine of creation *ex nihilo* is a metaphysical doctrine about the origin of being, not a hermeneutical doctrine about scripture; although with it in place it no doubt bears upon questions of how we ought to interpret scripture.

Now, if the doctrine of creation *ex nihilo* is true, then it places severe restrictions upon how we characterize God's activity in relation to the world. The first restriction is that we cannot treat God's causal activity as simply another form of the kind of mundane causality we study within the natural sciences. The natural causes studied within the natural sciences presuppose objects upon which they act; they change the various states of those objects. They exchange energy with one another in accord with the first and second laws of thermodynamics, Maxwell's equations, Heisenberg's uncertainty principle, and so on. But in exercising their causality they do not produce their effects *ex nihilo*, that is, from absolutely nothing other than themselves. They always work *upon* some kind of preexisting matter or energy state; natural causes rearrange and make the matter or energy state upon which they work into something else or some other state.

But creation, insofar as it is genuinely *ex nihilo*, presupposes nothing upon which it acts. It is not a making *of* something into something else. It is not an exchange of energy. It is not an interaction with something, or an intervention in an already presupposed nexus, matrix, or receptacle (Aquinas 1941, Ia. 44–47). God's causality in creation is not a causality that competes with natural causes, or even cooperates with them. It is best thought of as enabling natural causes to be what they are. In this respect a creator god is quite different from Plato's demiurge of the *Timaeus* or the world soul of book 10 of the *Laws*. The demiurge acts upon a preexisting receptacle with elements already existing but in a formless or chaotic state. He forms the world according to the Forms, in particular the Form of the World, which is the best possible world—He brings a preexisting *logos* to an equally preexisting *chaos*. The demiurge just looks like a greater or even greatest possible natural cause. The demiurge acts as one element or part of and within a presupposed setting or context, and in that sense is a natural agent within it, even if not a physical or terrestrial agent. By contrast, a creator in the sense of RC is responsible for the very existence *in toto* of the setting or context and so is neither a part or element of that context nor a natural agent of any sort. So it is important to clearly distinguish creation from natural causation by saying that natural causation is always a making of something from something, while creation is not a making of anything. Using the words "create" and "make" in semitechnical senses, we say that God creates, He does not make, while natural causes make, they do not, indeed they cannot create.[3]

Now one thing that follows from this claim about creation *ex ni-hilo* is that it is not a process existing in time, a process that begins, endures through some temporal span, and ends at some time. Time measures the natural causal relations and energy exchanges between things in this world. Insofar as time can be considered to be real, and to be something other than God, it too must be a creature (in the appropriate sense) of God.[4] It follows that there is no time at which a creature exists and is not being created by God. This is the doctrine of continuous creation—everything that exists at every moment it exists is sustained in being by God. God's act of creation did not take place at a moment in time in the past; neither is it a process in time that continues. Rather, it is the case that the world of temporally and spatially related causes, the world subject to the study of the natural sciences, is sustained in being by God. Even if one holds as a matter of RC that the universe had a first temporal moment, the doctrine of creation itself does not entail such a first moment. And the doctrine of creation would be true even if there were no first temporal moment in the existence of the universe.[5] In other words, even if the past history of the universe were unbounded and extended infinitely into the past, God would still be the creative cause of it.

Again, the natural sciences study causes that are effective and transpire according to spatiotemporal relations. Relativity theory and quantum mechanics indicate that at the cosmic and subatomic levels these spatiotemporal relations can become quite bizarre. Nonetheless they are spatiotemporal relations between existing material objects. So it follows that God's creative activity as such is not subject to scientific investigation. It is true that one can say that the natural sciences study the *result* of God's creation, describe it, attempt to understand it, and so on. But God's creative act is presupposed by what the natural sciences study; it is not itself amenable to the techniques of scientific inquiry into natural causes. Similarly, if the result of God's creative activity is intelligible to us, that is because it is an expression of God's practical intelligence (Aquinas, *De veritate*, Q. II-III, in Aquinas 1949). The natural sciences do not prove this intelligibility—they presuppose it. The doctrine of creation *ex nihilo* helps us to understand why this presupposition by natural science of the intelligibility of the world is rationally justified, as opposed to those who say that even among scientists the intelligibility of the world is a mere matter of faith, even if not religious faith (Clark 2010).

Of course a quick objection at this point would be that something that is not amenable to the techniques of scientific inquiry into natural causes is not something that can be known—this seems to be the attitude of some cultural popularizers of science, particularly some of the more atheistic of them. Thus, the objection would continue, I seem to have protected against the possibility of a conflict between NET and RC by making God unknowable as such. On the contrary, there are many objections to any such restriction on our ability to know to the confines of natural science, the restriction that raises this objection. Mathematical relations are presupposed in the techniques of modern natural science; they are not amenable to them—"the googolroot of the googolplex is 10" is a knowable mathematical truth that is not amenable to the techniques of scientific inquiry into natural causes to discover or prove it. That I am wondering right now whether my next example of this point will be effective is a knowable truth not amenable to the techniques of scientific inquiry into natural causes. I know that I am so wondering, and you know it because I wrote it; I simply manifested it to you who are capable of reading. You do not understand how science works if you think that manifestation is amenable to scientific investigation by anything we call science nowadays. Your knowing it in virtue of my writing it is not grounded in a scientific technique of inquiry or discovery unless the objector makes "science" cover every mode of knowing, in which case he grants my response and loses his objection. And of course the assertion itself, namely, *something that is not amenable to the techniques of scientific inquiry into natural causes is not something that can be known*, purports to be knowable, indeed purports to be true; and yet it is not itself amenable to the techniques of scientific inquiry into natural causes. If it is so amenable, just try explaining what scientific techniques you would use to know it and show it to be true. The assertion undercuts its coherence in the assertion of it. So we need not conclude from what I have said so far that I have only protected against a conflict between NET and RC by making God unknowable as such.

The second restriction on talk about God's creative activity follows closely upon the first. We should not think of the knowledge we gain of God's existence from His creation as something akin to the affirmation of a scientific theory, one hypothesis of which is God's existence. So when Aquinas for one gives the "Five Ways" (proofs of God's existence) his arguments are not to be understood to be scientific theories. Only a certain scientism or residual logical positivism would claim that only scientific theories can make meaningful existence claims. Again, I am not making a scientific claim when I claim that my mother exists or that mathematical functions exist. However, some religious believers looking at particular failures of natural scientific explanation, the gaps in NET for instance, can be tempted to posit the existence of a god, or an "intelligent designer" to explain those things that natural science fails to explain, as if they were proposing an alternative scientific explanation, and so on. But this approach to the existence of God is mistaken in at least four ways.

First, it fails to recognize that God's creative activity is not akin to the activity displayed by the natural causes that are the subject of the theories of natural science. If A and B plausibly compete to explain C, they must be subject to roughly the same context of evaluation as to the warrant of their competing causal claims. But God's causality isn't subject to anything like the same context of evaluation as natural causes are, because God does not transform presupposed matter or energy as natural causes do. Second, insofar as God is brought in to explain what the natural sciences fail to explain, it seems to suggest that the existence of God is not implicated in those aspects of reality that the natural sciences *do* explain. Again, insofar as positing the existence of a god or an intelligent designer is understood along the lines of a competing scientific theory, it must be subject to relatively the same criteria of evaluation as NET and other naturalistic theories in whatever realm of nature we are trying to understand. So when NET or these other theories succeed in explanation, it is most rational to reject those competing theories, including the "theory" that God did it—God is banished, so to speak, to the gaps of understanding, the realm of the apparently irrational and lawless, with the result that the realms of reality that the natural sciences *do* successfully explain appear to be that much more Godless, a result that plays too easily into the hands of some cultural popularizers who wish to use science to argue for atheism.

On the contrary, RC holds that the existence of God is just as much implicated, indeed *more so* in the success of natural science than in its failures. This fact is known both philosophically and by virtue of revelation. It is known philosophically from the implications of the doctrine of creation *ex nihilo* and from revelation when the Gospel of John opens by asserting:

> In the beginning was the Word [*Logos*], and the Word was with God, and the Word was God. He was in the beginning with God. All things came to be through him, and without him nothing came to be. (*NABRE* 1:1–3)

Again it is revealed when St. Paul asserts that the invisible things of God are made manifest by the visible things of this world. The first text suggests that there is nothing of this world that does not manifest the *logos* of God. The Catechism of the Catholic Church summarizes this by saying, "Our human understanding, which shares in the light of the divine intellect, can understand what God tells us by means of his creation, though not without great effort and only in a spirit of humility and respect before the Creator and his work" (Vatican 2012, 4. IV: #299). But thinking of knowledge of the existence of God as a theory in competition with the theories of natural science forces us to abandon the thought that the *logos* of God is manifest in those areas of reality successfully understood by natural science. And the assertion of Paul provides both conceptually and as a matter of history the impetus to study the world more closely to see how the *logos* of God is manifested in its workings, that is, provides an impetus for advocates of RC to pursue more, not less scientific understanding of the natural world.

Indeed, I think an argument can be made that efforts to posit God, or an intelligent designer, as a competing scientific hypothesis to naturalistic explanations requires something like a heretical notion of God, heretical, that is, from the perspective of RC. Essentially the argument goes like this. (1) For the existence of God to be like a scientific theory proposed as an alternative hypothesis or explanation of some natural phenomenon, it must at least be falsifiable. (2) To be falsifiable there must be something conceivable that would count as evidence falsifying the God hypothesis. (3) That something would conceivably be some phenomenon from which it followed that God is not causally involved in that phenomenon. (4) But the RC doctrine of providence requires that God be causally involved in anything whatsoever that exists (Vatican 2012, 4. IV-V: #298–304). (5) Therefore, the phenomenon that would falsify the hypothesis that God is responsible for it would in turn falsify the RC doctrine of providence. (6) Therefore, the claim that it is a scientific hypothesis or theory that God is responsible for some natural phenomenon is heretical as implying the denial of an orthodox RC doctrine on the condition of the success of natural science. There will be some who object to the strong notion of "falsifiability" employed in the argument above. But it is just for illustrative purposes. The argument could be modified appropriately for whatever criteria one prefers to distinguish scientific theories from one another as in some sense competing with one another, and subject to being ruled out in preference for another according to similar conditions of warrant and evaluation. If to preserve orthodoxy and the doctrine of providence the religious believer moves to an intelligent designer rather than God, an intelligent designer distinct from God whom God causes to act, the believer's thesis either collapses into a position in which no intelligent designer is necessary because God as creator is causally involved (i.e., the position I am defending above), or he again butts up against heresy in claiming that the intelligent designer is necessary for God to be causally involved in the acts of His creatures.

Third, scientific theories being what they are, if the existence of God is thought to be a competing scientific hypothesis established by the present failures of natural science, and if the natural sciences develop over time in such a way that they eventually come to explain what had heretofore been explained by the competing explanation involving God, then it seems that we have reason to abandon the existence of God as rationally demonstrable from the things of this world (Romans 1:20). The existence of Ptolemaic epicycles was rejected, as it should have been, with the rejection of Ptolemaic cosmology in favor of the simpler Copernicanism, and eventually Kepler-Newtonian cosmology, which did away with epicycles altogether. Similarly, when a scientific theory competing with the hypothesis of God comes to explain what it heretofore had not, that is just one more reason to reject the hypothesis of God. So, paradoxically, treating the existence of God as an element of a competing scientific explanation or theory promotes agnosticism and atheism, insofar as natural science succeeds, setting natural science generally at odds with Christian belief.

Fourth, and finally, as a matter of history from the perspective of RC such an approach to God suggests a reversion to a pre-Judeo-Christian pagan conception of divinity, in which the divine is understood to be more powerful, perhaps even infinitely more powerful than the mundane causes we are ordinarily confronted with, but, nonetheless, is still understood to be operating within the same broad context within which natural causes operate. We can take our pick between the polytheism of the gods of Mount Olympus as described in Hesiod's *Theogony*, or the monotheism of Plato's demiurge of the *Timaeus* and *Laws*, book 10. These gods or the demiurge operate upon an already existing order of being, even if only a formless and chaotic matter and a receptacle, an order of things that has its own intelligible necessities apart from and presupposed by the divine activity. They interact with it and influence it precisely because it can be conceived of as a stuff that stands over and apart from them but within the same overarching matrix or context of activity and passivity. In short they mess with stuff and with us; but their messing with stuff and with us is determined by necessities they themselves are not responsible for and within which we all exist. They do not create *ex nihilo* the order of things that has its intelligible necessities as a result of that creation.

In either case, with the gods of Mount Olympus or the demiurge, we return to a non-Judeo-Christian, that is, pagan conception of the divine, poetic-mythological or philosophic-mythological, that does not conceive of God who is the cause of all being other than Himself. Such a creator God does not operate upon beings in creating them, does not interact with, influence, or "mess" with them, but causes them to be, sustains them in being, and gives to them the order of necessity and contingency within which they operate. Aquinas writes, "Divine Providence imposes necessity upon some things, but not on others. . . . [It] prepares certain effects to happen from necessary causes, and they happen necessarily, while [It] has prepared others to happen from contingent causes, and they happen contingently, according to the condition of proximate causes" (Aquinas 1941, 1.22.4). The "god" or "intelligent designer" of scientific hypothesis is a rejection of this providential God of whom Aquinas for one speaks. Given the history of Christianity, it is a return to the gnostic god against which the church developed the doctrine of creation *ex nihilo*.

Indeed, it follows from what I said about creation presupposing nothing that the intelligible causal structures that are amenable to the techniques of natural science are the product of God's creative sustaining activity. Put another way, natural science investigates the intelligible structure of the world as it is displayed in natural causal relations, but it does not demonstrate the very existence of that intelligible structure; natural science presupposes it. In that respect, one can say that the intelligible structure of the world studied by natural science, insofar as it is created and sustained in being by God, is the product of a divine act, and thus the expression of divine intelligence—not the demiurge, but more akin to what the Greeks called *logos*, but a *logos* that is more than the intelligible structure of reality, a *logos* that acts and is thus a person communicating intelligibility. It is also a *logos* that does not stand over against an uncreated *chaos*, but is prior to all things. "In the beginning was the Logos, and the Logos was with God, and the Logos was God. He was in the beginning with God. All things came into being through him, and without him not one thing came into being." The *logos* that finds expression within creation comes to further elaboration and expression within our own minds through the theories of the natural sciences that are formulated and judged as adequate or inadequate to the reality of creation, that is, judged to be true or false.

To give an image, the created order is the expression of intelligence and stands, as it were, between two intellects, the divine and ours. The divine intellect expressing itself creatively and *ex nihilo* gives existence to creation and a rule or measure to it that is expressed within it. That same rule or measure expressed within creation then becomes the rule or measure of our intellect as we attempt to understand creation. And yet it is better to say that our intellects stand not on the other side of creation from God, but within that creation as well, being informed by it, and learning more about it from within through the study of the natural sciences. As St. Augustine said echoing St. Paul, creation and ourselves within it are like a cloudy mirror within which we glimpse ever so faintly the *Logos* of God. Aquinas later adds that creation itself comes to self-understanding in the human mind, precisely because it pertains to the perfection of creation that it contain creatures capable of understanding it (Aquinas, *De veritate*, II.2. *respon.*, in Aquinas 1949). And so from the perspective of RC, scientific endeavor and its success is not to be feared or avoided by Christians but, rather, pursued as a perfection and the glory of being rational creatures of God, made from the beginning in His image and likeness—in a way we are all as rational creatures called to be scientists to the extent possible.

But it is because the intelligible structure of creation that is studied by the natural sciences, including NET, is an expression of the divine *logos* that it is impossible for there to be a conflict between NET and RC. We have every reason to listen to the biologists when they tell us that NET is the best theory of the natural causal processes by which the living species that confront us came to be. Like every scientific theory, it likely contains some falsehoods subject to revision through further investigation. Nevertheless, it appears to be the closest thing to the natural scientific truth about the origin of species that we have. But the doctrine of creation forces us to recognize that the natural causal factors operative in evolution are only operative insofar as God renders them so by creating and sustaining them.

What this emphasis upon the proper notion of creation forces upon us is an understanding of natural causality as embedded within a larger context of divine causality, in which the natural causes are *enabled* to be genuinely causal because God is causing them to be so. But again, we want to avoid the conceptual trap of placing this relation of embedded causality within some common framework, where we fear that the more God does, the less the natural cause does, and the more the natural cause does, the less God does. When we think of the operations of natural causes we often and correctly think that the more one does the less another does, even if they are both involved in bringing about the effect. So, for instance, the more I lift the less you do when carrying a trunk. Or we think we are less free in acting to the extent that we think some other agent or cause is responsible for some features of what we claim to do—raising our arms, for instance. We do not think of the other cause as enabling us, but, rather, as competing with us, indeed often inhibiting and threatening our freedom of action.

However, even within the natural order of causality there are examples where it is better to speak of the one cause enabling the other, rather than speaking of it as competing with it—I enable my pen to participate causally in my writing, I do not compete with it for responsibility as to what is written; it makes no sense to speak as if the more it does the less I do, and the more I do the less it does. And in the case of God's causality in creation, it is a matter of God actively and presently enabling causes to cause, not competing with them. So to take my arm again, I can say God caused me to freely raise my arm—and the modality of the act as "free" is itself within the scope of God's act, a fortiori for non-free causes; it is free because God causes it to be and to be so. That is the sense in which God does not interact with the world, or mess with it, since its very possibility of acting is the result of God's intimate causal presence to it. Mutatis mutandis, we can think of Augustine's assertion that God is more interior to me than I am to myself. And the structure of embedded causality can be described as one of primary causality, God, to secondary causality, creatures.

One direct result of this emphasis upon the structure of primary and secondary causality as descriptive of God's action in nature and natural causality is the conception of miracles it forces upon us. Commonly we may think of miracles as a kind of divine intervention or interference in the course of nature, where God takes the place of the natural cause—if no natural cause did it, then God did it, and it's a miracle. And insofar as we think of the natural order as involving laws, we think of a miracle as a violation of the laws of nature. But we let God off the hook by saying that He made the laws ages ago, and so can't really be said to violate them, since He made them, and then generally and for the most part lets them run on. But, insofar as He is omniscient, when He did that ages ago, He always knew when and where He would occasionally intervene. This picture lends itself to the intervening, interacting, messing around god of the pagans and intelligent design. This god is an alternative explanation of what happens in nature, when nature fails. On this understanding of miracles, when the church, for example, examines a purported case of a miracle, it is involved in a scientific enterprise of determining which of two alternative and mutually exclusive scientific hypotheses is true—the natural one or the supernatural one. Who or what did it?

On the contrary, the problem with this picture from the perspective I have been sketching is that it presupposes that God isn't causally responsible for the things natural causes *do* bring about. But the structure of primary and secondary causality denies just that presupposition. The natural cause did it, and so did God by enabling and sustaining the natural cause to act in just the way it did when it did. And so a miracle must be conceived of as God bringing about some effect that He ordinarily brings about as primary cause simultaneously employing secondary causes, but that in the particular case under consideration He has refrained from employing those secondary causes. God always does it, with or without secondary causes. When He does it without, that is a miracle. And the church in examining a miracle is not determining which did it—God or a natural cause. God did it as primary cause, just as He is responsible for the effects of nature as primary cause. The church is simply determining whether a natural cause was involved as secondary cause in what it already knows God did as primary cause. In other words, God does it all.

Thus, insofar as NET appeals to natural causes to give an account of the origin of species and human origins, there can be no conflict between its account and the claim that God creates those species, since it is God who is causing those natural causes to operate in the ways they do, and sustains in being the effects or results of those natural causes.

But here another objection is obvious. NET appeals not simply to natural causes, but to chance or random variations in the genetic code—variations that are then subject to natural causes in the process of natural selection. Insofar as it appeals to chance, God is not involved in such processes, as neither are any other causes. But this misconceives chance or randomness, as if they meant "uncaused," and it reifies "chance" as if it were itself a cause without a cause. But they don't, and it isn't. If, digging a garden in my back yard, I come across a safe filled with cash that you buried there, that is a chance event. But it is not uncaused. It is caused by your burying the safe there and my digging there. It is just that neither causal line alone is responsible for it—my digging is uncorrelated with your burying. If a meteor strikes the Earth causing the extinction of the dinosaurs, and thus allowing for the development of other competing species that would otherwise be selected out by the dinosaurs, the impact of the meteor is a random event from the perspective of the evolutionary biologist, but it is certainly not uncaused or random for the astronomer. And if a gene mutates under the influence of a cosmic ray or radioactive particle, that is not uncaused. It is just not caused by the preexisting biological conditions of the organism and its environment taken into account by the biologist, as cosmic rays and radioactive decay are not correlated with those biological conditions.

But if you insist that there are genuine uncaused events in nature, for example, the radioactive decay of atoms,[6] and so on, and they enter into the process of evolution bringing about random events, I respond that that possibility still does not exclude a primary causality employing those uncaused events to bring about an effect. Let us say that the result of throwing dice in the game of craps is in each instance random in the sense we are talking about here. Nonetheless, the house always wins, which is why one ought to invest in a casino rather than play in one. I want to take all your money. That is my purpose. So I build a casino in which craps is the only game played, and you must bet at least a dollar. But I know given the rules of craps that I can run a game of craps as the house, and win on average 51 times out of 100, since the odds of a shooter winning on a turn are very nearly 49 in 100. And over time, even with long stretches where you are up, you will lose all your money if you keep playing. Even if you quit, there will always be another player to take your place, and continue the series that ends with my accumulation of the money bet. (And keep in mind that nature cannot quit, as the shooter might while she is ahead.) Thus my purpose in enabling the game will be fulfilled. That is why casinos never go out of business. A casino owner's purpose, his goal, is to build successful gambling houses employing games in which each turn is a random event uncorrelated with the events that precede it and the events that follow it. And there is no need for the house to intervene by loading the dice, for example, to correct the process to guarantee the outcome, because the rules of the game enabling it determine the outcome of the process, not the individual events within the process. All the house needs to do is sustain the rules of the game. It does not need to cheat or break the law—it can legally take all of one's money simply by inducing one to stay at the table. The rolls of the dice do not give intelligibility to the rules; the rules give intelligibility to the rolls, even as every role is uncorrelated with every other roll. He who controls the rules controls the game. And the house always wins.

It simply does not follow that randomness in a process, even a process in which every event is random and uncorrelated with every other event, entails that the process has no point. It does not follow that if you roll the clock back and then go forward again with the game that you will get a different result; you will simply get a different path to that same result—the house always wins. And it is striking to consider the fact that *every* event in a game of craps is uncorrelated with every other event in the game. But the utter lack of correlation characteristic of a game of craps does not hold of the biological processes studied by NET; no matter how many uncorrelated events take place within the process of biological evolution, by and large most of the reproductive events involved in evolution are correlated from generation to generation, the evidence of which is the relative stability of species populations over time and the length of time necessary for a new species to develop through random mutations and natural selection. If the events of evolution were as uncorrelated as those in a game of craps, by and large one would not get horses from horses, pigs from pigs, and fruit flies from fruit flies.

Or suppose we say that the half-life of carbon-14 is 5730 years plus or minus 40 years. Suppose I build a "doomsday bomb" destructive of all life on the planet and attach it to a Carbon-14 trigger that will trigger when at least half the amount of carbon has decayed radioactively. Suppose we say that each radioactive decay event is "random." And yet I guarantee you that in 100,000 years there will be no life on the planet, provided all the other operative causes continue to work as they do. And yet every individual carbon decay event is random in your absolute sense. Still, my purpose in building the bomb will be fulfilled despite the fact that the process involved in accomplishing my purpose involved random, uncorrelated, and even perhaps uncaused events. The conclusion then is that randomness is not in fact opposed to order and even purpose, and does not in general undermine ordered even purposive processes within which the randomness occurs. That's not an argument that the process of biological evolution actually has a purpose or point, just that randomness within it is not an argument that it *does not* have a purpose or point.

Finally, even here, because causing as a creator so transcends the conditions of causing as a natural cause, we ought to refrain from denying that God can directly cause a random event to be "random" even in this sense that I have allowed for the sake of argument, just as I asserted earlier that only God can cause my act to be free. The natural event may well be random against the background of natural causes, and yet not be so as caused by God. As incipient pagans, we are forever tempted to try to place God within the ways of natural causality according to some sort of common nexus. But God's ways are not our ways. There is no conflict between NET and RC.

THE SECOND THESIS

This brings me to my second thesis, namely, that there can be a conflict between RC and certain claims made *about* NET and human origins. The conflict arises from a particular doctrine concerning the human soul that is undeniably orthodox—that God directly creates from nothing the human soul without the mediation of secondary causes (Vatican 2012, 6. II: #366). Notice, this creation of the soul *ex nihilo* would not be a miracle in the sense that I described above, since the claim is that natural causes could never be involved as secondary causes in causing the soul to exist, while miraculous events as described above involve types of events that are ordinarily caused by natural causes acting as secondary causes.

Now I don't want to leave this claim simply as a dogmatic theological assertion. There are plausible philosophical arguments that can be made for the thesis, apart from the theological reflection of the church upon the data of revelation. Such arguments in general would proceed to the conclusion that the soul has a certain immaterial mode of existence. Some of these arguments are familiar in the history of philosophy, as for example the argument from human cognitive reflection upon one's purposes and the ability through such reflection to freely order one's actions according to a general understanding of the goods of distinctively human life. Another argument is from the capacity to have general intellectual knowledge of kinds of things, a general knowledge that transcends the particularities of the here and now in engaging the world like other animals do. There is another argument from the intentionality of our thoughts—the relation of a thought to its object is necessary, while the relation of bodily states to material objects is always contingent—my thought of the tree is necessarily related to the tree, while no bodily state is. There is the argument from moral experience—it is one thing to give an evolutionary causal account of how we may have come to assert as a survival mechanism that it is wrong to kill innocents, quite another to understand that it *is* wrong and to be able to give reasons why it is wrong. This is the difference between giving causes and giving reasons. Another argument is that from the failure of material states to represent truths, mathematical truths for example, except insofar as such material states are *used* by human beings to represent mathematical truths—what does one and zero (10) plus one and zero (10) equal?—twenty if the system of representation is decimal, four if the system is binary, and thirty-two if it is hexadecimal—suggesting that material states are not intrinsically representational, but take it on as used; since there can be no infinite regress of representations there must be a state that is not a representational state as physical states are. Another argument is the inability in general to give a naturalist/evolutionary account of truth and understanding in our beliefs—there is likely an evolutionary benefit to believing the sun moves around the earth, and yet that is false, we reject it, and we *understand* why, so having true beliefs does not seem to be related necessarily to beneficial evolutionary conditions so long as the right adaptive behavior is produced by false beliefs. What then of truth and understanding truth? The failure of naturalistic explanations to account for our understanding the *what* and *why* of things—there is likely an evolutionary account of why it is useful for me to perceive this large object approaching me as a threat, but there appears no account in general of the scientific endeavor of classifying it as a lion, and understanding what lions are, much less understanding their evolutionary history. There are more such arguments. Now each of them may be defeasible, and perhaps all of them are. I have my philosophical doubts about several of them. But it is fair to say that without an un-argued dogmatic assertion of physicalism akin to the dogmatic theological assertion of the direct creation of the soul,[7] these philosophical arguments remain plausible within philosophical discussion, or at least have not been shown definitively to be unsound.

So what? Well, if the human soul is caused to be and has an immaterial mode of existence, insofar as no one I know of argues that material causes can cause something with an immaterial mode of existence, it follows that the human soul cannot be caused by material causes. And it follows that evolutionary processes cannot causally result in the human soul.

These remarks about the soul are all very compressed, and would need much greater elaboration had I but world enough and time. But if these are granted as they are, the history of Christianity has variously adopted two broadly different and mutually exclusive attitudes toward the soul, which I will call in an unscholarly fashion the "platonic" and the "aristotelian." On the platonic view the soul is a thing fundamentally distinct from the human body, even if it happens to be closely connected with a particular human body; it uses a human body, perhaps even has a special care and concern for one body over another, but is not identical with it. On the platonic view one can say of a dead body that it is a human body because in life it is associated with a human soul. It is characteristic of the platonic view that the human person is identified with the soul, not the body, or at best the union of the two different things, soul and body, with soul being taken to be the essential person. And the soul moves the body like a kind of agent cause, much the way a child would move a ball, or a marionette a puppet; this is the sense of the soul in which Plato argues for the existence of a world soul.

On the aristotelian view, the soul is the living principle as substantial form of the body; it makes the body to be the kind of body it is; it is determinative of the distinctive acts of the living body as the kind of living thing it is; and it is indeed fundamentally identical to the body as the manifestation of the kind of life it is. It is not *united with* the body, but is, rather, the unity of the living acts of the body, nutritive, reproductive, motor, and in the case of the human, rational. It is not an agent cause moving the body in certain ways. It is the characteristic form of the movement of the body. Although he was not speaking of this himself, Wittgenstein captures this aristotelian notion of soul in saying that "my attitude toward him is an attitude towards a soul. I am not of the opinion that he has a soul. And the best picture of the human soul is the human body" (Wittgenstein 1953, II.4). On this aristotelian view, a dead body is not and cannot be a human body, for a human body is a living ensouled material object. The dead body at best can be called the remains of a human body. And the person is identified with the living animal body. The soul is manifest in the life of the body. On the platonic view the soul and the body have fundamentally distinct modes of existence, and we have to infer the hidden soul from the ways it pulls the strings of the body it is associated with. On the aristotelian view, by contrast, the mode of existence of the body is identical to the mode of existence of the soul, and we do not infer it—it is manifest (Aquinas 1941, I.76, esp. a.1). Again, this is all very compressed.

Now allow me to claim that as far as RC goes, it by and large adopts the aristotelian view. Indeed, at the Council of Vienne (1311–12), it was defined dogmatically as *de fide* that the rational soul is per se the essential form of the body. Of course it might be objected that this was not an ecumenical council and not binding upon Christians generally. But I am making a claim about RC, not Christianity generally. The Council of Vienne was in fact weighing in on a broad theological controversy of the thirteenth and fourteenth centuries, precisely the question of the unity of soul and body in the context of what is known as the Plurality of Substantial Forms controversy. The theological result of such a weighing in on the side of the "essential form of the body" view is that it follows that the human person is the living body, not the soul alone or the soul with the body. Finally, it may be argued that this is not a dogmatic recognition of the aristotelian position; one could come up with some kind of exegesis of the text and a suitably complex account of the soul such that it is the essential form of the body and yet not the substantial form of the body, or the essential form of the body while being a distinct thing from the body. But while we wait for that analysis, it appears very difficult to say that it does not at least favor the aristotelian position, particularly given the context of its use, and represent a rejection of the platonic. And finally, consider this recent 2004 statement from the International Theological Commission on Evolution:

> Present-day theology is striving to overcome the influence of dualistic anthropologies that locate the imago Dei exclusively with reference to the spiritual aspect of human nature. Partly under the influence first of Platonic and later of Cartesian dualistic anthropologies, Christian theology itself tended to identify the imago Dei in human beings with what is the most specific characteristic of human nature, viz., mind or spirit. The recovery both of elements of biblical anthropology and of aspects of the Thomistic synthesis has contributed to the effort in important ways. (International Theological Commission 2004)

But it was by and large the Thomistic position against dualism and the plurality of forms in the context of theological discussions of the *imago dei* that is most plausibly read as the position affirmed at the Council of Vienne. So let's just say for the sake of argument, and despite elements of platonism in Catholic-Christianity, the aristotelian view of the soul is at least favored by RC even now.

So, consider this problem. I mentioned that on the aristotelian view, the soul as essential, that is, the substantial form of the body makes the body to be the kind of body it is. If we have some organization of matter, however much like some kind of body it may be, if it lacks the kind of substantial form appropriate to a particular kind of substance, it is not a body of that kind. In particular, however much some organization of matter may be like a human body, even some living organization of matter, if it lacks a human soul, it is not a human body. But we have already seen that no natural secondary causes can be involved in causing the existence of a human soul. From the perspective of RC then, a human soul must be directly created by God as the substantial form of some body. Whatever natural processes may precede that creation of the soul as the substantial form of a human body, they do not bring about the existence of a human body. At best they could be said to prepare matter to a certain state, perhaps even a living state of some organism, at which time a soul is created by God as a substantial form of the matter, thus transforming the matter from whatever it was into a human body.

But clearly this is a challenge for any claim that evolutionary processes constitute a fully sufficient natural explanation in terms of secondary causes of the origins of human life. It is not that NET explanations happen to fail where they might succeed, and for that reason we must hold that God is the cause of human origins. We have already seen that God is involved in all evolutionary processes as primary cause to secondary. On the contrary, the claim is that no evolutionary account could possibly give a sufficient account of human origins properly speaking, if what we mean by that is a sufficient explanation of the actual existence of the human species. At best we could say that evolutionary processes act to prepare matter to a certain stage of development of human-like beings, such that at the appropriate stage of development or in the next generation in reproduction, God creates human souls as new substantial forms for those human-like organisms, in the process transforming them such that they cease to be what they were and become something new in creation, properly human beings. So this conflict only arises if one claims about evolutionary theory that in principle it sufficiently explains every aspect of human origins. But this is all very speculative.

Still, we might think at this point we have a number of options for dealing with this apparent conflict between RC and a claim *about* the ability of NET to give a sufficient account of human origins. One option would be to simply abandon talk of the soul as in any way useful within theology given the pressure of evolutionary accounts. In other words, modern natural science has shown the concept of the soul to be no longer useful for explaining human life. I have seen this suggested, and I understand that it is at least informally favored by some theologians.

However, I think that there are two problems with it. First, the philosophical suggestions that I have seen for abandoning it tend to misconceive the role of the soul in a platonic fashion, not recognizing the aristotelian sense I have been emphasizing. They tend once again to conceive of the soul acting like a kind of agent cause making the body do things, an agency that would be in competition with other natural causes, much like the account of God as a competing hypothesis eliminated by the advance of natural science—it is a "soul of the gaps" that is rejected. But the aristotelian account does not posit the soul as an agent cause of the body filling the gaps of explanation, but, rather, as specifying what something is. It posits what needs to be explained. To deny the soul in that sense is to deny that there is any "what" in human life. But that is to deny the very notion of *human* origins, and thus to undermine the claim about NET that it fully explains human origins, since there is no such thing as the *human* in that direction.

And while I am no theologian, I think it fair to say that the havoc caused for Christian belief by a cavalier abandonment of talk of the soul is worthy of serious reservation. It is even more dogmatically asserted by the church that the soul survives the death of the body than it is asserted that the soul is directly created by God. Are we to abandon that doctrine as well? Indeed, a colleague of mine concerned with this theological option gave a plenary address to the American Catholic Philosophical Association entitled, "Good News, Your Soul

Hasn't Died Quite Yet," arguing that any such move is unwarranted (Freddoso 2002). Incidentally, Aquinas for one accepts the implication of the aristotelian view that even if the soul survives the death of the person, the soul is no person, and certainly not the individual with that soul. "*Anima mea non est ego*" (Aquinas 2012).[8] Thus there is no reason to take the survival of the soul to imply that something like the platonic position is to be adopted. Indeed, the survival of the soul but not the person is cause for Aquinas to emphasize even more the fittingness of the centrality of the doctrine of the resurrection of the body according to RC which is affirmed in the Nicene Creed—the resurrection of the body is the resurrection of the person.

However, another option for us is to recognize the limited nature of NET in explaining human life generally. It is a philosophical assertion, and in contemporary philosophy more like a philosophical dogma, that evolutionary theory in particular and physical theories in general must account for all features of reality including all features of human life. Indeed it is a philosophical dogma that often strikes one as serving little more than the purpose of avoiding theological considerations at all costs. The best of physicalist philosophers recognize that this is a philosophical attitude one takes toward science, while many of the scientific popularizers misconceive it as a result of science itself and make a philosophical thesis into pseudo-science. NET is science; this dogma about NET is not.

But why must we insist that NET fails if it does not account for every feature of human life, in particular the properly human, and thus does not completely account for human origins? Suppose we held that NET provides an in principle adequate account of many features of human life, in particular, those bodily features of human life inherited by human beings from the prehuman biological species that was transformed into the properly human when God created human souls as the substantial forms of the members of a transformed and new species. What is certainly lost is the philosophical thesis that NET explains absolutely everything.

But we already knew that claim about NET to be false. It is one thing to give an account of the evolutionary benefit of believing and acting upon the belief that 2+2=4; it is another thing entirely to prove mathematically that 2+2=4, and mathematical proof is not a physical procedure or algorithm, or a matter of natural selection. NET does not explain that 2+2=4, even if it explains why we believe it. NET presupposes replication or reproduction—it does not explain it as a result of natural selection. Indeed, NET presupposes many physical conditions, like the inverse square form of the universal law of gravitation, without explaining them, and thus cannot be a theory of absolutely everything. But if it cannot be a theory of everything, why should we be shocked if it is not a theory of everything human? To insist that it must be a theory of everything human and of all human origins looks simply like a prejudice against or fear of acknowledging the action of God as creator and religious reflection upon it. It is to be engaged in cultural politics, not science, a cultural politics that in its fear of God constitutes an abuse of genuine science.

In any case we have already seen that fear to be unwarranted, since the very possibility of the natural causes operative in random mutation and natural selection are only enabled to be causes by God's ever present activity of creation. The existence of God is as implicated in what NET does explain as it is in what it does not explain. So banishing God by insisting that NET explains more than it is capable of explaining does not banish God at all; all such banishment does is harm genuine science by making it serve cultural politics, a servitude to cultural politics that quite often leads to the counter assertion that science is just one more mode of faith no better than any other (Clark 2010). Thus, it would seem better to simply let NET do its own job and not substitute it as an alternative pseudo-scientific-pseudo-metaphysics-pseudo-religion. As Catholic Christians, we may just have to rest content with the thought that human beings, however much they are conditioned by the material world around them, and however much there is a history that can be told of how the world was prepared for their arrival, nonetheless as animals, not as souls, simply transcend the conditions of life of other animals. In other words, there may be much to be said scientifically about how God prepared the world for the origin of human life, and yet reserved that origin properly speaking to Himself. If religious people are tempted to think platonistically that it is in virtue of the soul or spirit that human beings transcend the body and the life of animals, they might be better off thinking that it is in virtue of the soul directly created by God as substantial form of the body that as living animal bodies they in some measure transcend the ordinary conditions of matter studied by the natural sciences and NET. It is not as souls that human beings are transcendent—the paradox of RC is that it is as living animals that they are transcendent. And it is as transcendent embodied creatures of God that they are in fact ordered in a dynamic movement of their living bodies as *imago dei* to union with God Himself.

In conclusion, it is perhaps worthwhile pointing out that the sketch I have just given of an RC approach to NET and human origins was condemned by the church. Well, not really condemned. It is by and large the account proposed by the nineteenth century French Dominican Dalmace Leroy in *The Evolution of Organic Species* (*L'Evolution restreinte aux espèces organiques*) (Leroy 1891). He too suggested an evolutionary account of the development of the physical organism that came to be the human body upon the direct creation by God of human souls as substantial forms of those heretofore human-like bodies. When it was reported to the Congregation for the Index of Forbidden Books, the Congregation moved to prohibit the book and demand a retraction from Leroy, who, in a spirit of religious obedience, straightaway complied. But they did not actually place the book on the Index. And in any case, the Congregation for the Index never had the authority actually to *condemn* a theological or philosophical position—that belonged to the Holy Office, which never condemned LeRoy's position (Artigas, Glick, and Martinez 2006, 270–81). And so it is slightly ironic, given the evolution of the church's thought, that as far as I can tell the view I sketched above is pretty much the position that the church now holds as expressed in John Paul II's statement to the Pontifical Academy of Science in 1996 (John Paul II 1996), and the 2004 statement of the International Theological Commission cited above, "Communion and Stewardship: Human Persons Created in the Image of God."

So let us finish by considering the words of a poet: "the world is charged with the grandeur of God, it will flame out, like shining from shook foil; it gathers to a greatness, like the ooze of oil . . . nature is never spent; there lives the dearest freshness deep down things" (Hopkins 1877). Catholic faith seeking understanding does not find God in the failures of science, but, rather, it finds Him in the marvelous successes of science, including the success of NET. But Catholic faith also recognizes that God's action in creation is not bound like Prometheus to the rock of natural causes. It transcends those natural causes in giving the warm flame of life to human beings, as responsible for creating a kind of animal who is yet different as an animal. The saint says, "question the beauty of the earth, question the beauty of the sea, question the beauty of the air distending and diffusing itself, question the beauty of the sky . . . question all these realities. All respond: 'See, we are beautiful.' Their beauty is a profession. These beauties are subject to change. Who made *them if not the Beautiful One who is not subject to change*?" (Augustine 2012, III/7 Sermon #241.2). The marvelous growth of the natural sciences since St. Augustine wrote those lines fifteen hundred years ago, far from conflicting with or threatening RC, simply enriches and enlivens its profession. But even more than the natural causes studied within the natural sciences testify to the glory of God, it is human beings who sing the glory of God as they are the only animals we know of who are given the particular dignity of being made transcendent animals, as Genesis tells us, made in the image and likeness of God, commanded to be fruitful and multiply. The mark of their dignity is that they consent.

NOTES

1. This paper will employ a distinction between what belongs to NET as such, that is, as a natural science concerned with the development of species, and the cultural and popular use of NET beyond that study. Claims pertaining to the latter are no doubt about NET in some sense. But I will rely upon the assumption that they are not part of the *content* of NET. Philosophers of science may struggle with determining criteria that establish neat and tidy boundaries of a natural science. Nonetheless, I will assume that some sense can be made of distinguishing claims *about* NET and its relationship to other intellectual endeavors such as ethics, politics, and metaphysics on the one hand and claims *of* or *within* NET on the other hand. With that

rough distinction in mind it is important to recognize that even evolutionary biologists, particularly leading figures responsible for the advance of NET, can at times also engage in making claims *about* NET that are not thereby claims of NET, and should not be evaluated as such or treated as having the same authority as a claim of NET. Perhaps it will have greater authority, but just as likely it may have less. Socrates in the *Apology* explored the various ways in which a society may grant an unwarranted authority in one area to various classes of people on the basis of their success in another area.

2. One must distinguish "literal interpretation" as it is now commonly described from the earlier and much richer sense of the "literal" given by Augustine in *On the Literal Interpretation of Genesis* and Thomas Aquinas in Iae.1.8–10 of the *Summa Theologiae*. Nowadays it commonly means something like the plain or ordinary sense of a passage excluding the use of images or metaphors. So "Rommel was a fox," would "literally" mean that Rommel had four legs, fur, and might be subject to being hunted by dogs in a common English pastime in the countryside; and it would be false. But then one might nowadays say that it employs an image or metaphor, and so in some other nonliteral or "spiritual" sense it is true. Augustine on the other hand does not think that literal truth excludes the use of metaphors to say something true about the world; the literal sense of a text consists of the facts about the world the author intends to convey, however she intends to convey them. So "Rommel was a fox" can be literally true for Augustine, since the author intends to say that Rommel's battle plans were very clever, or some such. It was precisely to combat the "spiritualist" readings of scripture offered by the gnostics in the setting in which orthodox Christianity was developing the doctrine of creation *ex nihilo* that Augustine insisted upon this account of what "literal" interpretation is. Why we now use "literal" in a different way from Augustine and Aquinas after him is explored in Carroll (2001).

3. In Aquinas's case, it is significant that his discussion rejects the view that there are preexisting Forms to creation, or that one can make sense of "the best possible world." Since there is no maximal combination of preexisting Forms, there could always be a better world. See Aquinas (1949, Q.1, 3).

4. It is interesting to consider in this respect the concurrence of general relativity with this particular position on God and temporal relations. Insofar as general relativity would hold that temporal relations are inseparably bound up with spatial relations, it would seem that a religious believer who thinks that God is not subject to spatial relations, and is in some sense creatively responsible *ex nihilo* for those spatial relations, ought also to consider that God would not be subject to temporal relations, and would be in some sense creatively responsible for them *ex nihilo*. Conversely, a religious believer who held that God is in some sense bound by temporal relations, and thus not creatively responsible for them, ought also to hold that God is

in some sense bound by spatial relations and not creatively responsible for them. This is admittedly highly speculative on my part, but nonetheless worthy of serious reflection for religious believers.

5. It is important to note that in the work of Thomas Aquinas, it is not necessary for creation to have an initial moment of time in order for it to be created *ex nihilo*. Aquinas argued with many of his contemporaries that even if creation had had no first moment of time, it would still be created *ex nihilo*, since the latter signifies total dependence in existence, not a temporal starting to exist. For this reason, even if there were temporal facts that precede what is popularly called "the Big Bang," that would not disprove the fact of creation as at least Aquinas understood it. And one can take Aquinas's view on creation as having something of a normative weight on this particular issue in RC.

6. I am allowing this for the sake of argument.

7. Jaegwon Kim (1996) is the most straightforward contemporary physicalist in recognizing and admitting that physicalism is not a conclusion in philosophy; it is rather a position adopted in order to think philosophically about natural science and its results.

8. See also Aquinas (1941, Ia.75.2–4) on the human soul being subsistent, but not a substance in its own right.

REFERENCES

Aquinas, T. (1941). *Summa Theologiae*. Ottawa: College Dominicain d'Ottawa.

———. (1949). *Quaestiones Disputatae Dei*. Turin: Marietti.

———. (2012). *Super primam epistolam B. Pauli ad corinthios lectura*, Cap.15, lect.2. Available at http://www.corpusthomisticum.org/c1v.html#87682.

Artigas, M., T. F. Glick, and R. A. Martinez. (2006). *Negotiating Darwin: The Vatican Confronts Evolution, 1877–1902*. Baltimore, MD: Johns Hopkins University Press.

Augustine, St. (2012). *The Works of Saint Augustine: Sermons on the Liturgical Seasons*. Trans. E. Hill, O.P. New Rochelle: New City Press. 2nd electronic release available through http://pm.nlx.com.

Carroll, W. (2001). "Galileo and Biblical Exegesis." In *Largo Campo di Filosofare: Eurosymposium Galileo 2001*, ed. J. Montesinos and C. Solís, 677–91. Canarias: Fundación Canaria Orotava de Historia de la Ciencia.

Clark, S. R. L. (2010). Review of *Science and Spirituality: Making Room for Faith in the Age of Science*, by Michael Ruse. *Notre Dame Philosophical Reviews*, 2010. Available at http://ndpr.nd.edu/news/24456-science-and-spirituality-making-room-for-faith-in-the-age-of-science/.

Clement, Pope. (1975). *Liturgy of the Hours*. Translated by the International Commission on English in the Liturgy. New York: Catholic Book Publishing.

Freddoso, A. (2002). "Good News, Your Soul Hasn't Died Quite Yet." In *Person, Soul, and Immortality*, ed. M. Baur, 99–120. Proceedings of the American Catholic Philosophical Association 75. New York: American Catholic Philosophical Association.

Hopkins, G. M. (1877). "God's Grandeur." Available at http://www.poetry foundation.org/poem/173660.

International Theological Commission. (2004). "Communion and Stewardship: Human Persons Created in the Image of God." Available through http://www.vatican.va.

John Paul II, Pope. (1996). Address delivered to the Pontifical Academy of Sciences, October 22, 1996. Printed in the *Quarterly Review of Biology* 72:381–83 and available at http://www.jstor.org/stable/3037603.

Kim, J. (1996). *Philosophy of Mind*. Boulder, CO: Westview Press.

Leroy, D. (1891). *L'Evolution restreinte aux espèces organiques*. Paris: Delhomme et Briguet.

May, G. (1994). *Creatio ex nihilo*. London: T. T. and Clark International.

Ott, L. (1960). *Fundamentals of Catholic Dogma*. Translated by Patrick Lynch. Rockford, IL: Tan Books.

Sedley, D. (2007). *Creation and Its Critics in Antiquity*. Los Angeles: University of California Press.

Sokolowski, R. (1982). *The God of Faith and Reason*. Notre Dame, IN: University of Notre Dame Press.

Vatican. (2012). *Catechism of the Roman Catholic Church*. Part 1: The Profession of Faith, Section 2.1: The Creeds, Chapter 1, Article 1. Available at http://www.vatican.va/archive/ENG0015/__P16.HTM.

Wittgenstein, L. (1953). *Philosophical Investigations*. Oxford: Blackwell.

TWELVE

AFTER DARWIN, AQUINAS
A Universe Created and Evolving

William E. Carroll

At the 2000 Jubilee Session for scientists, held at the Vatican in May
of that year, Archbishop Józef Życiński offered an eloquent assessment of contemporary discourse on the relationship between the
natural sciences and theology. He ended his address with the comment that what is needed today is a new Thomas Aquinas. I remember remarking to him after his speech that I thought that the old
Thomas Aquinas would do just fine! At least the principles that inform Thomas's philosophy and theology remain of enduring value as
we seek to probe ever more deeply into our understanding of nature,
human nature, and God. Professor John Haught has written that
after the life and work of Charles Darwin "any thoughts we may
have about God can hardly remain the same as before." As Haught
observes, "Evolutionary science has changed our understanding of
the world dramatically, and so any sense we may have of a God who
creates and cares for this world must take into account what Darwin
and his followers have told us about it" (Haught 2000, 6). Although
evolutionary science has significantly changed our view of the world
and of ourselves, I am not persuaded that our thoughts about God
need to undergo a radical revision. Of course, it depends on the particular thoughts about God to which one is referring. "As long as

we think of God," Haught writes, "only in terms of 'order' or 'design,' the 'atheism' of many evolutionists will seem appropriate." Rather than accept the conclusions of the "new atheism," Haught and others urge a "new theism" consistent with the evolving universe disclosed by contemporary science (Haught 2000, 6).

More generally, the claim is that the novelty, dynamism, chance, and self-organizing principles in nature are not consistent with an omnipotent, omniscient, and timeless God, especially as described in categories of Aristotelian philosophy. Additional reasons offered for rejecting the traditional conception of God include the claim that the Thomistic distinction between God as Primary Cause and the whole array of secondary causes is incoherent and that the god of Aristotle and Aquinas is religiously objectionable since such a god is not the God of love, who suffers for us, as depicted in the Bible.

Less sophisticated than Haught's approach is the proposal of Robert Wright, reminiscent of early modern deism, that any reconciliation between evolution and theology requires that "modern theology . . . bite the bullet and accept the fact that God did his work remotely—that his role in the creative process ended when he unleashed the algorithm of natural selection (whether by dropping it into the primordial ooze or writing its eventual emergence into the initial conditions of the universe or whatever)." Admitting that his "theo-biological scenario," according to which God initiated natural selection with "some confidence that it would lead to a morally rich and reflective species," is only speculative, Wright concludes that "these speculations are compatible with the standard scientific theory of human creation." Furthermore, if believers would accept them, "it would end any conflict between religion and the teaching of evolutionary biology." Were theology to follow this path it would, according to Wright, author of *The Evolution of God*, do what it has done before: "evolve—adapt its conception of God to advancing knowledge and to sheer logic" (Wright 2009).[1]

The conceit of my title, "After Darwin, Aquinas," is meant to suggest that the challenges that evolutionary biology presents to theology do not so much demand a "new theism" (much less a return to a form of deism, as Wright proposes) as they offer us the opportunity to reappropriate insights of Thomas Aquinas, especially concerning the doctrine of creation, God's transcendence, and God's action in the world. Evolutionary biology surely challenges the conception that each of the various types of living things is the result of some *special* divine act, some kind of special creation, or that the order and design in nature must be the result of a type of divine manipulation with little or no reference to natural causes themselves. Too often creation had been seen as the bestowing of order, and thus, if order could be explained by biological processes, it would seem that there was no need for a creator. The importance of God as designer and orderer had been emphasized by the physicotheologians (especially in England) in the seventeenth and eighteenth centuries; it was a view of God strongly criticized by David Hume.[2] Indeed, various forms of modern atheism have their origins in the rejection of the God set forth by physicotheology.[3]

In important ways, evolutionary biology has helped us to see the inadequacies of what had become generally accepted understandings of creation and of divine agency, but these understandings are not those of Thomas Aquinas. The god who, after Darwin, must be seen in new terms is not really the God described by Thomas. My point is that after Darwin's challenge we have new reasons for returning to the thought of Thomas Aquinas on these and related topics. We do not need to appeal, as many do, to a kenotic theology of creation, according to which God limits His power in order to allow for the vibrant causal agency in nature which evolution and the other sciences now describe. Nor must we embrace versions of process theology in which God and nature are evolving together. As we shall see, the autonomy and integrity of nature, so important for contemporary science, are part of Thomas's created universe: creation, for him, is the causing of existence, it ought not to be identified essentially as the ordering of reality. The path to God as Creator, which Darwin has helped to prepare by rejecting certain established views of god as the grand designer, can lead us back to Thomas's view of creation and science.

In fact, a return to Thomas would help to defuse much of the confusion in contemporary discourse about evolution, a discourse that can easily become obscured in broader political, social, and philosophical contexts. Indeed, evolution and creation have taken on cultural connotations, serve as ideological markers, with the result that each has come to stand for a competing worldview. For some, to embrace evolution is to affirm an exclusively secular and atheistic view of reality, and evolution is accordingly either welcomed or rejected on such grounds. Too often "creation" is confused with various forms of "creationism," which embrace either a literalistic reading of the Bible or think that creation *must* mean a kind of divine intervention in cosmic history with God's directly creating each individual species of living things.

The choice for many seems to be between a purely natural explanation of the origin and development of life, an explanation in terms of common descent, genetic mutations, and natural selection, on the one hand, and, on the other hand, an explanation that sees divine agency as the source of life in all its diversity and holds that human beings, created in the image and likeness of God, have a special place in the universe. The difference *appears* stark: either Darwin or God.

Often scientific developments have been used to support a kind of "totalizing naturalism" according to which the universe and the processes within it need no explanation beyond the categories of the natural sciences.[4] Jean-Michel Maldamé identifies such a conception in the following terms:

> Nature is understood as self-creating, this term connoting that the classic notion of creation has become useless. Nature—and it is proper to write this word with a capital letter—is self-sufficient to produce not only its effects, but to produce itself. . . . The notion of creation disappears in this perspective of reflection. (Maldamé 2006, 153)[5]

This view of Nature as completely self-sufficient supports the philosophical claim that existence is a "brute fact" that does not call for any explanation beyond itself.[6] Thus, only the emergence of new things from already existing realities, or their going out of existence, or other varieties of change, need to be explained; what does not need to be explained, so this position contends, is the mere existence of that which changes.[7] The argument is that the natural sciences are fully sufficient, at least in principle, to account for all that needs to be accounted for in the universe. Whether we speak of explanations of the Big Bang itself (such as quantum tunneling from nothing) or of some version of a multiverse hypothesis, or of self-organizing principles in biological change (including, at times, appeals to randomness and chance as ultimate explanations), the conclusion that seems inescapable to many is that there is no need to appeal to a creator, that is, to any cause that is outside the natural order.

Stuart Kauffman, famous for his work on information systems and biocomplexity, argues that we are "reinventing the sacred" as a result of a new view of science. This new view involves a rejection of reductionism and an affirmation of the emergent properties of a dynamic universe of "ceaseless creativity." As he observes, "life has emerged in the universe without requiring special intervention from a Creator God. . . . All, I claim, arose without a Creator God. . . . Is not this view, a view based on an expanded science, God enough? Is not nature itself creativity enough? What more do we really need of God . . . ?" (Kauffman 2008, 71, 229, 283).

As I have suggested, the universe described by contemporary science is often viewed as a self-contained universe, exhaustively understood in terms of the principles and laws of science. In such a universe there would seem to be little if any need for the God of Jewish, Christian, or Muslim revelation. The traditional doctrine of creation appears obsolete in the face of the recent advances of science. The role of a creator has, so it seems, been rendered superfluous; a creator represents, at best, an intellectual artifact from a less enlightened age. One source of confusion in such an analysis, as we have seen, is to view God's creative act essentially as the explanation for order and design in nature, that is, to identify creation with the causing of order and design. As evolutionary biology, for example, claims to be able to explain order and design without an appeal to an orderer or designer, but exclusively on the basis of natural processes, it appears to many that there is no longer a role for God to play. It is an error, as we shall see, to think that an explanation of order is an explanation of existence.

In contemporary biology, there have been important discussions about understanding living things in terms of "self-organization." As reductionism and mechanism are being replaced by appeals to dynamic, intrinsic, organizing principles, the conclusion often reached is that changes in nature are *exhaustively* based on principles and entities in the natural world, and that there is no need for any external "interference" to explain the change. Previously, when nature was conceived in exclusively inert and mechanistic terms, there were appeals to a source of activity beyond nature, although such appeals would often never be more than the affirmation of a kind of deism: to see God as only getting things started, so to speak, although there may be times when he has to tinker with the mechanism he has produced. Furthermore, as I have already indicated, for many thinkers today there is no category beyond change and the particular behavior of individual things that requires an explanation. We need to distinguish, however, between the specific claims that evolutionary biology makes about the development and diversity of living things, explanations that are properly in the domain of the natural sciences, and philosophical claims concerning whether or not additional explanations of these realities are necessary. An important point here is that to defend the competence of the natural sciences to describe what happens in nature ought not to be equated with a denial of creation. Although many contemporary theologians see the poverty of arguments that absolutize the natural order, they tend to think that the challenges of science require us to reformulate what it means for God to create.

The extent to which biologists, when they speak about self-organization, move beyond the domain of biology to make broad claims about "self-creation" and that, accordingly, there is no need to appeal to a source of existence of living things, is the extent to which their claims are broadly metaphysical. An important feature of these philosophical claims, namely, that "self-creation" and "self-sufficiency" evident in the natural order eliminate the need to appeal to God, involves conceptions of God and creation that, even if shared by some believers, are really not the same as those found in traditional philosophy and theology.

For many, there seem to be only three general theological responses to developments in evolutionary biology: (1) reject any reference to God as Creator in order to defend the autonomy and integrity of nature; (2) deny conclusions in biology in defense of a "creationist" reading of scripture, according to which, for example, God directly creates each new species; or (3) argue for a radical change in what it means for God to create, or even of what it means to be God. None of these alternatives will be defended here.

A RETURN TO THOMAS AQUINAS

In this essay, I want to argue for a Thomistic analysis of creation and the relative self-sufficiency of nature against both the "new atheism" and the various theological attempts to alter fundamentally what we mean by God as creator. After Darwin, we should no longer accept those notions of creation and divine agency that are incompatible with an evolving universe in which there is real novelty and in which the processes of development and the emergence of new species can be explained by principles in the universe. Thomas Aquinas remains an excellent guide for coming to terms with "God after Darwin." As Thomas would argue, the very processes that evolutionary biology explains depend upon God's creative act. The intelligibility of evolution itself depends upon a source that transcends the processes of nature. In fact, I want to argue that without a robust understanding of what creation means—indeed, without the very fact that all that is is completely dependent upon God as cause—there would be no evolution at all. Furthermore, for Thomas nature contains intrinsic principles of dynamic activity, an integrity that is not challenged by a robust notion of divine omnipotence but is made possible by this omnipotence.

Medieval discussions about creation (especially the intelligibility of *creatio ex nihilo*), divine agency, and the autonomy of nature, and ultimately the very possibility of the natural sciences' discovering real causes in nature, provide a rich source of insights for us today. What Avicenna, Maimonides, and Thomas Aquinas, for example, saw so clearly—that creation is an account of the existence of things, not of changes in and among things—allows us to conclude that there is no contradiction between creation, so understood, and any conclusion in the natural sciences.

The key to Thomas's analysis is the distinction he draws between creation and change, or, as he often remarked: *creatio non est mutatio*. Creation, as a metaphysical and theological notion, affirms that all that is, in whatever way or ways it is, depends upon God as cause. The natural sciences, whether Aristotelian (with which Thomas was primarily concerned) or those of our own day, have as their subject the world of changing things: from subatomic particles to acorns to galaxies. Whenever there is a change there must be something that changes. Whether the changes are biological or cosmological, without beginning or end, or temporally finite, they remain processes. Creation, on the other hand, is the radical causing of the whole existence of whatever exists. To cause completely something to exist is not to produce a change in something, is not to work on or with some existing material. If, in producing something new, an agent were to use something already existing, the agent would not be the *complete* cause of the new thing. But such complete causing is precisely what creation is. To create is to cause existence, and all things are totally dependent upon the Creator for the very fact that they are. As Thomas remarks in his treatise *De substantiis separatis* [c. 9]: "Over and above the mode of becoming by which something comes to be through change or motion, there must be a mode of becoming or origin of things without any mutation or motion, through the influx of being" (Aquinas 2013a).

Modern arguments about order and design (like medieval arguments about motion and an unmoved mover) are arguments in natural philosophy and not in metaphysics. To the extent that "to create" is susceptible to rational examination, it is a topic in metaphysics and not in natural philosophy, nor in the individual empirical sciences. Furthermore, for Thomas, creation is not primarily some distant event; rather, it is the ongoing complete causing of the existence of all that is. At this very moment, were God not causing all that is to exist—from quantum processes to the color of the sky, to our own thoughts, hopes, and dreams—were God not to be causing everything that is, there would be nothing at all.

Creatures are what they are (including those that are free), precisely because God is present to them as cause. Were God to withdraw, all that exists would cease to be. Creaturely agency and the integrity of nature, in general, are guaranteed by God's creative causality. Here is how Thomas expresses this view in the *Summa theologiae*:

> Some have understood God to work in every agent in such a way that no created power has any effect in things, but that God alone is the ultimate cause of everything wrought; for instance, that it is not fire that gives heat, but God in the fire, and so forth. But this is impossible. First, because the order of cause and effect would be taken away from created things, and this would imply lack of power in the Creator, for it is due to the power of the cause, that it bestows active power on its effect. Secondly, because the active powers which are seen to exist in things, would be bestowed on things to no purpose, if these wrought nothing through them. Indeed, all things created would seem, in a way, to be purposeless, if they lacked an operation proper to them, since the purpose of everything is its operation. . . . We must therefore understand that God works in things in such a manner that things have their proper operation. (Aquinas 2012e, I, Q.105. a.5)

God is so powerful that His causal agency also produces the modality of its effect: the effect is assimilated to God's will in every way so that not only what happens occurs because God wills it to happen, but it happens "in that way which God wills it to happen" (Aquinas 2012a, *De Veritate* Q. 23, a. 5). God's will transcends and constitutes the whole hierarchy of created causes, both causes that always and necessarily produce their effects and causes that at times fail to produce their effects.[8] We can say that God causes chance events to be chance events. The role of chance mutations at the genetic level, so important in current evolutionary theory, does not call into question God's creative act.

God is the cause of being as such—and to cause being as such is precisely what to create means. God's causation does not compete with the causation of creatures, but rather supports and grounds it.[9] Since it is characteristic of the causes in nature precisely to be causes, God's causal determination of them is not such as to deny their proper autonomy.[10] God as "primary cause" is not at the summit of a chain of causes. God is a kind of ultimate or foundational cause of all that is. Nor is God an entity among other entities.[11] Contrary to the view of some contemporary theologians (Murphy 1995), God does not need a metaphysical indeterminacy in nature so that His actions would not collide, so to speak, with other causes.[12] Often commentators fail to recognize that to speak of God as cause is to use cause in an analogical sense. John Yates makes this point well when he discusses the causality of free creatures and divine causality: "If the Thomist solution to the reconcilability of finite free action and divine causal power is to work . . . God cannot be inserted into the world's causal chains, the divine causal influence, as *ex nihilo*, cannot and must not be thought of as univocal with other causes. As in all other things, God is not to be conceived of as a 'cause' in the categorical sense; He does not belong to any categories precisely because He is the 'cause' of them all" (Yates 1990, 252–56). In a sense, of course, for Thomas we do not really know what it means for God to be "cause," as is the case with any characterization we predicate of God.[13]

Notice how Thomas's analysis of creation does not challenge what evolutionary theories describe as the source of order and design in terms of the natural selection of organisms "subject to the vagaries of genetic mutation, environmental challenge, and past history." Francisco Ayala points out that "the *scientific* account of these events does not necessitate recourse to a preordained plan, whether imprinted from the beginning or through successive interventions by an omniscient and almighty Designer" (Ayala 2007, 76, italics added). Ayala discusses chance occurrences at the level of random genetic mutations, but adds that the process of natural selection results in outcomes that are not fortuitous but are "determined by their [the traits that organisms acquire] functional utility to the organisms," and thus these traits are "designed, as it were" to serve the life needs of the organisms. Thus, Ayala thinks we ought to recognize that some natural processes (*viz.*, natural selection) at work in evolution are "not random, but oriented and able to generate order or 'create'" (Ayala 2007, 77). Although Ayala puts "create" in quotation marks, I think it would be better not to use the term at all in this context.

Ayala has often emphasized the revolutionary character of Darwinian evolution and its scientific progeny: "This is Darwin's fundamental discovery, that there is a process that is creative but not conscious. And this is the conceptual revolution that Darwin completed: the idea that the design of living organisms can be accounted for as the result of natural processes governed by natural laws. This is nothing if not a fundamental vision that has forever changed how mankind perceives itself and its place in the universe" (Ayala 2007, 202). However revolutionary evolution by natural selection is in our understanding processes in nature, it does not call into question the truth of a Thomistic metaphysical and theological understanding of creation.[14] At least, there is no incompatibility unless one erroneously thinks that to cause order or to design is the root sense of what it means to create or concludes that there is no need to seek an ultimate explanation of natural processes themselves.

KENOTIC CREATION AND THE METAPHYSICS OF HUMILITY

Many theologians today are attracted to what is sometimes referred to as a kenotic theology of creation. This is the view that, given the insights of chaos theory, complexity, and dynamic self-organization in nature, we need to see God as withdrawing, so to speak, in order to allow new order and new life "to unfold with spontaneity, freedom, and creativity" (Delio 2003, 336). John Haught speaks of a "metaphysics of humility" as the basis of divine action (Haught 1995, 41). The position often affirmed is that somehow God must be limited or self-limiting or "humble" in order for nature to be marked by the novelty and surprise that evolutionary biology discovers in the world. Novelty and surprise are incompatible with "programmed blueprints" so often associated with the view of God as designer. In a broader sense, such theologians think that traditional conceptions of God as omnipotent, omniscient, and beyond time, based as they are on a "metaphysics of being," need to be replaced as a result of a new understanding of reality, a "metaphysics of becoming."[15]

Keeping with his general theme that evolutionary biology offers a gift to theology, Haught thinks that modern science shows us that we live in a universe that in important ways is not yet created. There is no perfect created order from which the physical suffering in the world must be seen as a defect. If the world is developing, is in a sense unfinished, then "we cannot justifiably expect it yet to be perfect." The world "*inevitably* has a dark side." Haught insists that "the only kind of universe" a loving God could create is an evolving universe:

> For God's love of creation to be actualized, the beloved must be truly "other" than God. And an instantaneously finished universe, one from which our present condition of historical becoming and existential ambiguity could be envisaged as a subsequent estrangement, would in principle have been only an emanation or appendage of deity and not something truly other than God. A world that is not clearly distinct from God could not be the recipient of divine love. And an instantaneously completed world could never have established an independent existence vis-à-vis its creator. The idea of a world perfectly constituted *ab initio* would, in other words, be logically incompatible with any idea of a divine creation emerging from the depths of selfless love. (Haught 2003, 168–69)[16]

Here we see the frequent view that a God of love must be a self-limiting God in order to allow creatures to have an appropriate autonomy. For Thomas Aquinas, however, as we shall see again in the next section of this essay, one need not choose between a self-limiting God of love with a relative autonomy for nature, on the one hand, and an immutable, omnipotent God with a completed, static natural order, on the other. With a proper understanding of divine agency and divine transcendence, as we shall also see, there is no reason for God to limit His power and presence in order to provide space in which to create or space in which creatures can exercise their own agency (including the agency of evolutionary change that results in real novelty in nature). We can accept an important feature of Haught's analysis without rejecting, as he does, a traditional view of God as Creator.

Thomas Aquinas's understanding of creation is perfectly compatible with an evolving universe. The "creation" that is, according to Haught, not yet finished is the world of changing things, ever open to new and unexpected varieties of things. Such a changing world, with all its novelty, is being created by God; Aquinas never thought that creation is essentially some once-and-for-all distant event.

Keith Ward, formerly Regius Professor of Divinity at Oxford, thinks that the traditional attempt to make God the "efficient cause of all things, without compromising the simplicity and unchangeability which are characteristics of the Aristotelian picture of God" was "an heroic failure," since it "could not account for the contingency of the universe." This is so because "[t]hat which is wholly necessary can only produce that which is necessary. A contingent universe can only be accounted for if one makes free creativity a characteristic of the First Mover, which entails placing change and contingency within the First Mover itself." According to Ward, God's omniscience "is the capacity to know everything that becomes actual, *whenever* it does so. . . . The classical hypothesis [of a God who does not change] does not . . . seem compelling" (Ward 1996, 202, 188, italics added). In a sense, God must wait to know what is actual since there is an inherent contingency in nature itself, and as actualities change so does God's knowledge. Other critics of Thomas claim that, if God really knows each object and is immanent in each, God must in some way "contain temporality" and must be affected by the knowledge He has: to claim that such a God is immutable, without potentiality and atemporal is "metaphysically incoherent" (Stoeger 2008, 233, citing O'Donnell 1983, 17–21). Since modern science, especially evolutionary theory, discloses a fundamental contingency in nature, the classical conception of God, inconsistent with nature so understood, must be jettisoned. So, at least, is the argument of many contemporary theologians who think that in rejecting classical theism they are honoring the insights of science. Some argue, in addition, that the view of God-in-time, changing as the world and man change, neither omniscient nor omnipotent in the classical sense, is far more consistent with the core of biblical revelation than the God described by Thomas Aquinas and others.

Denying divine omnipotence, immutability, and the like is surely more appealing than the crude atheism of Dawkins, Dennett, Hitchens, and others, but, I think, it too suffers from a failure to take seriously the insights of Thomas Aquinas. I think that Haught, and certainly process theologians in general, underestimate the resources of traditional philosophy and theology to respond to the evidence of modern biology and cosmology.

There are some theologians who have adopted a theory known as "panentheism," according to which the world is somehow located within the divine, and that patterns of emergence in nature are grounded in the divine order and that God is continually responding to the evolutionary process. "Panentheists seek to formulate a single ontological vision rather than sharply separating the becoming of the world from the timelessness and aseity of the divine being. . . . [They reject] purely atemporal understandings of the God-world relationship, *in so far as* such views tend to underestimate the importance of time, process, and pervasive change within the natural world. . . . [Panentheism] also at least indirectly undercuts static views of the divine nature, for it would be surprising, though not impossible, that a natural reality characterized by ubiquitous process and interconnection would be the result of a creator whose nature is essentially non-relational and non-responsive" (Clayton 2006, 319–20, italics added).[17]

Thomas's analysis of God as cause of all that is is, I think, immune to such criticism. As I have already indicated, for Thomas, the notion of "cause" is predicated of God in an analogical sense such that it is radically different from all other uses of the term. Nor is God a super-entity among other entities. The incoherence that critics find in Thomas's approach to God as creator is the result, in part, of not recognizing the profound understanding Thomas has of the way in which terms are predicated of God and of not seeing clearly the distinctions he draws between metaphysics and natural philosophy, especially his understanding that creation is a subject for metaphysics and theology, not for the natural sciences. Natural philosophy, for Thomas, is a more general science of nature that considers topics such as the nature of change, time, and the like; unlike metaphysics, it does not address the questions of what it means for things to exist and why there is something rather than nothing.

In a thorough-going critique of a kenotic approach to creation and divine agency, Taede Smedes thinks that many of the proposals to embrace a kenotic understanding of God's action in creating the world are misguided attempts because they subject our understanding of God to categories of explanation in the world.[18] His criticism, finally, is theological: such a limited God raises questions about God's providence (in particular about God's power to bring about eschatological ends) and also about whether such a God is worthy of worship. Denials of divine omniscience and atemporality are also theologically suspect since they result in God's being just another agent in a world of many agents: "God is reduced to an ignorant, impotent, helpless, and reckless entity which has let things get out of control" (Smedes 2004, 187).[19] A fundamental error underlying the kenotic approach, Smedes thinks, is the failure to recognize that the natural sciences themselves "cannot tell us anything about divine action" (226).

DIVINE AGENCY AND THE SELF-SUFFICIENCY OF NATURE

The problem that those who defend a self-sufficiency in nature and its processes often see is that any appeal to a cause outside of nature is either superfluous or contradictory to the very claim that nature is the domain of self-organizing activities. There is a confusion here, however, about different orders or levels of explanation. If we ask, for example, why wood is heated in the presence of fire, we can explain the phenomenon in terms of the characteristics of both wood and fire. Thomas Aquinas remarks that if a person answers the question of why the wood is heated by saying that God wills it, the person "answers appropriately, provided he intends to take the question back to a first cause; but not appropriately, if he means to exclude all other causes" (Aquinas 2013b, III, c. 94). For Thomas, there is no question that there are real causes in the natural order: "if effects are not produced by the action of created things, but only by the action of God, it is impossible for the power of any created cause to be manifested through its effects." If no created things really produced effects, then "no nature of anything would ever be known through its effect, and thus all the knowledge of natural science is taken away from us"

(c. 69). Thomas thinks that to defend the fact that creatures are real causes, far from challenging divine omnipotence, is a powerful argument for divine omnipotence. As he says, to deny the power of creatures to be the causes of things is to detract from the perfection of creatures and, thus, "to detract from the perfection of divine power" (c. 69).

God is immediately active in all things and, in an important sense, God is more intimate to each creature than a creature is to itself (see Aquinas 2013d, I, 8, 1, ad 1). God, as the cause of each creature's being, is present at the very center of each creature's being. He is more interior to things than they are to themselves: not as an intrinsic principle entering into their constitution, but as the *abiding* cause of their existence (Aquinas 2013d, I, 37, 1, ad 1).[20] As Simon Tugwell has written: "The fact that things exist and act in their own right is the most telling indication that God is existing and acting in them" (Tugwell 1988, 213).[21]

Thomas shows us how to distinguish between the being or existence of creatures and the operations they perform. God causes creatures to exist in such a way that they are the real causes of their own operations. For Thomas, God is at work in every operation of nature, but the autonomy of nature is not an indication of some reduction in God's power or activity; rather, it is an indication of His goodness. It is important to recognize that, for Thomas, divine causality and creaturely causality function at fundamentally different levels.[22] In the *Summa contra Gentiles*, Thomas remarks that "the same effect is not attributed to a natural cause and to divine power in such a way that it is partly done by God, and partly by the natural agent; rather, it is wholly done by both, according to a different way" (Aquinas 2013b, III, c. 70, 8). It is not the case of partial or co-causes with each contributing a separate element to produce the effect.[23] God, as Creator, transcends the order of created causes in such a way that He is their enabling origin. For Thomas "the differing metaphysical levels of primary and secondary causation require us to say that any created effect comes totally and immediately from God as the transcendent primary cause and totally and immediately from the creature as secondary cause" (Shanley 1998, 100).[24] In response to the objection that it is superfluous for effects to flow from natural causes since they could just as well be directly caused by God alone, Thomas writes that the existence of real secondary causes "is not the result of the inadequacy of divine power, but of the immensity of God's goodness." God wills to communicate His likeness to things, "not only that they might exist, but also that they might be causes for other things" (Aquinas 2013b, III, c. 70).[25] To ascribe to God (as first cause) *all* causal agency "eliminates the order of the universe, which is woven together through the order and connection of causes. For the first cause lends from the eminence of its goodness not only to other things that they are, but also that they are causes" (Aquinas 2013a, *De veritate* 11,1; see also 2013e, I, q. 23, a. 8, ad. 2).

Here I should like to probe a little more deeply into how everything that exists depends completely upon God as cause, yet what is caused has its own proper stability in being. Of its own nature—that is, left completely to itself—the creature is non-being rather than being, and it must be caused by God continuously lest it return to the non-being which it properly is. It is true to say that the creature is literally nothing without the creative causality of God. Nevertheless, we must remember that the being of creatures, far from being an accident or a characteristic of things, is the ultimate perfection or actuality of the creature (Aquinas 2013d, I, 8, 1, a. 3). Most profoundly, in the depths of any creature is its being; a creature is nothing so much as its own being. The creature, thus, far from being an insubstantial, quasi-nothing, is a real something, existing, in a sense, on its own. In giving being to the creature, God does not merely make the creature to be an extension of Himself; rather He gives the creature an inherent stability in being, that is, a tendency to exist. God gives being in such a way that the tendency of the given being is not to lapse into non-being but precisely to remain in being.[26]

A. D. Sertillanges offers a particularly good account of what we might call the created autonomy of creatures:

> Thomas is so firmly convinced that creatures have an autonomy grounded in their creation itself that, unlike many other doctors, he stoutly refuses the position that creatures tend towards nothingness and require God's constant support to keep them from

doing so. In this respect, Thomas would reject Descartes' creatio continua. . . . Created being not only does not tend towards nothingness, it positively tends toward the growth and perfection of its own being and, in this way, to be always more and better. . . . Inasmuch as the creature is autonomous thanks to God, inasmuch as it exists in itself, albeit not per se, the creature tends to be and to increase. This tendency is the primary property of its very being. . . . We can say that the creature tends towards nothingness only if we consider the abstract notion of creatureliness. Concretely, the autonomy granted the creature by the very nature of its origin is prolonged in its consequences, which orient it towards the increase and betterment of its being according to the full measure of its power. (Sertillanges 1945, 62, in Petrosino 2001)

In a way, to speak of the creature's dependence upon God tends to suggest too weak a notion of dependence, since dependence signifies something added to being, whereas in the case of creation a thing depends upon the Creator for its very being. There is a kind of oscillation in the idea of creation: on the one hand, there is an emphasis on the creature's very autonomy as such, and, on the other hand, an emphasis on the derivation of the creature's very being from God.[27]

On this point concerning what I have termed creaturely autonomy, it is useful to compare Thomas's position to that of his contemporary St. Bonaventure, who, like Thomas, does not hold that created beings have a tendency toward non-existence, but who, unlike Thomas, thinks that since creatures are temporal they need to be maintained in their being, to be conserved in existence, and that this conservation in being is different from their being created in the first place. It is true for both Thomas and Bonaventure that creatures will cease to exist should God cease to cause their existence. For Thomas, however, God gives being, and no other act is required in order to keep creatures in existence. For Bonaventure, on the other hand, God must perform two different acts (different, at least, with respect to the creature): He gives being initially and, since the creature cannot naturally maintain its own existence, He conserves the creature in existence. In other words, according to Bonaventure, if we look at the natural principles of creation, form and matter, the creature is not mutable into non-being. If, however, we look at the fact that creatures are made out of nothing, we find an inherent emptiness (*vanitas*), instability (*instabilitas*), and mutability (*vertibilitas*). Hence, by nature creatures are mutable into non-being, but by God's grace they are conserved in being (Bonaventure 1934, *In 2 Sent.* 37.1.2. sol.; *In 1 Sent.* 8, 1.2.2. sol., and ad 7–8). For Thomas, on the other hand, being belongs really to the creature: "in the whole of created nature, there is no potency through which it is possible for something to tend into nothing" (Aquinas 2013a, *De potentia Dei*, q. 5, a. 3, sol.). It is true that material bodies tend to corrupt, but matter itself, prime matter, is incorruptible. The whole of the universe, considered in itself, has its own being and tends to continue in being. Of itself, it has no potency or tendency to non-being. However true it may be to say that a creature would be absolutely nothing without the creative causality of God, still, the creature really has its own being: its own autonomy, and, in a way, its own self-sufficiency.

God does not only give being to things when they first begin to exist, He also causes being in them so long as they exist. He not only causes the operative powers to exist in things when these things come into being, He always causes these powers in things (Aquinas 2013b, III, c. 67). Thus, if God's creative act were to cease, every operation would cease; every operation of a thing has God as its ultimate cause. As we have seen, Thomas does not think that such an affirmation of divine omnipotence eliminates the role of real created causes. The self-sufficiency of nature, the dynamism of natural processes that science discovers, does not mean that God is superfluous, because He is the cause of nature itself; He is a cause in such a way that nature has its own integrity, its own self-organizing principles. Nor must God "withdraw," so to speak, in order for there to be such intrinsic dynamism and novelty in nature.

Niels Gregersen has highlighted this approach in a series of articles, speaking, for example, of the "grace of self-organization." The "splendor of God" is disclosed in the self-productive powers intrinsic to the created order. Gregersen cites an apt parable from the New Testament in which Jesus refers to the coming of the reign of God in the midst of the created order: "The earth produces of itself [Greek: *automatiké*] first the stalk, then the head, then full grain in the head" (Mark 4:28). Gregersen contends: "The more self-productive the creatures are, the more God is at work in and through these creatures. Even though God cannot be confined to specific informational structures in matter, God should not be thought as working as an external agent in relation to such local patterns of information" (Gregersen 2006, 25–26).[28] God, so conceived, is the underlying source of evolutionary processes and of the order and design (especially the patterns of information) that result from these processes.

Thomas's emphasis on the ongoing character of God's creative act and that God is the source of the dynamic principles and potentialities found in nature resonates with what the evolutionary sciences disclose. As William Stoeger points out: the "evolutionary, emergent, and unfinished character of creation [that is, of the created order] revealed by the sciences serves to emphasize the continuing character of God's action through the regularities, processes, and relationships God sustains." The transience of entities and of groups of entities within nature that constitute the dynamic processes of evolution and which, for Stoeger and others, is part of a kind of directionality of the whole of nature, "is consonant with what . . . Christians have come to understand about God's presence and action within creation" (Stoeger 2008, 231). After Darwin, we have new encouragement to see God's creative and providential act as a continuing reality, something that Thomas always emphasized.

Creation, for Thomas, concerns first of all the origin of the universe, not its temporal beginning. Indeed, it is important to recognize his distinction between origin and beginning. To speak of origin is to affirm the complete, continuing dependence of all that is on God as cause. Thomas does, of course, believe that the universe is temporally finite: a truth revealed only in scripture. But, even if the universe were eternal it still would depend upon God for its existence; it still would be created.

For Thomas, to say that the world is created has no effect on the principles of the natural sciences, unless, of course, one would make the metaphysical claim that the principles of the natural sciences are the ultimate principles of reality. Whether the world is, as contemporary atheists would have, a "brute fact" and hence uncreated, or whether the world is created is irrelevant to how you describe it in scientific terms. As Denys Turner comments, "the difference between a created and uncreated world is no difference at all so far as concerns how you describe" that world: "the only difference it makes is all the difference to everything" (Turner 2004, 257–58).

POSTMODERN THEOLOGY AND THE FLIGHT FROM CAUSALITY

There is a response to traditional conceptions of creation, and of God, far more radical than that of process thought and panentheism, to which I have already referred. There are some contemporary theologians who reject what they consider to be outmoded notions of God as Creator and cause of all that is. Although they wish to affirm a role for God in the world, they do not think that speaking of God as cause is appropriate. In this rejection of divine causality they share many of the presuppositions of those who think that because nature is self-sufficient any appeal to the activity of a god is superfluous. Writing in *Theology and Science*, the Lutheran theologian and Swedish bishop Antje Jackelén thinks that, as she puts it, theistic discourse is wedded to an *ontology* that presupposes too neat a distinction between transcendence and immanence (Jackelén 2007, 153).[29] This defective discourse is built on the God of onto-theology whom Heidegger called *causa sui* and about whom he said: "One can neither pray nor sacrifice to this god. Before the *causa sui*, one can neither fall to one's knees in awe nor can one play music and dance before this God" (Heidegger 1969, 72). Theism, Jackelén argues, "is inextricably linked to causation," and "like deism, it gives the impression that causation is the primary category for an adequate understanding of God and Nature" (Jackelén 2007, 154–55). She cites approvingly the work of an Eastern Orthodox theologian, Sergei Bulgakov, for whom the entire doctrine of primary and secondary causality, and particularly the "doctrine of God as the *cause* of the world . . . is only a monstrous misunderstanding, a theological temptation." Bulgakov thinks that:

the idea of the Creator and creation does not need to be translated into the language of mechanical causality, for it has another category, its proper one, that of *co-imagedness*, since the creature contains the living *image* of the Creator and is correlated with Him. . . . The world does not have a cause, since it is created; and *God is not the cause of the world and not a cause in the world, but its Creator and Provider.* God's creative act is not the mechanical causation through Himself of the world's being, but His going out of Himself in creation through the positing of the world as the creaturely Sophia. (Bulgakov 2002, 220, 221–222, in Jackelén 2007, 154–55)

Note that Bulgakov refers to God's "mechanical causation" of the world. This is hardly Thomas's understanding of what it means for God to be the cause of all that is, but the notion of causality that informs Bulgakov's view is characteristic of the narrower notions of cause that have been employed since the seventeenth century. For Thomas, God's causality is radically different from any causality in the created order. Indeed, as we have seen, it is only through analogical predication that one can speak of God as a cause at all. The criticism of the theism associated with what is called onto-theology may very well be accurate, but such criticism does not really apply to the conceptions of God, creation, and causality central to the thought of Thomas Aquinas. For Thomas, God is not *causa sui*; He does not cause Himself to exist. In fact, Jean-Luc Marion, famous for his criticism of onto-theology in books like *God beyond Being*,[30] has come to recognize that Thomas is immune to such criticism.

DIVINE TRANSCENDENCE: AN UNDERLYING ASSUMPTION

The source of most of the difficulties in grasping an adequate understanding of the relationship between the created order and God is the failure to understand divine transcendence. It is God's very transcendence, a transcendence beyond any contrast with immanence, that enables God to be intimately present in the world as cause. God is not transcendent in such a way that He is "outside" or "above" or "beyond" the world. God is not different from creatures in the way in which creatures differ from one another. We might say that God "differs differently" from the created order.[31] "God operates immanently in nature in such a way that He sets nature, so to speak, free in its own operation. . . . Thomas [sees] . . . God as a cause which by its transcending immanence constitutes the causality of nature in its own order" (te Velde 1995, 164). Modern theism, with its "domesticated transcendence" (Placher 1996), too often sees "transcendence" and "immanence" as contrasting categories such that one necessarily excludes the other. This shift from a Thomistic understanding of divine transcendence has resulted in rejecting traditional attributes of God (e.g., omnipotence, immutability, and timelessness) in order to affirm the integrity of processes in nature. Also, as Charles Taylor has noted, a crucial characteristic of our secular age is the possibility of describing the cosmos in wholly immanent terms (especially when immanence is contrasted with transcendence) without any reference to a transcendent source of existence and meaning. This "immanent frame," as Taylor calls it,

> is the sense of an absence; it is the sense that all, order, meaning comes from us. We encounter no echo outside. In the world read this way, as so many of our contemporaries live it, that natural/ supernatural distinction is no mere intellectual abstraction. A race of humans has arisen which has managed to experience its world entirely as immanent. In some respects we may judge this achievement a victory of darkness, but it is a remarkable achievement nonetheless. (Taylor 2007, 376)[32]

Creatures of pure immanence still may be haunted, according to Taylor, by a longing for the transcendence rejected, but part of the problem, I think, is the notion that one must choose between transcendent and immanent sources of meaning. As we have seen, to view God's agency in the world essentially as designer or orderer has provided intellectual support for the view that we can just as well do without any reference to God, or to a god so conceived. As I have argued, however, the god so rejected is hardly the God about whom Thomas speaks.

If we follow Thomas's lead, we can see that there is no need to choose between a robust view of creation as the constant exercise of divine omnipotence and the causes disclosed by the natural sciences. God's creative power is exercised throughout the entire course of cosmic history, in whatever ways that history has unfolded. No matter how random one thinks evolutionary change is, for example; no matter how much one thinks that natural selection is the master mechanism of change in the world of living things; the role of God as Creator, as continuing cause of the whole reality of all that is, is not challenged. We need to remember Thomas's fundamental point that creation is not a change, and thus there is no possibility of conflict between the explanatory domain of the natural sciences—the world of change—and that of creation.

In a 2008 speech before the Pontifical Academy of Sciences, Pope Benedict XVI emphasized the continuing importance of Thomas's analysis of creation:

> To state that the foundation of the cosmos and its developments is the provident wisdom of the Creator is not to say that creation has only to do with the beginning of the history of the world and of life. It implies, rather, that the Creator founds these developments and supports them, underpins them and sustains them continuously. Thomas Aquinas taught that the notion of creation must transcend the horizontal origin of the unfolding of events, which is history, and consequently all our purely naturalistic ways of thinking and speaking about the evolution of the world. Thomas observed that creation is neither a movement nor a mutation. It is instead the foundational and continuing relationship that links the creature to the Creator, for he is the cause of every being and all becoming. (Benedict XVI 2008)

CREATION AND GENESIS

It is appropriate here to make a brief comment about Thomas's approach to the account of creation in the opening of the Book of Genesis. As is well known, some defenders as well as critics of evolutionary biology think that Genesis is incompatible with an evolutionary account of the origin and development of living things. This conclusion rests, in part, on the view that an essential part of the Genesis story is that God has directly intervened in the unfolding of nature to produce different distinct species. This direct divine intervention—direct, in the sense of not involving any natural causes—is often called "special creation." Viewing Genesis as an account in natural history leads to a variety of contrasts with the natural history that has become part of evolutionary biology. Thus, we have the conclusion that one must choose between seemingly competing historical accounts of how the great diversity of living things has come to be. After Darwin, so the argument goes, we can no longer accept the unscientific, indeed false, view that denies the origin of diversity in terms of evolutionary processes (without any need to appeal to special divine action). Thomas offers a particularly useful insight in how to approach what the Bible says about creation, an approach that is not challenged by developments in the natural sciences (either in his own times or now).

In his *Commentary on the Sentences of Peter Lombard*, Thomas sketches the debate between two traditions, one favored by Albert the Great, the other by Bonaventure, on "whether all things were created simultaneously and as distinct species." In his reply, he observes:

> There are some things that are by their very nature the substance of faith (*substantia fidei*), as to say of God that He is three and one, and other similar things, about which it is forbidden for anyone to think otherwise. . . . There are other things that relate to the faith only incidentally . . . and, with respect to these, Christian authors have different opinions, interpreting the Sacred Scripture in various ways. Thus with respect to the origin of the world (*circa mundi principium*), there is one point that is of the substance of faith, *viz.*, to know that it began by creation (*mundum incepisse creatum*), on which all authors in question are in agreement. But the manner and the order according to which creation took place concerns the faith only incidentally (*non pertinet ad fidem nisi per accidens*), in so far as it has been

recorded in Scripture, and of these things aforementioned authors, safeguarding the truth by their various interpretations, have reported different things. (Aquinas 2013d, II, 12, 1, a. 2)

Thomas notes that although the interpretation regarding successive creation, or what we might call "episodic creation," is "more common, and seems superficially to be more in accord with the letter [of the biblical text]," still that of simultaneous creation is "more conformed to reason and better adapted to preserve Sacred Scripture from the mockery of infidels" (Aquinas 2013d, II, 12, 3, a. 1). His firm adherence to the truth of scripture without falling into the trap of literalistic readings of the text offers valuable correction for exegesis of the Bible that concludes that one must choose between the literal interpretation of the Bible and modern science. For Thomas, the literal meaning of the Bible is what God, its ultimate author, intended the words to mean. The literal sense of the text includes metaphors, similes, and other figures of speech useful to accommodate the truth of the Bible to the understanding of its readers. A common example of such accommodation is that when one reads in the Bible of God's stretching out His hand, one ought not to think that God has a hand. The literal meaning of such passages concerns God's power, not His anatomy. Nor ought one to think that the six days at the beginning of Genesis literally refer to God's acting in time, for God's creative act is outside of time.

As biblical exegete, Thomas is in many ways representative of medieval biblical hermeneutics, which do indeed differ from contemporary approaches to scriptural interpretation. An analysis of Thomas, biblical scholar, is well beyond the scope of this essay. The point I wish to make here is the distinction Thomas draws between "the fact of creation" and the "manner and order" of the formation of the world; this is a distinction often lost among those who argue for an irreconcilability between creation in Genesis and evolutionary biology.[33]

Thomas, following the lead of Augustine, thinks that the natural sciences serve as a kind of veto in biblical interpretation. Augustine observed that when discussing passages of the Bible that refer, or seem to refer, to the natural phenomena one should defer to the authority of the sciences, when available, to show what the texts cannot mean. In examining, for example, whether the light spoken of in the opening of Genesis (before the creation of the Sun and the Moon) is physical light, Augustine says that if physicists show us that there cannot be physical light without a luminous source then we know that this particular passage does not refer to physical light (Augustine 1982, 1:38–39). The Bible cannot authentically be understood as affirming as true what the natural sciences teach us is false.

CONCLUSION

For Thomas Aquinas, the complete dependence of all that is on God does not challenge an appropriate autonomy of natural causation; God is not a competing cause in a world of other causes. In fact, God's causality is such that he causes creatures to be the kind of causal agents that they are. In an important sense, there would be no autonomy to the natural order were God not causing it to be so: there would be no evolutionary processes at all were God not creating them to be and to be what they are. Traditional conceptions of God as Creator certainly need not be abandoned in order to embrace an evolving universe in which real novelty and contingency are characteristic features of nature. For Thomas, the natural sciences, philosophy, and theology discover complementary, not competing, truths about nature, human nature, and God. The account he offers of divine agency and the autonomy and integrity of nature is not merely an artifact from the past, but an enduring legacy. Questions concerning the creation of all that is are distinct from those concerning design, order, and chance in the universe. Not only is there no contradiction between creation and evolution, without creation there would be no evolution. After Darwin, we are able to see, perhaps more clearly than before, Thomas's point that creation is not a change, and that as a result of the ongoing creative act of an omnipotent God nature possesses a dynamism and self-sufficiency that have been ever more evident to science.

NOTES

1. Wright thinks that, for the "grand bargain" to work, proponents of evolution would have to accept that "some notions of 'higher purpose' are compatible with scientific materialism." He cites the remarks of Steven Pinker: "There may be a sense in which some moral statements aren't just . . . artifacts of a particular brain wiring but are part of the reality of the universe, even if you can't touch them and weigh them" (Wright 2009, 9).

2. For an excellent discussion of the Stoic features of the view of creation as the bestowal of order, and how this differs from the traditional Christian notion of creation as the bestowal of existence, see Sloan (1985).

3. This is one of the important claims made by Buckley (1987).

4. "The great conflicts of the past between science and religion, first over Copernicanism, and later over Darwinism, have involved what have seemed to be insoluble conflicts between two competing explanations of the same body of phenomena—the motions of the heavens and earth; the origin, distribution and development of living beings in relation to the history of the earth. I would suggest, however, that the issues presented by contemporary life science are not of the same character as represented by these classic cases. In our present context, a new level of conflict between theology and science is being generated not by any single issue or theory—it can be argued that molecular biology is not even governed by a unifying idealizing theory—but by the convergence of a wide range of inquiries—evolutionary biology, molecular genetics, reductive physiology, naturalistic scientific cosmology, and cognitive neuroscience—in a totalizing naturalistic world view that claims to give a comprehensive explanation of all aspects of existence" (Sloan 2000, 25).

5. "La nature est comprise comme auto-créatrice, ce terme connotant que la notion classique de création est devenue inutile. La Nature—et il convient d'écrire le mot avec un majuscule—est autosuffisante pour produire non seulement ses effets, mais pour se produire. . . . La notion de création disparaît dans cette perspective de la réflexion."

6. In an important sense, once one admits that the question "why is there anything at all?" is a meaningful question (as distinct from simply affirming the brute fact of existence), one is already a theist. Here is how Denys Turner explains it: "For since any question which is not merely idle must have an answer, you have conceded, in conceding that the question is intelligible, that there is an answer: the world is created out of nothing. For if it is a valid question—that is to say, if nothing in the nature of the question itself places it beyond the bounds of sense—then human reason by the very fact of asking it has already been placed outside the universe of what there is, *whatever* there is: reason is, as it were, displaced, forced out of its natural, intra-mundane situatedness, forced by this question to confront the mystery

that there is anything at all. What the question's legitimacy expresses, therefore, is a sense of the world's radical contingency—there might have been nothing at all, so its existence must have been brought about. But to get at the precise form of that contingency which forces us to conclude that the world has been brought about, we should note that our everyday notions of contingency cannot capture the sense of the *world's* contingency as such, for they are . . . intra-mundane, they concern the contingency of things and events *in* the world. And even in that intra-mundane connection contingency is a pluriform concept. For how things 'might have been otherwise' can come in kinds and degrees" (Turner 2004, 242–43, italics in original).

7. Throughout this paper I leave aside questions such as the role of God as Unmoved Mover being a cause of all motion and change. My concern is the metaphysical topic of God as cause of existence.

8. "God's will is to be thought of as existing outside the realm of existents, as a cause from which pours forth everything that exists in all its variant forms. Now what can be and must be are variants of being, so that it is from God's will itself that things derive whether they must be or may or may not be and the distinction of the two according to the nature of their immediate causes. For He prepares causes that must be for those effects that He wills must be, and causes that might cause but might fail to cause for those effects that He wills might or might not be. And it is because of the nature of these causes that these effects are said to be effects that must be and those effects that need not be, although all depend upon God's will as primary cause, a cause which transcends the distinction between must and might not. But the same cannot be said of human will or any other cause, since every other cause exists within the realm of must and might not. So of every other cause it must be said either that it can fail to cause, or that its effect must be and cannot not be; God's will however cannot fail, and yet not all His effects must be, but some can be or not be" (Aquinas *Commentary on Aristotle's 'De Interpretatione'*, Book I, lectio 14, in McDermott 1993, 282).

9. Harm Goris notes that the distinction between divine causality and creaturely causality is based on the distinction between divine being and creaturely being: "Aquinas distinguishes the being of the Creator from the being of the creature not in terms of necessary being versus contingent being but more radically in terms of being versus non-being, while God causes the either necessary or contingent being of the creature. Likewise divine causation differs from creaturely causation as being differs from non-being. Without God's causation there is no creaturely causation at all" (Goris 1996, 299).

10. In discussing how the human will is free to choose, and yet caused to be so by God, Thomas notes that the autonomy of the will does not require that it be the "first cause" of its activity: "Not every principle is a first principle. . . . [A]lthough it is essential to the voluntary act that its principle

be *within* the agent, nevertheless it is not contrary to the nature of a voluntary act that this principle be caused or moved by an extrinsic principle: because it is not essential to the voluntary act that its intrinsic principle be a first principle" (Aquinas 2013e, I-II, q. 6, a. 1, ad 1). In a later question, Thomas writes: "God is the first cause of both natural causes and voluntary agents. And just as His moving natural causes does not prevent their acts from being natural, so also His moving voluntary agents does not prevent them from acting voluntarily, but rather makes it be just that, for He works in each according to its nature" (Aquinas 2013e, I, q. 83, a. 1, ad 3). Indeed, "every movement either of will or of nature proceeds from God as the First Mover" (ibid.). What is crucial for Thomas, however, is that we recognize that both natural and voluntary movements proceed from an *intrinsic* principle, but that need not, indeed cannot, be the truly first principle of action.

11. William Stoeger, S.J., suggests that "God is more like a verb, a continuing action in which everything else participates, but participates according to its own individuality. God's primary causality does not substitute for nor interfere with nor countermand the integrity and adequacy of the (secondary) causal structures of nature or of history—despite being their ultimate foundation or source" (Stoeger 2008, 229).

12. I have in mind here those who think that quantum indeterminism or chaos theory provide a new kind of "metaphysical space" in the world in which God can act without interfering with natural processes.

13. As Denys Turner puts it, Thomas's apophaticism "begins with the proposition that God can be demonstrated to exist, but that what such inference to God succeeds in showing is precisely the unknowability of the God thus shown." Turner compares Kant with Thomas on this topic: "It is a conflict in which the opposing sides occupy some common ground. Thomas and Kant contest the common territory of the unknowability of God. For both, God could not be the cause of all that is in the sense in which anything in the world is a cause. For both, then, what a cause in the world explains could not in the same sense of 'explanation' be what God's existence explains. . . . For both, what reason knows is all the world needs by way of explanation as to *how* it is, and God is not something known in any of the ways in which the world needs to be known; except for Thomas, the mystery *that* it is at all compels upon reason an acknowledgement that its deficiency is already theological: but not for Kant. . . . For Kant, speculative reason's falling short of God consists in the impossibility that the transcendental conditions of human knowledge and agency—the conditions of the possibility of our knowing the world and of acting as free agents within it—could themselves be an objective of our knowledge and agency in the world. Hence, they cannot be an object of knowledge; not one arrived at, therefore, even by inference, whether from the nature of things, or from the fact of the existence of things rather than of nothing" (Turner 2004, 254). On this very point, Kant

parts company from Thomas. Turner thinks that what is called the "Derridean dilemma" in postmodern thought remains fundamentally a Kantian dilemma: "it has seemed that proof and unknowability work against each other: that proof might be had at the price of an 'onto-theological', and so idolatrous, theism, or else that resistance might be made to 'onto-theology', but only at the price of abandoning the possibility of proof. Or . . . the dilemma is whether to say that the 'there is' in 'there is a God' lies 'within' language, or 'outside' it, either answer having unacceptable consequences. For if 'there is' lies within language, and so retains its connective tissue unbroken with our ordinary sense of 'there is', then this would appear to buy into a 'Scotist' and onto-theological univocity; whereas if we seek to evade this horn of the dilemma by saying that 'there is' lies on the other side of language, then we become impaled on the dilemma's other horn. For in breaking the tissue of connection with our ordinary meanings of 'there is', the existence of God is placed beyond the reach of any possible proof precisely because it is placed beyond the reach of language" (Turner 2004, 255–56). For Turner: "the means of escape from the 'Derridean dilemma' is *through* its horns, as the classical logicians used to say. For if the argument-strategy consists in the justification principally of a *question*—the question of 'Why anything?'—then we can say that it is the question which lies on the 'inside' of language, and so of reason, and so of logic, and it is the answer which must lie on the other side of all three. Hence, while the question retains its lines of continuity with our ordinary causal questions, the answer does not and could not do so. In short, the existence of God is in the nature of a demonstrated unknowability. *Et hoc omnes dicunt Deum*" (Turner 2004, 256).

14. I leave aside here discussions in Thomistic natural philosophy concerning species, natural substances, change, the distinction between living and non-living beings, and the like, and their relationship to contemporary science. On these subjects one should consult Connell (1988) and Wallace (1996).

15. John Haught thinks that evolution and cosmology disclose an unfinished universe such that one ought not so much to speak of *esse*, or being, in some timeless sense: "being must in some way mean the still-to-come; *esse est advenire*" (Haught 2003, 170). In an earlier work, Haught distinguished among a "metaphysics of the past" (that of the materialists), a metaphysics of an eternal present (which he identifies as a metaphysics of *esse* [being]), and a "metaphysics of the future"; the latter is the position he adopts (Haught 2000, 84). Haught, following the analysis of Gerhard von Rad (1965), thinks that the doctrine of creation is subordinate to soteriology. One should view it as one of God's saving acts, a kind of consummation of His saving work, and thus locate it as a future goal. Thus, this "soteriologization of creation" offers a biblical warrant that the best metaphysics for interpreting creation is a "metaphysics of the future" (Haught 2000, 84–93).

16. Jerry Korsmeyer (1998) and Denis Edwards (1999) argue that the kind of autonomy the universe possesses is only possible if God is "self-limiting." Edwards argues that a loving God must limit Himself in order to provide the space for creation and thus respect "the integrity of nature, its processes, and its laws" (Edwards 1999, 42). God persuades, He does not coerce.

17. As Clayton continues: "It seems to me, finally, that emergence theory tends to undercut dogmatic knowledge claims about the nature of God. Such claims tend to presuppose that one can have timeless knowledge, a view that implicitly lifts the epistemic agent above and hence out of the flow of history in which she is immersed. If emergence is right, our epistemic situation is constantly changing, in so far as we are products of a pervasive process of biological and cultural evolution. Acknowledging this fact should make one suspicious of any knowledge claims that imply, however tacitly, that the knower stands above the march of history and has direct and immediate access to timeless truths" (Clayton 2006, 320). Denis Edwards provides an example of panentheism when he argues that "the universe can be understood as unfolding 'within' the trinitarian relations of mutual love." Creation is the self-expression of the Trinity, which is divine Persons-in-Relation. All of this suggests, according to Edwards, "a worldview in which relations and praxis are primary." For Edwards, evolutionary biology discloses the primacy of relations and systems, a reality that has its foundation in the life of the Trinity (Edwards 1999, 30, 33).

18. This is a point made as well by Ted Peters and Martinez Hewlett: "The fallacy in the divine self-limitation approach is that it presupposes a conflict between divine power and creature power; whereas the classic Christian view, we contend, emphasizes that God's power empowers and thereby liberates God's creatures. The fallacy presupposes a fixed pie of power. According to the fixed pie image, if God gets a big slice then creation gets a proportionally smaller slice" (Peters and Hewlett 2003, 143). See also Gregersen (2006).

19. Smedes argues that those who think that quantum indeterminacy or chaos theory can provide a kind of metaphysical space in which God can act suggest that God's action "competes with the laws of nature [and is] on the same ontological level as the workings of the natural order" (Smedes 2004, 198). On this latter point, see Carroll (2008).

20. Thomas draws an analogy from the sun. Just as the air is lighted as long as it is illuminated by the sun, and falls into darkness when the sun does not shine at night, so creatures are caused to be by the creative diffusion of God's goodness. If God were to withdraw His presence all creatures would fall into non-being (Aquinas 2013e, I, q. 104, a. 1). Aquinas is influenced in this analysis by the works of Pseudo-Dionysius. Referring to Thomas's

analogy of the sun's illumination of light, Fran O'Rourke observes: "As the sun is naturally luminous, while air is lighted by sharing in the light of the sun although it does not partake of its nature, so also God alone is by his essence Being, while every creature is being through participation since its essence is not identical with its *esse*. Beings do not share in divine essence but in the illuminative effusion of divine Being which emanates from him. . . . God is present in all things not according to his essence but through a participation of his created likeness. . . . Divine similitude is not just a gift bestowed upon beings, but is their very being itself. . . . Creatures participate in God's presence but God is not participated. Beings share in the similitude of God while God in no manner resembles them" (O'Rourke 1992, 257–58).

21. As Denys Turner aptly notes: "God's intimacy to the world as Creator is the foundation of that ultimate intimacy of God to creation which is the incarnation" (Turner 2004, 258).

22. In an essay in the *Revue thomiste,* Jean-Michel Maldamé notes that a proper understanding of the relationship between evolution and creation requires that we recognize two distinct but complementary orders of causality, and that it is the failure to recognize the existence of these two orders that results in confusion on all sides of the current debate. "On doit donc reconnaître deux ordres de causalité: d'une part, l'ordre de la causalité des phénomènes décrits par les faits et les lois qui les régissent, et, d'autre part, l'ordre de la causalité première qui rend raison de leur sens" . . . . [L]'intervention de Dieu n'est pas une falsification des processus naturels. Si les phénomènes élémentaires sont aléatoires, ils restent aléatoires" (Maldamé 2007, 553, 554). Aquinas: "it would be contrary to the character of divine providence if nothing were to be fortuitous and a matter of chance in things" (Aquinas 2013b, III, c. 74). "If some things did not occur in rare instances, all things would happen by necessity. Indeed, things that are contingent in most cases differ from necessary things only in this: they can fail to happen, in a few cases. But it would be contrary to the essential character of divine providence if all things occurred by necessity, as we showed" (c. 72). As Maldamé continues: "le refus par certains de la théorie de l'évolution à cause du caractère aléatoire des phénomènes élémentaires est dû à leur incapacité de comprendre la distinction qui doit être faite entre la manière d'agir et la capacité d'agir" (Maldamé 2007, 557). As Aquinas writes about the divine will—there are some things on which the divine will confers necessity and others on which it does not.

23. "God and creatures are not two causes collaborating on the same level to produce a joint effect. God causes on the transcendental level and He thereby constitutes the creatures' causation on the categorical level" (Goris 1996, 301).

24. Shanley argues that no real explanation of exactly *how* God's causality functions is possible, since God transcends the mundane world of causation.

Michael Miller has argued that Bernard Lonergan, following in the tradition of Thomas, provides a more philosophically satisfying account of divine causation without sacrificing divine transcendence (Miller 1999). David Burrell observes that the "terms 'primary' and 'secondary' [causality] come into play when we are faced with the situation where one thing is by virtue of the other. So each can properly be said to be a cause, yet what makes one secondary is the intrinsic dependence on the one which is primary. This stipulation clearly distinguishes a secondary cause from an instrument, which is *not* a cause in its own right: it is not the hammer which drives the nails but the carpenter using it" (Burrell 1993, 97). See also Carroll (1999).

25. The dignity of being a cause God imparts to His creatures "not on account of any defect in God's power, but by reason of the abundance of His goodness" (Aquinas 2013e, q. 22, a. 3).

26. God so constitutes the being of creatures that they tend to exist and not to fall into nothingness. "The natures of creatures manifest that no creatures are degenerating into nothing, either because they are immaterial beings, in which there is no potency to non-being, or because they are material beings, and these remain in existence, at least in their matter, which is incorruptible" (Aquinas 2013e, I, 2. 104, a. 4, sol). See also *De potentia Dei*, q. 5, a. 4 (Aquinas 2013a). An illustration of the fact that in Aquinas's doctrine being belongs essentially to the creature can be found in *De potentia Dei* (Aquinas 2013a, q. 5, a. 3), where he asks whether God can return the creature to nothing. When Thomas answers this question he rejects the view of Avicenna, who had argued that the essence of the creature is of itself a pure possibility toward either being or non-being. Thomas agrees with Averroes in thinking that some creatures, such as immaterial substances and heavenly bodies, have an inherent necessity for existing, for there is in them no possibility for corruption. Thomas, however, carries Averroes' point further, and argues that no creature, whether material or immaterial, has any sort of potency for non-being: "in the whole of created nature, there is no potency through which it is possible for something to tend into nothing" (Aquinas 2013a, *De potentia Dei*, q. 5, a. 3, sol). It is true that material bodies tend to corrupt, but matter itself, prime matter, is incorruptible. The whole of the universe, considered in itself, has its own being and tends to continue in being. Of itself, it has no potency, or tendency, to non-being. However true it may be to say that the creature would be absolutely nothing without the creative causality of God, still, the creature really has its very own being. Thus, since creatures do have their own being, they are able to be true, autonomous causes.

27. "This oscillation runs throughout philosophy both in the East and in the West. On the one extreme, we have pure naturalism, where the creature is so autonomous that it no longer has any link with the divine whatsoever. On the other extreme, we have pantheism, where God absorbs every-

thing and is no longer distinct from what emanates from himself" (Sertillanges 1945, 62).

28. See also Gregersen (2000). Gregersen (2006, 25) offers the Lutheran eucharistic theology of consubstantiation as a model, as well as the Lutheran maxim: *finitum capax infinitum* [the finite is capable of hosting the infinite].

29. Jackelén rejects the term "theistic evolution" because, among other things, it involves notions of causality and creation that she thinks contemporary theology needs to avoid.

30. In *Dieu sans l'être* (Marion 1982), Marion argued that "by reversing the Pseudo-Dionysian priority of the good over being in his doctrine of the divine names, Aquinas had moved fatally away from the God of revelation and faith, who is fundamentally Love, towards the construction of the metaphysical idol of 'God' who would come to dominate modern thought" (qtd. in Shanley 1996, 617).

31. Kathryn Tanner, who has written persuasively on this subject, observes: "This non-competitive relation between creatures and God is possible, it seems, only if God is the fecund provider of *all* that the creature is in itself. . . . This relationship of total giver to total gift is possible, in turn, only if God and creatures are on different levels of being, and different planes of causality" (Tanner 2001, 3–4). For an excellent discussion of the transition between a Thomistic understanding of divine transcendence and a modern sense, especially beginning with Suarez, see Placher (1996).

32. Taylor thinks that various forms of what he calls "exclusive humanism" are attempts to fill the gap and offer only a kind of *ersatz* transcendence. He claims that "exclusive humanism closes the transcendent window, as though there were nothing beyond. More, as though it weren't an irrepressible need of the human heart to open that window, and first look, then go beyond. As though feeling this need were the result of a mistake, an erroneous world-view, bad conditioning, or worse, some pathology" (Taylor 2007, 638).

33. Throughout his commentary on Genesis, Thomas adheres to the following principle: there is a distinction between primary and secondary material in the Bible. When writing about codifying the articles of faith in a creedal statement, Thomas responds to the objection that "all things in Holy Scripture are matters of faith" and because of their multitude "cannot be reduced to a certain number." He writes: "[O]f things to be believed some of them belong to the faith, whereas others are purely subsidiary, for, as happens in any branch of knowledge, some matters are its essential interest, while it touches on others only to make the first matters clear. Now because faith is chiefly about things we hope to see in heaven, 'for faith is the substance of things hoped for,' [Hebrews 11:1] it follows that those things which order us directly to eternal life essentially belong to faith; such as three Persons of

almighty God, the mystery of Christ's incarnation, and other like truths. . . . Some things, however, are proposed in Holy Scripture, not as being the main matters of faith, but to bring them out; for instance, that Abraham had two sons, that a dead man came to life at the touch of Elisha's bones, and other like matters narrated in Scripture to disclose God's majesty or Christ's incarnation" (Aquinas 2013e, II-II, q. 1, a. 6, ad 1).

REFERENCES

Aquinas, T. (2013a). *S. Thomae de Aquino Opera Omnia.* Available at http://www.corpusthomisticum.org/iopera.html.

———. (2013b). *Summa contra Gentiles.* In Aquinas 2013a.

———. (2013c). *Commentary on Aristotle's De Interpretatione.* In Aquinas 2013a.

———. (2013d). *Commentary on the Sentences of Peter Lombard.* In Aquinas 2013a.

———. (2013e). *Summa theologiae.* In Aquinas 2013a.

Augustine, St. (1982). *On the Literal Meaning of Genesis.* 2 vols. Translated by J. H. Taylor. New York: Newman Books.

Ayala, F. (2007). *Darwin's Gift to Science and Religion.* Washington DC: Joseph Henry Press.

Benedict XVI, Pope. (2008). "Address to the Plenary Meeting of the Pontifical Academy of Sciences, 31 October 2008." Available at http://www.vatican.va.

Bonaventure, St. (1934). *Opera theologica selecta.* 4 vols. Edited by L. M. Bello. Florence: Quaracchi.

Buckley, M., S.J. (1987). *At the Origins of Modern Atheism.* New Haven, CT: Yale University Press.

Bulgakov, S. (2002). *The Bride of the Lamb.* Translated by Boris Jakim. Grand Rapids, MI: Eerdmans.

Burrell, D. (1993). *Freedom and Creation in Three Traditions.* Notre Dame, IN: University of Notre Dame Press.

Carroll, W. E. (1999). "Aquinas on Creation and the Metaphysical Foundations of Science." *Sapientia* 54:69–91.

———. (2008). "Divine Agency, Contemporary Physics, and the Autonomy of Nature." *Heythrop Journal* 49:582–602.

———. (2010). "Creation and the Foundations of Evolution." *Angelicum* 87:45–60.

Clayton, P. (2006). "Emergence from Quantum Physics to Religion: A Critical Appraisal." In *Re-Emergence of Emergence*, ed. P. Clayton and P. Davies, 303–22. Oxford: Oxford University Press.

Connell, R. (1988). *Substance and Modern Science*. Houston, TX: Center for Thomistic Studies.

Delio, I. (2003). "Does God 'Act' in Creation? A Bonaventurian Response." *Heythrop Journal* 44:328–44.

Dodds, M., O.P. (2012). *Unlocking Divine Action*. Washington DC: Catholic University of America Press.

Edwards, D. (1999). *The God of Evolution: A Trinitarian Theology*. Mahwah, NJ: Paulist Press.

———. (2010). *How God Acts: Creation, Redemption, and Special Divine Action*. Minneapolis, MN: Fortress Press.

Goris, H. J. M. J. (1996). *Free Creatures of an Eternal God: Thomas Aquinas on God's Infallible Foreknowledge and Irresistible Will*. Nijmegen: Stichting Thomasfonds.

Gregersen, N. H. (2000). "God: The Creator of Creativity." In *Evolution and Creativity: A New Dialogue between Faith and Knowledge*, ed. C. W. du Toit, 25–56. Pretoria: University of South Africa Press.

———. (2006). "The Complexity of Nature: Supplementing the Neo-Darwinian Paradigm?" *Theology and Science* 4:5–31.

Haught, J. (1995). *Science and Religion: From Conflict to Conversation*. New York: Paulist Press.

———. (2000). *God After Darwin. A Theology of Evolution*. Boulder, CO: Westview Press.

———. (2003). *Deeper Than Darwin: The Prospect of Religion in the Age of Evolution*. Boulder, CO: Westview Press.

———. (2010). *Making Sense of Evolution: Darwin, God, and the Drama of Life*. Louisville, KY: Westminster John Knox Press.

Heidegger, M. (1969). *Identity and Difference*. Translated by Joan Stambaugh. New York: Harper and Row.

Jackelén, A. (2007). "A Critical View of 'Theistic Evolution.'" *Theology and Science* 5:151–65.

Kauffman, S. (2008). *Reinventing the Sacred: A New View of Science, Reason, and Religion*. New York: Basic Books.

Korsmeyer, J. (1998). *Evolution and Eden: Balancing Original Sin and Contemporary Science*. Mahwah, NJ: Paulist Press.

Maldamé, J-M. (2006). *Création et Providence: Bible, science et philosophie*. Paris: Cerf.

———. (2007). "L'évolution et la question de Dieu." *Revue thomiste* 107:531–60.

Marion, J.-L. (1982). *Dieu sans l'être*. Paris: Fayard.

McDermott, T., ed. (1993). *Aquinas: Selected Philosophical Writings*. Oxford: Oxford University Press.

Miller, M. (1999). "Transcendence and Divine Causality." *American Catholic Philosophical Quarterly* 73:537–54.

Murphy, N. (1995). "Divine Action in the Natural Sciences." In *Chaos and Complexity: Scientific Perspectives on Divine Action*, ed. R. J. Russell, N. Murphy, and A. Peacocke, 324–57. Vatican City: Vatican Observatory Publications.

Nagel, T. (2012). *Mind and Cosmos: Why the Materialist Neo-Darwinian Conception of Nature Is Almost Certainly False.* Oxford: Oxford University Press.

O'Donnell, J. J., S. J. (1983). *Trinity and Temporality: The Christian Doctrine of God in the Light of Process Theology and the Theology of Hope.* Oxford: Oxford University Press.

O'Rourke, F. (1992). *Pseudo-Dionysius and the Metaphysics of Aquinas.* London: E. J. Brill.

Peters, T., and M. Hewlett. (2003). *Evolution from Creation to New Creation: Conflict, Conversation, and Convergence.* Nashville, TN: Abingdon Press.

Petrosino, S. (2001). "Is Creation a Negation?" *Communio* 28:311–23.

Placher, W. (1996). *The Domestication of Transcendence.* Louisville, KY: Westminster Press.

Rad, G. (1965). "The Theological Problem of the Old Testament Doctrine of Creation." In *The Problem of the Hexateuch and Other Essays,* trans. E. W. Trueman Dicken, 131–43. New York: McGraw-Hill.

Sertillanges, A. D. (1945). *L'idée de création et ses retentissements en philosophie.* Paris: Aubier.

Shanley, B. J., O.P. (1996). "St. Thomas Aquinas, Onto-theology, and Marion." *The Thomist* 60:617–25.

———. (1998). "Divine Causation and Human Freedom in Aquinas." *American Catholic Philosophical Quarterly* 72:99–122.

Sloan, P. R. (1985). "The Question of Natural Purpose." In *Evolution and Creation*, ed. E. McMullin, 121–50. Notre Dame, IN: University of Notre Dame Press.

———. (2000). "Introductory Essay: Completing the Tree of Descartes." In *Controlling Our Destinies: Historical, Philosophical, Ethical, and Theological Perspectives on the Human Genome Project*, ed. P. R. Sloan, 1–26. Notre Dame, IN: University of Notre Dame Press.

Smedes, T. A. (2004). *Chaos, Complexity, and God: Divine Action and Scientism.* Leuven: Peeters.

Stoeger, W. (2008). "Conceiving Divine Action in a Dynamic Universe." In *Scientific Perspectives on Divine Action: Twenty Years of Challenge and Progress*, ed. R. J. Russell, N. Murphy, and W. R. Stoeger, S. J., 225–47. Berkeley, CA: Center for Theology and the Natural Sciences.

Tanner, K. (2001). *Jesus, Humanity and the Trinity.* London: Continuum.

Taylor, C. (2007). *A Secular Age.* Cambridge, MA: Belknap Press of Harvard University Press.

te Velde, R. A. (1995). *Participation and Substantiality in Thomas Aquinas.* Studien und Texte zur Geistesgeschichte des Mittelalters 46. Leiden: Brill.

Tugwell, S. (1988). *Albert and Aquinas: Selected Writings.* New York: Paulist Press.

Turner, D. (2004). *Faith, Reason and the Existence of God.* Cambridge: Cambridge University Press.

Wallace, W. (1996). *The Modeling of Nature.* Washington DC: Catholic University of America Press.

Ward, K. (1996). *Religion and Creation.* Oxford: Clarendon Press.

Wright, R. (2009). "Week in Review: A Grand Bargain Over Evolution." *New York Times,* August 23. Available at http://www.nytimes.com/2009/08/23/opinion/23wright.html.

Yates, J. C. (1990). *The Timelessness of God.* Lanham, MD: University Press of America.

THIRTEEN

EVOLUTIONARY THEISM AND THE EMERGENT UNIVERSE

Józef Życiński

The 150th anniversary of the publication of Charles Darwin's *The Origin of Species* has been celebrated in the context of an animated debate concerning both scientific and philosophical issues implied by the theory of evolution.[1] One finds a deep diversity of attitudes, both methodological and semantic, in the current debates on evolutionary theory. Semantic differences in the understanding of such basic terms as "chance," "necessity," "directedness," and "design" are emphatically exemplified in publications in which the expression "by chance" is used as a synonym of "by Darwinian processes" (Alexander 2008, 297, 305). As a result, in the radical critique from the Darwinian theory, "blind chance" is supposed to replace the omnipotent God guiding the evolutionary processes. In this chapter I will argue that such an interpretation is both scientifically and philosophically groundless.

POST-DARWINIAN DILEMMAS

All exaggeration causes troubles, whether it may refer to philosophical, theological, or scientific theories. When describing "quasi-religious enthusiasms of ultra-Darwinists" in the area of the theory of evolution,

Simon Conway Morris observes that their radical interpretations, their self-confidence, and maybe their pretension to offer an alternative "secular" religion are based on a very weak understanding of the complexity of reality. Thus, they disregard in fact a true scientific approach, and when talking about theology they combine "ignorance and derision" by referring to irrational clichés of so-called scientific creationism (Conway Morris 2003, 316) instead of serious theological reflections. Sometimes these creationist clichés are presented as ground-breaking discoveries in science. This was the case with Michael Behe, who in 1996 proposed his version of the hypothesis of the so-called intelligent design. He claimed that the hypothesis was one of the greatest achievements in the history of science that rivaled the discoveries of Newton, Einstein, Lavoisier, Schrödinger, Pasteur, and Darwin (Behe 1996, 232 ff). This opinion was supported in 2004 by William Dembski who, in *The Design Revolution*, identified the evolutionary designer with God (Dembski 2004).

Behe himself was much more cautious when he admitted that the cosmic design of evolution might be the work of satanic powers or even of mysterious forces described in the New Age ideology. In *The Edge of Evolution*, published in 2007, he proposed a radical revision in his earlier views (Behe 2007). After revising his initial interpretation, he finally declared readiness to accept as true the fundamental biological principles of the theory of evolution except for the stochastic interpretation of genetic mutations.

In the newest version of his argument, Behe claims that genetic mutations are not stochastic in nature but are guided by the evolutionary Designer. Some defenders of such an interpretation, who claim at the same time that our Earth is less than ten thousand years old, represent the Catholic Kolbe Center for the Study of Creation, founded in Woodstock, Virginia, in 2000. It is a form of intellectual drama that while Fr. Maximilian Kolbe himself tried to use all new achievements in science and technology to spread the Gospel of Jesus Christ, his name is being used by Hugh Owen, the founder of the Kolbe Center, as a symbol of religious opposition to modern scientific discoveries. The most painful component of the present practice is manifested in introducing false conflict between the world of science and the

Christian interpretation of nature. Such a praxis has been evidently inconsistent with John Paul II's teaching expressed in his famous letter to George Coyne, S J When the pope emphasized that the dialogue between science and faith should continue and grow in depth, he argued that these two important domains of the human search for truth "should continue to enrich, nourish and challenge the other to be more fully what it can be and to contribute to our vision of who we are and who we are becoming" (John Paul II 1988, 1711).

Nicholas Rescher seems closer to the methodology suggested by the pope when he rejects the opposition between evolutionary theory and the hypothesis of intelligent design. In one of his recent papers, Rescher argues that it would be "a profound error" to regard evolutionary theory and the design hypothesis as "conflicting and incompatible" (Rescher 2009, 101). One could try to reconcile evolution and design on the philosophical level. Nonetheless, in the epistemological perspective of scientific research, such a synthesis cannot be accepted due to the conflict with the principle of methodological positivism, which forbids scientists to refer to extra-natural phenomena to explain properties that occur in the world of nature.

When one tries to accept the concept of "design" on the level of philosophical explanation, many problems reappear that were discussed already in the time of William Paley. The basic questions concerning chance, teleology, and directedness were already criticized in the past as anthropomorphism that brought metaphors closer to poetic description of the world than to its rational explanation.

BEYOND PALEY'S PARADIGM

Contemporary attempts to identify mechanisms governing evolution are invariably dominated by an influence of the mentality in which mechanics was regarded as the highest achievement of science. The paradigm of mechanics is expressed by comparing physical relations to clockwork, as well as in publications inspired by the idea of *The Blind Watchmaker* developed in Richard Dawkins's famous bestseller (Dawkins 1986). Since 1802, however, when Paley in his *Natural Theology* presented the original version of the argument comparing nature to clockwork, not only has the concept of time been relativized, but also time measurement techniques have undergone deep changes. Paley's contemporaries did not understand that a clock must not necessarily be a set of cogged wheels but may be based on geological stratification or radioactive decay. Geological deposits may be a source of more precise information on the age of an environment than a system of cogged wheels, of interest to watchmakers. Thus the watchmaker metaphor may seem to be psychologically suggestive; its informative value, however, disappears when we propose the metaphor of a geologist or radioisotope expert (see Pratchett, Stewart, and Cohen 2005).

There are no substantial reasons to ascribe to the clockwork metaphor proposed by Paley any explanatory role. After the eighteenth-century fascination with physical mechanisms, in the twentieth century the perception of the universe as an organism emerged in philosophy. A classic example of such practice is Alfred North Whitehead's process philosophy. If we were to consider, following Whitehead, the universe as a living organism, Paley's problem of the cosmic watchmaker would naturally be replaced by the problem of the cosmic womb and quite different philosophical questions would emerge (Davies 1992, 201).

Texts dominated with poetic metaphors and analogies need to be complemented with reflection on the substantial content revealed by the universal and constant relations determined by the laws of nature. Regardless of the controversies accompanying the question of what the laws of nature are, it is possible to detect at different levels of evolution the structural relations that determine the nature of evolutionary mechanisms. However, where in the past a simple opposition of "teleology or determinism" was applied, nowadays emerges a vast spectrum of explanation patterns that has not previously been considered. One of the possibilities is created by admitting the existence of "open teleonomic systems" (Stoeger 1998, 183) while recognizing the importance of deterministic chaos, in other words, of deterministic (nonrandom) systems that behave as seemingly random ones. Many terms in the vocabulary of evolutionary concepts—terms previously believed fundamental—show their relative nature. For instance, Richard Dawkins admits that his bestseller *The Selfish Gene*, which started extensive debates on the role of the "selfish gene" in biological evolution, may have as well been entitled *The Cooperative Gene* (Coyne 2009, 247).

BEYOND CHANCE AND NECESSITY

In the texts by both the critics and the supporters of the theory of evolution, there are strong claims about blind chance, the blind forces of nature, or even the blind laws governing evolution. Charles Darwin unsuccessfully wrestled with those issues in his philosophical commentaries. He was intrigued by a deep split between the level of purposeful actions of man, the order of mathematizable laws of nature, and the so-called game of chance, whose inherent logic of events is impossible to determine. In his letters, he expressed astonishment at the fact that electricity worked according to the laws of physics, trees grew arousing our admiration, and all that originated from blind brute forces. Thus in a letter to William Graham, the author of *The Creed of Science,* Darwin thanked him for expressing the view that the universe was not the work of chance (Darwin to Graham, 3 July 1881, in Darwin 1887, 1:315). Nowadays, those issues are the subject of numerous controversies in which an important role is ascribed to random events. Their ultimate effects are not unequivocally determined by the very course of processes. Considered on a small scale, in a small dose, individual processes appear unpredictable and can thus make one inclined to speak of blind chance. However, having taken into account a greater number of events, it is possible to observe the *probabilistic compressibility of the universe*; the processes that earlier appeared accidental, uncoordinated, "blind," on a greater scale begin to reveal internal relations described in the language of probability theory. What earlier seemed a manifestation of physical disarray is compressible to the general formulae of the probability calculus. An apparent game of "blind" chance ceases to be blind when the applied formalism also reveals the *algorithmic compressibility* of nature,[2] making it possible to foresee its future states. Were the universe not algorithmically compressible, its every state would require a separate description, and it would be impossible to use the generalized expressions that would also refer to future, as yet unrealized processes.

We have become reconciled with the lack of algorithmic compressibility of nature in many other areas of life. In the field of aesthetics, we do not expect general laws of creating masterpieces, and in psychology we do not look for identical principles explaining reactions of all geniuses. The philosophical significance of the compressibility of phenomena, however, tends to be consistently ignored in philosophical commentaries on the theory of biological evolution. Their authors disregard an important element of rationality and mathematizability of the universe, an element that is a no less interesting philosophical phenomenon than the manifestations of the biological order that so greatly fascinated William Paley in his famous argument.

Considering the complexity of physical situations to which the concept of chance is applied, one must observe that processes believed to be accidental *in relation to* a given law of physics may, in a different system of principles, be due to the nonaccidental operation of the laws of physics. The laws of physics do not function in mutual isolation but constitute an overall (dynamic) structure called the nomic, or nomological, structure of nature. In the cosmic network of interactions, the necessary elements concur with the accidental ones; chance and necessity complement each other, also in a terminological sense, because what appears an accidental coincidence from one cognitive perspective is a necessary process in the light of knowledge of different laws.

Considering also metaphysical and theological aspects of the problem, it should be noted that, from the vantage point of theology, speaking of the role of "blind chance" in evolution does not make sense. The expression is deprived of theological meaning as it seems to suggest that God the Creator loses control over a certain class of accidental, random events; God, however, is present in the same way as in strictly deterministic and probabilistic laws. Coming into existence of particular conditions may be unlikely, but this does not justify the claim that the processes that produce them are "blind."

Not only does including stochastic processes and probable events in the process of biological evolution not destroy the great philosophical questions concerning the ultimate conditionings of the order of nature, but it also gives those questions a more sophisticated form. Witnessing the actualization of merely probable sequences of events that happen in nature, we become observers of the great process of creation that lasts, revealing the mysterious rationality of the universe in the sequences of events earlier considered "blind" and accidental. Innovative studies are being conducted in many fields revealing new manifestations of the cosmic order generated by stochastic relations that, until recently, have been believed symptoms of disorder. John Barrow, Michał Heller, and Joseph Silk seriously consider the possibility that the probabilistic relations at the deep level of nature may underlie the mathematical symmetries and universal laws that physics describes at the higher levels (Heller 2005, 59; Barrow and Silk 1983, 213). From this perspective, the fundamental "law of nature" would be the cosmic game of probabilities.

The mutual relationship between "accidentality" and contingency—the latter concept being very much discussed in the Christian tradition of metaphysics—leads to philosophically interesting consequences, justifying the claims about the nomic structure of the universe where deterministic and quasi-teleological processes are mutually complementary. The so-called bottom-up relations are expressed both in the deterministic and strictly deterministic processes, while the so-called top-down influence can be described in terms of the analogy of the attractor. Thus, the teleologically understood power of the Creator manifests itself in the laws of nature, constituting an overall structure, and not in gaps in our knowledge. The very concept of design does not seem to play any explanatory role in contemporary philosophical debates. In place of it I would like to propose the concept of emergence as a basic one.

EVOLUTION IN THE EMERGENT UNIVERSE

Roughly speaking, the term "emergence" recognizes that in evolving physical systems at each level of complexity new and unpredictable qualities emerge that cannot be explained by reference to the properties of the lower level of the system. The essence of philosophical emergentism, conceived of very often as a denial of reductionism, in its most compact version can be expressed in the words: *if emergence exists, absolute reductionism fails* (Rotschild 2006, 156).

The emergent structures are more than the sum of their parts because the emergent order will not arise if the various parts are simply coexisting; the interaction of these parts is central. The new qualities emergent in the evolving systems are irreducible to the system's constituent parts. Evolutionary emergence implies the successive appearance of organized systems that display properties nonexistent in the earlier stages of cosmic history.

In explaining the very nature of emergence, an important role is played by the concepts of supervenience and subvenience. According to McLaughlin, one can define the emergent properties of the evolving systems in the following way: "If P is a property of w, then P is emergent if and only if (1) P supervenes with nomological necessity, but not with logical necessity, on properties the parts of w have taken separately or in other combinations; and (2) some of the supervenience principles linking properties of the parts of w with w's having P are fundamental laws" (McLaughlin 1997, 39).

The subvenient level of being is a necessary, but insufficient, condition for the occurrence of the properties recognized as supervenient. The relation of supervenience is treated as asymmetric and transitive (Heil 1992, 65). Supervenience is necessary, but not sufficient for emergence. If M emerges from $N_2, \ldots N_k$, then M supervenes on N_s but is not reducible to N_s, nor is it explainable in terms of N_s.

In the evolutionary process of cosmic growth we observe the particular stages of cosmic and biological evolution: the emergence of physical particles from vacuum fluctuations, the emergence of stellar structures, the emergence of carbon forms of life, the emergence of human psychism. These processes could be described in ontologically neutral terms of philosophical emergentism. In the framework of philosophical interpretations where the principle of methodological naturalism, basic for the natural sciences, does not function, one could refer to a transcendent factor explaining the process of evolutionary emergence. On this level of reflection one could repeat John Haught's question: "Is cosmic emergence . . . an empty plunge toward final nothingness, or a response to something like an invitation?" (Haught 2006, 79).

There are two possible answers to the basic questions brought by emergentism: either (1) we will accept positivistically that emergence is nothing but emergence and further debate will be regarded as meaningless; or (2) we will accentuate the claim that to explain it philosophically requires a new element emerging on more complicated levels of evolution. That is, one has to refer to the Divine Logos that is immanently present in the laws of evolving nature. To defend the latter approach we need a new terminological basis in which outdated finalistic categories would be abandoned and novel categories provided by the natural sciences would be philosophically interpreted.

TOP-DOWN DETERMINISTIC INTERACTIONS

Within the framework of evolutionary emergentism, a great deal of attention is currently paid to the so-called top-down deterministic interactions in which the integrally understood structure of mutual interactions at a higher level of a system's development makes it possible to determine the direction of the development of lower constituent structures. The flow of information between different levels of structures is essential to the distinction between the top-down (i.e., downward) and the bottom-up (i.e., upward) interactions in the evolutionary process. The study of this relation is enabled by the fact that properties absent at the lower levels of structural complexity emerge at the higher levels; their emergence is not liable to reductionistic explanation referring to their components, as this type of explanation includes only bottom-up interactions.

When determinism and reductionism dominated the scientific paradigm, the mentioned relations were believed to be one of the paradoxes of contemporary science, and penetrating questions concerning the relationship between biology and transcendence, free will and physical determinism, and so forth, were formulated. However, with the progress of research, the importance of both types of interactions has been more and more appreciated, along with the recognition of the relative autonomy of structures that were interpreted reductionistically at the earlier stages of knowledge development. The need to step beyond the reductionist interpretative scheme was earlier admitted by those natural scientists who, on the one hand, did not want to reintroduce purpose-related categories into science and, on the other, who believed in the insufficiency of explanations in which one referred only to the principles of a reductionistic, bottom-up approach. To overcome the simplifications of the reductionist explanatory schemes some authors used so-called teleonomic categories, also called quasi-finalistic or quasi-purposive ones. These applied to the situation when the processes of change in the studied system proceeded *as if* they were oriented toward an as yet unachieved end that would physically come into existence only at later stages of evolution. Such an understanding of quasi-purposiveness did not merely express a linguistic purism; it was also of a heuristic value, since it enabled the selection, from the set of alternative scientific explanations, of those particularly valuable for science.

Among the authors who devoted their particular attention to the top-down interactions were such natural scientists as G. F. R. Ellis, Paul C. Davies, A. R. Peacocke, D. T. Campbell, F. Ayala, and T. Dobzhansky (see Davies 2006; Campbell 1974). They emphasize that only the holistic approach covering the complete structure of mutual interdependencies may satisfactorily explain the phenomena that occurred at the lower stages of evolution. Confining oneself only to the bottom-up approach would mean leaving off some important information (Foerst 2002, 40).

Sugar fermentation, nucleosynthesis in the cosmos, an influence of the holistic structures on their components, an influence of the mind on numerous functions of the body are given as classic examples of the top-down influence. The fact is that at the level of the human psyche, there are relationships that cannot be reduced to the level of physical-biological determinants.

It is important for the understanding of the structure of such processes that transformations occurring in nature depend not only on the events that have already happened, but also—although in a different way—on the events that will happen, leading to the formation of the complete structure. In the studied system, there are not only "bottom-up" determinants, which are traditionally of interest for physicists, but also "top-down" integral influences of the system on its individual components. It changes its character along with the development of emerging structures, and—as Jacques Monod expressed it—"everything happens as if living things were structured, organized and conditioned with a view to an end" (Monod 1968, 9).

Adopting the above-mentioned cognitive perspective, G. F. R. Ellis believes reductionism to be a symptom of a kind of fundamentalism, the former being an attempt to reduce human personality, morality, culture to the level of physical-biological phenomena (Ellis 2006, 104). The holistic approach investigating the top-down influence on the whole evolving system is still criticized by those physicists whose cognitive attention is absorbed by the functioning of strictly deterministic laws.

CONVERGENCE IN THE EMERGENT UNIVERSE

In the contemporary debate over biological evolution, there is a deep difference of opinion concerning so-called convergence. Taking a polemical stance against the works of S. J. Gould, Simon Conway Morris recognizes an essential role of convergence in the process of biological development. He convincingly argues that different evolutionary processes will follow the same pathways of growth and result in the repetition of the same effects. Referring to numerous examples from paleontology, Conway Morris claims that in evolutionary processes distant in space, time, and on the scale of belonging to a different genealogical tree, there are structural similarities expressed in the preference of certain paths of development. Thus there is a close dependence between the laws of physics and the conditions controlling biological evolution.

The laws of physics have a decisive influence on the fact that the structure of hyperspace is not a result of chance combinatorial possibilities. To determine this structure, it is necessary to consider not only the bottom-up influence of the laws of physics on the progressing organization of the system, but also a top-down influence of the emerging structures on their less organized components. Conway Morris expresses the essence of his convictions in stating that "convergence tells us two things: firstly—evolutionary trends are real, secondly—adaptation is not an accidental component of an organic structure, but is central to explaining the emergence of man" (Conway Morris 2003, xv).

Convergence is an expression of the structural directedness at the level of living organisms. Accepting it, Conway Morris does not proclaim pandeterminism. Neither does he attempt to subordinate billions of years of the cosmic development to one simple principle. However, he gives many examples of large-scale processes that, regardless of the profound differences in physical conditions, led to the same biological consequences at the level of organ function and to the presence of characteristic properties in certain living organisms. From this perspective, the process of evolution is not a simple game of chance and necessity, since some processes are preferred structurally and play a decisive role in the subsequent stages of evolution. The emergence of the same evolutionary structures appears in spite of differences in initial conditions.

DIVINE LOGOS AND THE EVOLVING UNIVERSE

If we bear in mind that an important form of God's action in nature is that the laws of nature determine its potential development, we must recognize God's presence not only in the strictly deterministic or stochastic processes, but also in those occurring as if they were directed toward as yet unrealized physical attractors.[3] The claim that God, understood immanently, "attracts" evolving systems is an expression of the same idea that Teilhard de Chardin expressed in his global characteristic of all evolutionary processes, speaking of the Omega Point of evolution.

"The combination of contingency and knowability makes us look for new unexpected forms of the rational order" (Barbour 1988, 34). This conviction about the fundamental character of the thesis about the rationality of the world was expressed with particular force in a lecture held by Benedict XVI in Regensburg on 12 September 2006. The Pope develops there an argument that action contradictory to the principles of rationality is inconsistent with the nature of God (Benedict XVI 2006). From such a perspective, evolutionary processes occur despite discontinuity, bifurcation, and stochastic processes.

In the evolving universe, a new horizon of hope, so close to Christianity, can be found. It makes us overcome determinisms of the past and gives a new system of reference in which our being and our culture are grasped in the perspective of the harmony and order introduced into the evolving nature by the Divine Logos. Then, the tension between the natural and the religious dimensions of events disappears, nature becoming one of the important places where the immanent God is present in the laws of nature. Theological reflection, consistently developed, makes us aware of the importance of the truth that nature is the arena of the history of salvation. In consequence—*extra naturam nulla salus*—there is no salvation outside of the created world. An evolutionary descendant of an amoeba has been endowed not only with an extraordinary gift of reflection, but also with a possibility to participate in God's life.

KENOSIS OF GOD IN EVOLVING NATURE

The suffering of human beings in evolving nature brings problems no less important than the issue of the mechanism of evolutionary processes. How could we reconcile the manifestations of the Divine Logos found in nature with the tragic pain experienced by so many creatures? In the model of the emergent universe one has to remember that God's immanence in physical laws accepts the drama of suffering as a natural consequence of interactions between the universal laws of nature and concrete physical conditions. Our anticipations of a world free from suffering are perhaps as unrealistic as were the pre-Gödelian dreams of logicians for a rich system of logic that would be both complete and consistent. The discovery of two incompleteness theorems in 1930 disclosed the unrealistic character of such expectations and shook the foundations of mathematics. Perhaps our yearning for the unrealized scenario of evolution free from suffering can be alleviated when we understand deeper the *kenosis* of God immanent in nature.

The very term *kenosis*, taken from the biblical Greek, means an impoverishment, denudation, or humiliation. From a theological perspective, the *kenosis* of Christ is based on the fact that He "emptied himself . . . humbled himself and became obedient unto death, even death on a cross" (Philippians 2:7–8). The *kenosis* of Christ, revealed not only in His passion but in His entire earthly life, was a result both of the acceptance of the limitations of human nature and of the cultural conditions in which Jesus systematically experienced the resistance of His contemporaries, who accepted a different hierarchy of values.

The significance of the *kenosis* of God is emphasized by John Paul II, who does this in theological formulations that shock our logic. In the final parts of the encyclical *Fides et Ratio*, he writes:

> the prime commitment of theology is seen to be the understanding of God's *kenosis*, a grand and mysterious truth for the human mind, which finds it inconceivable that suffering and death can express a love which gives itself and seeks nothing in return. . . . On this score, some problems have emerged in recent times, problems which are only partially new; and a coherent solution to them will not be found without philosophy's contribution. (John Paul II 1998, para. 93)

The question concerning the evolutionary explanation of the origin of suffering is only partially new; finding an answer to that question requires the introduction of a new explanatory perspective in which the concept of God's *kenosis* plays a very important role. Had Jesus chosen another way of fulfilling His mission, had He abandoned the *kenosis* that hid His divine traits, then probably the sphere of His influence would have been significantly greater. The *kenosis* of Christ's suffering at Golgotha and the *kenosis* of maximally simple means of action seems to be an important feature in Christ's message, the full significance of which we cannot yet fully comprehend.

At the present stage of knowledge, we cannot determine how alternative scenarios of evolution might have looked. We nevertheless know that God does not cease to love human beings and that, thanks to the undeserved gift of His grace, we can strive for that spiritual harmony that is not our natural state. The drama of sin is not an epilogue in evolutionary anthropology—it is the introduction to the truth about Divine grace, which opens before us a new horizon of existence, inaccessible at the purely natural level. The God kenotically hidden in nature engages human species in the process of evolution, which, on the level of our existence, manifests itself above all in the sphere of culture. In the emerging reality of cosmic growth, our pain is not eliminated but it receives a radically different meaning. The sense of rationality can be discovered both in the immanent divine Logos and in emerging structures of the evolving universe.

NOTES

1. Archbishop Życiński died unexpectedly on 10 February 2011. We are indebted to Msgr. Tomasz Trafny, head of the Science and Faith Department and executive director of the STOQ Project Pontifical Council for Culture, and a long-time associate of Archbishop Życiński, for final editing of this lecture manuscript. This volume is dedicated to his memory. See also the special issue of the journal *Logos* 15 (Winter 2012) devoted to his legacy in the science and theology dialogue.—*Ed. note*

2. Algorithmic compressibility is understood as the possibility of giving a simple economical description of complex physical phenomena by introducing a mathematical equation (algorithm) that describes important physical relations among the parameters in an evolving system. See Davies (1995, 258).

3. I have developed the conception of the evolutionary attractors in the study *God and Evolution* (Życiński 2006, 133 ff, 161–64).

REFERENCES

Alexander, D. (2008). *Creation or Evolution: Do We Have to Choose?* Oxford: Monarch Books.

Barbour, I. (1988). "Ways of Relating Science and Theology." In *Physics, Philosophy and Theology: A Common Quest for Understanding*, ed. R. J. Russell, W. Stoeger, and G. Coyne, 21–48. Vatican City: Vatican City Observatory.

Barrow, J., and J. Silk. (1983). *The Left Hand of Creation: The Origin and the Evolution of the Expanding Universe*. London: Unwin.

Behe, M. J. (1996). *Darwin's Black Box: The Biochemical Challenge to Evolution*. New York: Free Press.

———. (2007). *The Edge of Evolution: The Search for the Limits of Darwinism*. New York: Free Press.

Benedict XVI, Pope. (2006). "Faith, Reason and the University: Memories and Reflections." Meeting with the representatives of science in the Aula Magna of the University of Regensburg, in *Acta Apostolicae Sedis* 10 (XCVIII), 728–39. English translation available at: http://www.vatican .va/holy_father/benedict_xvi/speeches/2006/september/documents/hf_ ben-xvi_spe_20060912_university-regensburg_en.html.

Campbell, D. T. (1974). "Downward Causation in Hierarchically Organized Systems." In *Studies in the Philosophy of Biology: Reduction and Related Problems*, ed. F. Ayala and T. Dobzhansky, 179–86. London: Macmillan.

Clayton, P., and P. Davies, eds. (2006). *The Re-Emergence of Emergence: The Emergentist Hypothesis from Science to Religion*. Oxford: Oxford University Press.

Conway Morris, S. (2003). *Life's Solution: Inevitable Humans in a Lonely Universe*. Cambridge: Cambridge University Press.

Coyne, J. A. (2009). *Why Is Evolution True?* Oxford: Oxford University Press.

Darwin, F., ed. (1887). *The Life and Letters of Charles Darwin: Including an Autobiographical Chapter*. 3 vols. London: Murray.

Davies, P. (1992). *The Mind of God: Science and the Search for Ultimate Meaning*. London: Penguin Books.

———. (1995). "Algorithmic Compressibility, Fundamental and Phenomenological Laws." In *Laws of Nature: Essays on the Philosophical, Scientific and Historical Dimesions*, ed. F. Weinert, 248–67. Berlin: Walter de Gruyter.

———. (2006). "The Physics of Downward Causation." In Clayton and Davies 2006, 35–52.

Dawkins, R. (1986). *The Blind Watchmaker*. New York: Norton.

Dembski, W. (2004). *The Design Revolution: Answering the Toughest Questions about Intelligent Design*. Downers Grove, IL: Intervarsity Press.

Ellis, G. F. R. (2006). "On the Nature of Emergent Reality." In Clayton and Davies 2006, 79–110.

Foerst, A. (2002). "The Body of Christ: Embodied Robots and Theology." In *God, Life, Intelligence and the Universe*, ed. T. J. Kelly and W. D. Regan, 37–59. Adelaide: Australian Theological Forum.

Haught, J. F. (2006). *Is Nature Enough? Meaning and Truth in the Age of Science*. Cambridge: Cambridge University Press.

Heil, J. (1992). *The Nature of True Minds*. Cambridge: Cambridge University Press.

Heller, M. (2005). *Filozofia i wszech wiat: Wybór pism*. Kraków: Universitas.

Horn, S. O., and S. Wiedenhofer, eds. (2008). *Creation and Evolution: A Conference with Pope Benedict XVI in Castel Gandolfo*. Translated by M. J. Miller. San Francisco: Ignatius Press.

John Paul II, Pope. (1988). "Our Knowledge of God and Nature: Physics, Philosophy and Theology," Letter to the Director of the Vatican Observatory. Available at http://www.vatican.va/holy_father/john_paul_ii /letters/1988/documents/hf_jp-ii_let_19880601_padre-coyne_en.html.

———. (1998). *Encyclical Letter: Fides et Ratio*. In *Acta Apostolicae Sedis* 1 (XCI), 5–88. English translation available at http://www.vatican.va /holy_father/john_paul_ii/encyclicals/documents/hf_jp-ii_enc_ 15101998_fides-et-ratio_en.html.

McLaughlin, B. (1997). "Emergence and Supervenience." *Intellectica: Emergence and Explanation* 25, no. 2:25–43.

Monod, J. (1968). *Leçon inaugurale faite le vendredi 3 novembre 1967, Collège de France, Chaire de biologie moléculaire*. Paris: Collège de France.

Pratchett, T., I. Stewart, and J. Cohen. (2005). *The Science of Discworld III: Darwin's Watch*. London: Ebury Press.

Rescher, N. (2009). "On Evolution and Intelligent Design." In *Darwin and Catholicism: The Past and the Present Dynamics of a Cultural Encounter*, ed. L. Caruana, 95–106. London: T&T Clark.

Rotschild, L. J. (2006). "The Role of Emergence in Biology." In Clayton and Davies 2006, 151–65.

Stoeger, W. (1998). "The Immanent Directionality of the Evolutionary Process, and Its Relationship to Teleology." In *Evolutionary and Molecular Biology: Scientific Perspectives on Divine Action*, ed. R. J. Russell, 163–90. Vatican City: Vatican Observatory Press.

Życiński, J. (2006). *God and Evolution*. Washington DC: Catholic University of America Press.

FOURTEEN

BEYOND SEPARATION OR SYNTHESIS
Christ and Evolution as Theodrama

_____

Celia Deane-Drummond

The fervor with which popular discourse on science and religion has continued to bubble up in the anniversary year celebrating Darwin's achievements shows that the publically perceived conflict between science and religion will not go away. Academic discussion on such matters is therefore not just peripheral to cultural concerns but takes the form of public responsibility. Yet it is important to situate conflictual stances in an appropriate historical context. Biologist and theologian Arthur Peacocke is, I believe, correct to note that historically the conflict between Darwin's theory and Christian belief was not nearly as stark as subsequent biologists and popular perception have supposed (Peacocke 2005, 59–60). He suggests that the enmity that T. H. Huxley had towards Christianity was related to the desire to free professional biologists from any stranglehold of ecclesial control. Peacocke's reply, following Aubrey Moore, that Darwin is really a disguised friend, is one based on a profound belief that Darwin's key idea of evolution by natural selection forces religious believers to think more carefully about the meaning of the incarnation and develop a renewed sense of the sacramental presence of God understood as immanent or present in the world.[1] Those reading the most recent vitriolic outpourings of Richard Dawkins, such as his *The God Delusion*, might be forgiven for wondering what kind of friend Darwin might be for religious faith, unless it is seen in a paradoxical way of providing the opportunity to speak about what Christian belief in God as a Creator really entails.[2] This, of course, raises the issue of how far and to what extent Dawkins can speak either for Darwin or for Christian belief. Dawkins is particularly naïve in his portrayal of Christian theism as founded on claims that lack a rational basis, while denying the cultural ferment in which his own narrative is placed.

The fact that evolution is not inevitably atheistic is a matter that Darwin himself recognized, even though more evidence has come to light recently that he understood religion to be an *evolved* capacity, a view that he clearly understood as still being compatible with religious belief, even if it helped shift his own position towards agnosticism (Pleins 2013). David Pleins exposes the evolutionary naturalism that is inherent in Darwin's explanation of religion in common with many adherents of evolutionary psychology writing today. There are important distinctions that are worth mentioning. First, Pleins shows in a fascinating way how far Darwin's explanation of religious belief was still bound up with preconceived ideas about what "counts" as a mature religious belief and, perhaps not surprisingly, presents monotheism as the most advanced religious belief progressing from more "primitive" religious beliefs of other cultures. His Victorian view of the advancement of religion is naïve and culturally somewhat racist in its assumption that religious beliefs have progressed over time. Second, the evolutionary naturalism that Darwin and his successors were anxious to promote only becomes atheistic in as much as it insists on its exclusivity and materialism in its explanatory power. Darwin resisted taking this step himself, even though he tottered on the boundary of such a belief.

Terry Eagleton, writing from a nontheistic perspective, has exposed some of the glaring errors in more extreme versions of evolutionary naturalism represented by new atheist positions, such as Dawkins's portrayal of God, not least his flawed understanding that God intervenes directly in the natural world.[3] The creationist movement holds to such a view, and those creationists who support intelligent design seem to reserve such intervention by God or an intelligent mind for the appearance of, for example, complex organs in evolutionary history. The latter is seemingly an elaboration of the "God of the gaps argument," but the prospect that God intervenes directly is not faithful to the classical Christian tradition. This tradition, as elaborated clearly in William Carroll's and John O'Callaghan's contributions to this volume, envisages God not as a being among other beings, but the ground of all that exists as primary cause. While the threat of Darwinism for religious belief may well have fostered the birth of more fundamentalist versions of creationism in North America (Numbers 2006), it is a mistake to believe that Darwinism and theism are incompatible.

While Christian commentators have more often than not pointed to considerable flaws in Dawkins's understanding of theism, or even the cultural agenda behind his theories, they are less inclined to pay attention to the scientific or philosophical basis for his evolutionary arguments. Part of the problem with Dawkins's attack on religion is not just its flawed theological and scientific claims, but its representation of truth claims as an absolute mirror image of the very theism that he rejects. One way to respond to Dawkins might be to show up the flaws in his philosophical, theological, and scientific approach. Such a response to Dawkins's attack joins Christianity with all other monotheisms that express belief in God as Creator. Another approach is to suggest that both Dawkins and his creationist respondents have failed to address the heart of Christian belief by ignoring its central premise, namely, the incarnation.

The premise of this chapter is, therefore, that there is much more to be said about evolution and Christianity than simply *either* taking the path of friendship *or* hostility towards Darwin. Such debates regularly miss out or push to the background proper consideration of the central tenet in Christian theology, namely, an understanding of the place and significance of Christ, or Christology. On the other hand, perhaps in a paradoxical way, those who make it their profession as systematic theologians to reflect on the significance of Christ in different cultures have also largely failed to give proper attention to the significance of either science in general or evolutionary ideas, both in terms of biological science and wider cultural meanings.[4]

Evolution is significant in that it provides an interpretation of the biological history of the natural world and, more narrowly conceived, the history of living forms. Therefore, theological engagement with evolution is necessarily historical. However, there are various styles of rendering history theological that might be appropriate even within the narrative or storied nature of theological reasoning. When that narrative becomes an overarching grand narrative it takes on the flavor of an epic narrative that seems to have deterministic elements within it. An alternative approach to storytelling that at least attempts to avoid that temptation is *drama*, and this approach pays much closer attention to what might be termed mini-narratives in any given plot. This allows for a greater degree of flexibility as to the future outcome. In theological parlance, drama that includes taking account of the action of God in history is *theodrama*. Yet such dramatic accounts necessarily include contingent events in a way that make it more difficult to imagine when considering a grand narrative or epic approach. In as much as a theodrama is oriented to the future in God it is fundamentally eschatological in scope, and so maps out the shape of future hope in relation to, in this case, evolutionary history, including that of human beings. I will, therefore, argue here for a different way of appropriating evolutionary ideas into a theological discourse using theodrama, rather than *narrative* as a framework for discussion.[5]

CHRISTOLOGY AND EVOLUTION

Classical debates on the person and nature of Christ struggled to articulate the meaning of the human and divine natures of Christ while keeping within the Chalcedonian Council (451) framework of who Christ is, namely, one person and two natures. It is possible to lean in the more Alexandrian direction, where the emphasis is on the importance of the divinity of Christ and the Word made flesh, or in the Antiochene direction, where more emphasis is placed on the humanity of Christ and the human soul taken up by the Word from the moment of incarnation. If the first view tended to squeeze out the possibility of Christ having a human soul, by an overemphasis on Christ's divinity, the second ended up with two persons in Christ, the divine indwelling the human. Historically, however, the story was likely to have been rather more complicated than this account implies, with authors such as Cyril of Alexandria adopting some ideas on Christ's rational soul that seem closer to the Antiochene tradition (Crisp 2007, 38–40).

The point here is that such a framework then leads to further discussion about how one might consider the human nature that is assumed by Christ—is it an abstract universal, or does it only make sense in the particular?—along with related technical discussion about *anhypostasia* and *enhypostasia*, with the former putting more emphasis on the possibility of a human nature existing as an abstract universal human nature, and the latter emphasizing the particular human nature as pertains in Christ's person.[6] Other ways through the problem of relating the divine and human natures in Christ posit that the two are related through mutual indwelling, that is, *perichoresis*, so that each indwells the other in a manner analogous to the relationships of the Trinity.[7]

All these discussions are essentially *closed* insofar as they represent internal theological debates about what might be logically possible, given certain premises. They are of logical importance given the Chalcedonian definition, but they only make sense once faith in Christ as divine and human is accepted *in toto*. They seem to bear little or no relationship to evolutionary biology except inasmuch as the concept of two natures becomes incredible or difficult to understand. I have a suspicion that many biologists reading this paragraph will even consider the figure of Christ to be an evolved myth that has emerged as a further development of a symbolic form of supernatural agency arising in human consciousness. Theology is then an expression of humans as the symbolic species *par excellence*.

It is hardly surprising that many pioneers who were engaged in the dialogue between evolution and theology found liberal accounts of Christology more attractive compared with the traditional view represented above. In the liberal position Christ is portrayed as a man unique only in as much as he is uniquely obedient and open to God. Arthur Peacocke, for example, suggests that in his oneness to God, Jesus is an archetype, a chief exemplar of what it is for a human to be united in self-offering to God (Peacocke 1979, 248). In Jesus we find God's character as Love displayed, and this, in its turn, is expressed in the life, death, and resurrection of Jesus, where resurrection is interpreted in terms of what happens in the minds of the disciples. Drawing heavily on the theology of Geoffrey Lampe, Peacocke views Jesus as the evolutionary point where perfect humanity appears for the first time, but it is perfection in relationship. Jesus, here, is portrayed as one whose deity *emerges* as a result of carrying out the divine will, and is seen as in direct parallel to the normal workings of emergent reality in other spheres. Jesus becomes "the manifestation of what, or rather of the One who, is already in the world though not recognised or known" (Peacocke 2007, 37).[8] Jesus's humanity evolved into a form of "transcendence" that then is recognized by others as having divine cogency. Although Peacocke maintains, therefore, a "top-down" approach to the interaction of God and the world, analogous to evolutionary emergence of "higher" levels of interaction, his understanding of Christ is "bottom up," in as much as Jesus seems to become a fully God-informed subject rather than being endowed with divine subjectivity from the beginning. Of course, this is the only option he really has, for it would be hard to imagine, in a general sense, how Christ might acquire divine characteristics through general "top-down" providential activity.

Ian Barbour is similarly exercised against traditional notions of Christ's two natures, so that "what was unique about Christ, in other words, was his relationship to God, not his metaphysical 'substance'"; even though he had "two wills," he was also able to exercise human "freedom and personal responsibility" (Barbour 1990, 210). Also, similarly drawing on the theology of Lampe he suggests that Jesus is best understood through a Spirit Christology, where Christ emerges as the archetype and pattern of union between God as Spirit and the spirit of humanity, and moves towards the final goal of creation, where humanity will be formed in the likeness of Christ, the model "Adam." Christ represents, therefore, a new "stage" in evolution and a new stage in God's activity. But on what basis is humanity going to be conformed in an eschatological sense to the likeness of Christ if Christ has simply emerged through evolution? On this basis, redemption seems to be reduced to what happens to creation, expressed entirely in evolutionary terms. In the work of Barbour we find that the newness of Christ is related to, first, his personal relationship with God, second, his ideas, and third, the response by the community around him. But is this really sufficient? Would I really be inclined to worship Jesus as Lord, in the manner given in John's Gospel, if this is *all* that can be said about Christ? The challenge is how to retain something of the spirit behind the Chalcedonian definition, but link it in some way with evolutionary accounts rather than dispensing with it altogether.

Using process thought, Barbour also suggests that the difference between humanity and animals is similar to the difference between humanity and Christ, in that it is a difference of *degree*, rather than an absolute difference, so, in this way, "Christ is the distinctive, but not exclusive revelation of the power of God" (Barbour 1990, 213). I would agree with Barbour that the difference as far as Christ's humanity is concerned is one of degree, so that while Christ is perfect in his relationship to God, his humanity is fully grounded in evolutionary biology as far as his human personality is concerned; he suffers, is fearful, is angry and grieves, and is tempted like all of us. What is much weaker in Barbour's account is a sense of Christ as divine gift, in spite of the language of authors such as John Cobb or David Ray Griffin, where Christ becomes God's "supreme act," yet such an act still seems integral to evolutionary processes, so it is hard to understand what purpose it serves. Christ's uniqueness is reduced to "the content of God's aims for him and in his actualisation of those aims" (Barbour 1990, 235). Hence, the classic notion of Christ as God incarnate has virtually disappeared.

The influence of Jesuit priest and palaeontologist Pierre Teilhard de Chardin is palpable in such accounts.[9] He was, without doubt, one of the great pioneers relating theology to evolution, and Christ to evolution in particular. He therefore needs to be given credit for the originality of his thought and the boldness of his vision. While Teilhard did take on board Darwin's theory of evolution by natural selection, he stressed the significance of the human in a way that was still decidedly anthropocentric. He attributed the expansive influence of humans to an explosion of consciousness, a shift from what he called "direct psychism to reflective psychism" (Teilhard de Chardin 1966, 62). He called such a change an evolutionary breakthrough "convolution," in what he termed a radically new phase in evolutionary history, namely, the appearance of the "noosphere" (Teilhard de Chardin 1966, 79). David Sloan Wilson is not the only biologist who has adopted some of Teilhard's ideas about the "noosphere" for his most recent research project (Wilson 2011, 102–24). Yet Wilson is quite ready to admit that such ideas remain imaginative speculation and are not thus far supported by empirical science.

Of course in Teilhard's account of evolution there is an inherent push towards unification, a view that arguably stems just as much from ideals of progress gleaned from Herbert Spencer as from Darwinian branching theory of natural selection that in its strictest sense resisted grand ideals of progress, even though it is reasonable to suggest that Darwin was not always fully consistent himself in this respect, especially in his account of *The Descent of Man* (Darwin [1871] 2004). It is also in the light of this vision of progress that Teilhard situates the significance of Christ, for he proposes that such a center is necessary if the cosmos is to progress, and, in this way, Christ is recognized as the crossroads where "everything can be seen, can be felt, can be controlled, can be vitalised, can be in touch with everything else. Is not that an admirable place in which to position (or rather recognize) Christ?" (Teilhard de Chardin 1971, 87). He interprets the incarnation of God in Christ primarily in evolutionary terms, and the cross of Christ primarily on a cosmic stage first, rather than speaking of it in terms of human misdemeanors. It would, however, be incorrect to suggest that for Teilhard Christ is simply explained by evolution. Rather, Christ is also in some way emancipated from time and space, so that "in one of its aspects, different from that in which we are witnessing its formation, it has always been emerging above a world from which, seen from another angle, it is, at the same time, in the process of emergence" (Teilhard de Chardin 1966, 84). He also envisages the future as impinging on the present, so that "this is what renders the movement not only irreversible, but irresistible" (85). Before humankind emerges, such attraction is "received blindly"; afterwards it is partially conscious in reflective freedom, leading eventually to religious belief (86).[10]

Teilhard was undoubtedly a pioneer in the field of relating science and theology, and Christ and evolution in particular. However, question marks about adopting his approach relate to: (1) His understanding of evolution as progressive towards the evolution of more and more sophisticated forms, which seems to rest on metaphysical theory and cosmology rather than the eliminative evolutionary theory of natural selection of biological organisms according to Charles Darwin. The term "evolution" covers the cosmic process at the beginning of the universe through to the evolution of life by natural selection, even though the processes involved are very different. (2) An outdated anthropocentrically oriented theory of the evolution of life, in as much as his interpretation of its direction towards the human is not strictly accurate, even if the present landscape of the world in the geological era dubbed the Anthropocene shows *Homo sapiens sapiens* to be the most dominant species. Aspects of his thinking are characteristic of the knowledge available at the time when he was writing. For example, he was ignorant of the range of hominins that have since been discovered, and he was also unaware of the importance of other heritable factors in evolution, such as that of niche construction, genetic drift, and epigenesis. (3) His tight association of evolution and Christology, such that Christ becomes embedded in the process as such, and thus endorses that process, rather than being freely given by God to creation. (4) His Christomonism where Christ becomes somewhat remote from the historical Jesus. (5) His neutralization of evil through his treatment of it as a necessary part of the evolutionary process. And (6) his somewhat naïve and optimistic view of human progress and science in particular, whereby all its endeavors could necessarily be synthesized with the goals of the kingdom of God.[11]

THEODRAMA AND EVOLUTION

Many other authors in this field are, like Teilhard, rather more inclined to discuss the process of cosmological evolution rather than strictly biological evolution. This leads to his interweaving of a generalized theory of evolution and theology, using a hermeneutic of historical narrative, viewing the history of nature as a grand narrative or *story* to be told in a way that is comparable to the *human story*. During his life, Teilhard was caught up in a dramatic encounter with the established church, which resisted making his eclectic ideas on evolution and theology widely known. It is in this sense that Teilhard's work is highly dramatic, for his vision of the future of the universe sustained his faith in God in spite of all the obstacles that he faced in his lifetime. In one sense I agree heartily with his idea that the natural world can be interpreted as sharing historical features that run to some extent in tandem, or even in continuity with human history. Both appeal to the powerful concept of story as a way of imagining human history bound up with the history of other creaturely kinds, and the history of the very stuff of the universe. But what if a theology of history becomes even more vivid and, I suggest, truer to itself as theology, through a *different* reading of history, one that draws specifically on *drama*, rather than a narrative account of God's ways with the world?

Drama, in that it specifies particular agents, actions, contexts, and events, puts stress on the *contingent* aspects of a storied life history in a way that a grand narrative does not. Drama looks to a specific scene in the story of life, focusing on the particular contribution of the individual actors and the where and why and how a particular scene takes place. A grand narrative has connotations of inevitability in a way that drama does not; in drama it is possible for change to take place by the fraught decisions of individual players. Drama admits to the tragic and incorporates particular internal and external factors in a way that grand narrative tends to eschew. For example, in a grand narrative the internal states of players would not be considerable in a way that is illuminated beautifully in dramatic script. Quite simply, the play *could* go in a different direction, whereas in a grand narrative or epic players are caught up in a storied account of their existence that feels, to them at least, outside their control. The category of drama is more consistent with evolutionary theory by natural selection as it allows for both the contingent and the slow moving forward that is only really visible after the event. When evolution of life becomes a grand narrative, the contingency in that process is lost from view, it points to a determinism that is characteristic, ironically, of the selfish-gene-type grand narratives associated with Richard Dawkins. While it would theoretically be possible to use "mini-narratives" in order to overcome the latter, as in Matthew Ashley (2010), in the end I prefer the language of drama as it helps put more emphasis on the importance of *agency*, including that of other creaturely kinds. Such a view also fits in better with the evolutionary idea of niche construction, that humans and other biological agents are actively involved in making their world that itself feeds into the process of evolutionary change (Deane-Drummond 2014). *Theo*drama can also articulate a link to a central act in the drama, in this case, the life, passion, and resurrection of Jesus Christ, in a way that is rather more difficult, though not necessarily impossible, with the concept of mini-narrative.

John Haught has also drawn on the language of drama in order to situate his account of cosmic evolution (Haught 2010). Even while sharing a common language there is a significant difference from the approach I am developing here. In the first place, Haught takes his theme of drama primarily from the observations of the natural world. He then matches this dramatic account onto theological categories. The difficulty in his case arises because the dramatic and often tragic elements in the evolutionary story seem to become endorsed through a theological reading. By starting with the passion, death, and resurrection of Jesus Christ as the key drama, I hope to avoid such endorsements; rather, the drama of each act becomes interrogated through an encounter with the living Christ who shows human beings how to act.

Before developing these ideas further, however, the first and most basic question to be addressed is whether theology is compatible with history, given the influence of scientific understanding on modern historical scholarship; this is also indirectly a question about the compatibility of theology and science. Ben Quash suggests that theologians are people prepared,

> to see the dense, historical world as having an origin and an end in the creative purposing of God, a God who can relate personally to his creatures. People ready to acknowledge the idea that there can be revelation: a prevenient ground *for* our knowledge

and perception that is not the product *of* our knowledge and perception, which is neither accidental nor impersonal but which freely, and even lovingly, communicates itself. (Quash 2005, 2)

Such a reading gives history a key oriented to the future, and theodramatics is a way of thinking about eschatology and history together in their relationship with each other (Quash 2005, 2–3).[12] Drama is about human actions and particular events in particular contexts, and theodrama is that which is connected to God's purpose. A theodramatic approach will always be in one sense eschatological in orientation. Attention to drama draws out the specific significance of human agency, the particular context, and also the wider plot or time dimension. Consideration will therefore include that of the *subjects* themselves, the acting area in which they perform, or the *stage*, and the movement of the play, or *action*.

Another key issue that arises here is that of freedom, and what this means in the Christian life. If God is perceived as one who is in possession of divine freedom, this means that any narration of history cannot be simply an inevitable chain of events while preserving the sense of the importance and validity of individual freedom. The advantage of theodrama is that it envisages an *encounter* between God and creation, where the freedom of the creature is preserved without resorting to pantheistic interpretations of the relationship between God and creation, which surface in many renditions of evolution as narrative.[13] In this, theodrama resists making either subjects, through family genealogies, or structures, through a systematization in mechanistic or organic models, the most important keys to the interpretation of history.

But what if we allow theodrama to include not just human history but evolutionary history as well? Such an expansion has the advantage of viewing other evolved creatures *as more than* simply the stage in which human action and freedom is worked out. The ability to read evolution not just as science but also as history means that, through evolutionary accounts, nature as such becomes historical, a perspective that, according to some, is one of the most significant discoveries of science (Haught 1996, 57). The most common way of reading human history is that according to the activities of individual human subjects in genealogies or that according to the dynamics of historical change through systematic analysis in what can be termed a "grand narrative" approach (Quash 2005, 6). This is applied most commonly to a reading of the activity of human subjects, but in evolutionary science we find similar trends toward either tracing genealogies or constructing grand narratives such as evolutionary psychology, but also others such as Darwin's theory of natural selection, Mendel's laws, and so on. Ben Quash, drawing particularly on the work of Hans Urs von Balthasar, has argued for the recovery of theodramatic approaches to history that concern human actions in time in relation to God's purpose (Quash 2005, 3–4). For him, the concern of theodramatics is with the character of human agency, its necessary conditions in terms of place, and its relation to the wider plot over time. Of course, the degree of awareness of divine action will be different according to different levels of consciousness and capacity for decision, but by placing creatures in kinship with humanity, the evolution of life becomes an integral aspect of the drama between God and God's creatures. Moreover, in the light of such a dramatic reading, an overly systematized account of evolutionary process begins to look far too thin. Ben Quash suggests that a nondramatic reading of human history leads to "synchronic" principles that amount to a form of betrayal, for they "fail to give due attention to particulars, to the individuals, the exceptions to rules, the resistances to explanation and the densities of meaning that ask for recognition in a good description of historical reality" (Quash 2005, 7).

Yet I believe that such synchronic readings of evolutionary history are rife in secular and religious accounts; examples include the Christomonism of Teilhard; the aesthetic principle of process thought; even the organic model of James Lovelock's Gaia hypothesis, which envisages the sum total of the biota feeding back on the earth in order to render environmental conditions constant; or even more general ideas such as the "balance" of nature. The difficulty, of course, when it comes to the millions of years of evolutionary history, is that human imagination finds it hard to appreciate the dynamics of the particular in any given "scene" of the drama. Also, given that evolution takes place over a long period of time, the "play," if it is to do justice to the individual characters concerned, will find itself dealing with long epochs of history where such characters have come and gone in different scenes presented. In other words, the characters that may be picked out for discussion are selective and illustrative in as much as they represent just one small fraction of the overall evolutionary process. Sometimes it may prove preferable, therefore, to use a close examination of those creatures that we know, in order to provide an analogy of what earlier creatures may have been like.

A good example of this is the study of primates in order to give clues as to the life of early hominids.[14] Yet such study also helps open up the realization of human ignorance about the drama itself, by focusing on the so-called punctuated phases of evolution where improbable events came together in a way that mean only one lineage survived and not others.[15] Such events, which effectively wiped out myriads of species, many of which may not even yet be identified by present research, means that the tragic nature of the evolutionary drama comes into view. A theodramatic approach takes proper account of the tragic, one that is so vivid in terms of the evolutionary history of the earth, but now brings this into juxtaposition with an understanding of how God works in the tragic in human history. It therefore will resist any generalization of evil or attempt to wash over the contingency of events. In theodrama, the tragic is recognized fully rather than absorbed and neutralized in the manner that more often than not happens in an epic account. The tragic has been the pattern for the drama of evolutionary history for millennia, as witnessed in the paleontological record.

In order to highlight the advantages of a theodramatic approach, the tendency for narrative accounts to become *epic* is also worth considering. Evolutionary history, with its tremendously long time scale, is almost always drawn into such an interpretation. I do not believe that we can avoid at least some narrative description, but such description needs to be self-aware in as much as it recognizes the tendency for it to be taken over by the genre of the epic.

What do we mean by epic? In the second book of his trilogy, *Theodrama*, theologian Hans Urs von Balthasar considers whether there is some standpoint from which we can merely be observers to a sequence of events, including the events of Christ's death and resurrection. In such a view he suggests we "smooth out the folds and say that Jesus' suffering is past history; we can only speak of his continued suffering in an indirect sense, in so far as those who believe in him are referred to, metaphorically, as his members" (von Balthasar 1990, 54). At its worst, epic becomes the "genre of false objectification," and "reifies what is given to it to know. It substitutes monological narration for dialogue, without supposing that this is a loss for truth. And it tends towards determinism" (Quash 2005, 42). The idea of evolution as incorporating some sort of *necessity* is a typical reading of evolutionary history in some quarters as well.

On the other hand, Christian spirituality is more often than not expressed in terms of particular experiences of the individual, what von Balthasar terms the "lyric" mode, where all thoughts of universal significance found in epic thought falls from view. Theodrama avoids equally problematic mystical "lyric" accounts, where "the whole substance of an action is transposed into a highly volatile, highly individual; immediate and emotionally coloured mode of response and expression" (Quash 2005, 42). The councils of the church may frame their theological deliberations with prayer, but this does not deal with the problem that they are delivered in an epic mode. Even a theology that focuses on scripture can lead in the same direction, that is, if it sees itself as an objective (epic-narrative) account that has taken place and is now done. Reactions against such "arrogance" in theopraxy is, von Balthasar believes, simply a one sided reaction (von Balthasar 1990, 57).[16] Instead, the theodramatic considers the ongoing action of God in history, as witnessed particularly in the lives of the apostles and in the early church. Interpreting evolution in theodramatic terms has a further advantage in that anticipation still plays a vital role, but it does not take the form of resignation, as is characteristic of purely secular accounts of evolutionary science, but rather encourages further interpretation and engagement.

Ignatius of Loyola's sacramental view of the world allowed for a positive attitude towards all of creation, God's presence being found in all things. It is no surprise, therefore, that von Balthasar also perceived this dramatic account of God's action as extending back not just to the creation of Adam and Eve but also to the creation of the world, and forward to the future revelation of what is to come, expressed through apocalyptic literature. He suggests that:

> It so overarches everything, from beginning to end, that there is no standpoint from which we could observe and portray events as if we were uninvolved narrators of an epic. By wanting to find such an external standpoint, allegedly because it will enable us to evaluate the events objectively (*sine ira et studio*), we put ourselves outside the drama, which has already drawn all truth and all objectivity into itself. In this play, all the spectators must eventually become fellow actors, whether they wish to or not. (von Balthasar 1990, 58)

Such an approach has profound implications on how to envisage Christology. Now the incarnation is understood as kenotic, which implies a self-emptying, but it is a kenotic Christology that is less about God "giving up" particular attributes or divine and human essences and more about a theodrama expressed in a radical, deep incarnation of God assuming human, and thereby creaturely, being in Christ. Jesus's life and ministry are thereby connected with the cosmic role of Christ, for it is expressed in the theodrama of Trinitarian interrelationships. For von Balthasar, it is the filial response to the Father that unites both the human activity of the Son and the divine Son in Trinitarian relationships (von Balthasar 1989, 170–71). This is important in consideration of the significance of the cross of Christ, for the primacy of love comes first in any considerations of the atonement, especially that which relates to the self-giving of the inner kenotic movement of the Trinity, rather than kenosis as understood in a primary sense as that between Creator and creation. I have also extended von Balthasar's theodrama into the life of Christ as one who chose to take on the sins of the world by suggesting that this choice also embraced not just the negativity of human sin, but also sin more generally associated with creaturely being (Deane-Drummond 2009, 159–93).

We also need to ask another question, namely, is consideration of evolution in dramatic terms meaningful from an evolutionary perspective? Evolutionary biologist Jeffrey Schloss has noted such a possibility in describing evolution in terms of a play on an "ecological stage." He suggests, "The lines, the players and even the plot may change over evolutionary time, though they are ever constrained by the props and setting and choreographic syntax of the ecological moment" (Schloss 2002, 58).[17] While I agree with the analogy, I suggest that we can go even further than this, in that ecology is rather more dynamic than this view might suggest. Hence, ecology does not *just* represent the stage but also, for many nonhuman animals at least, includes the possibility of agency.

The profile of the dramatic includes indeterminacy and thus is inclusive of circumstance, compulsion, and decision, which most characterize human existence (Quash 2005, 35–37). Such indeterminacy is also, I suggest, characteristic of life in general and becomes most intense in animals that share the capacity for decision making. Drama also works through particular events, as well as showing a social dimension through including the audience as much as those on the stage. The possibility of sharing in a *performance*, it seems to me, makes for a more readily accommodated perception of inclusiveness with other finite creatures, compared with, for example, a simple portrayal of evolution in terms of a rational system of truth claims.

Jeffrey Schloss has argued more specifically that there is an analogy between "ideal" and "actualized" niches found in evolutionary history and that which can be expected in an eschatological future "heaven" (Schloss 2002, 65). Most species occupy theoretically "suboptimal" niches due to competitive displacement by other, more dominant species. Yet while some biologists resist any language of the "ideal"— the "optimal" niche always being a compromise in terms of physiology and availability of resources, competition, and so on—the meaning of "optimal" is equally under dispute. The use of terms like *optimal* and *ideal* implies teleology, a future to which species aspire. Yet Schloss is critical of those attempts to endorse evolutionary accounts through a developed teleology inasmuch as many of the assumptions in such stories do not appear to be true. For example, there is no real evidence for an increase in complexity over time, and ever since the Cambrian Period, morphological diversity has stayed more or less constant. There is also no evidence that parasites or other harmful organisms gradually become more mutualistic.[18] Such benign readings of nature characterize much natural theology, especially in the work of those theologians that seek to find a continuum between evolution and eschatology.

Schloss draws on Richard Dawkins's idea of the extended phenotype as operative in evolution, where the phenotype extends beyond the immediate "skin" of a given organism to all its fitness-enhancing influences on the environment (Dawkins 1986, 1999). Such a move breaks down an organism's identity, rendering it fluid, in such a way that there is an ambiguity over "whose" body a body is. Schloss uses this model as analogous to the resurrection, in that while some accounts are literal, others are less so. Yet this seems to assume that Dawkins's account of natural selection working primarily through genes is correct, a view that I call into question (Deane-Drummond 2009, chap. 2). Dawkins would no doubt shudder to see his views expounded in support of claims for the resurrection, but this is perhaps beside the point. In addition, if Christ becomes rendered through his resurrection as some kind of "super-organism" in the manner of Dawkins's extended phenotype, what might this mean for individual encounters? Perhaps a key insight here is that the boundary of what might be termed ontology of being is rather more fluid than used to be considered the case, but it is a kind of fluidity that is still bound by existing laws of nature rather than operative outside it.

In evolutionary terms, there are trends to be observed, such as increase in body temperature, increase in body size, episodic jumps in energy intensiveness, increases in developmental cascades, increases in hierarchical integration, as well as levels of organization. But there are also increases in competition, rather than evasion, and developments of powers of cognitive processing.[19] A future hope tied into such processes becomes problematic in that "a critique of eschatologies based on evolutionary aesthetics is that we have fashioned God after our own fallen image, at precisely the cultural moment when theology needs to recall us in eschatological hope to the renewal of his moral image within us" (Schloss 2002, 77).

Moving from an epic narrative to a dramatic approach has wider implications than those raised by the limited confines of this chapter, restricted to asking questions about how to approach Christology in an evolutionary world. While further questions that relate to a specific understanding of Christology require far more development, there are also wider implications raised by the specific methodology proposed in this chapter.[20] In concluding I will point to some of these implications as a way of opening up further discussion. In the first place, the methodology suggested here raises questions about how a narrative imagination might inform the way science is interpreted and filtered culturally. Matthew Ashley is similarly critical of grand narrative approaches to cosmic history, and has suggested that epic narratives might be deconstructed using a mosaic approach of mini-narratives woven together like a patchwork quilt (Ashley 2010).[21] It seems to me that representation of cosmic and human history as mini-narratives successfully avoids the problems associated with grand narratives and bears some similarity to a dramatic approach. However, a dramatic understanding deliberately draws attention to elements of a particular scene in a drama set in the context of a stage, inviting audience participation.

An eschatological reading of scripture may also require a more dramatic reading in some cases, such as apocalyptic texts of, for example, the book of Revelation. This book, which speaks in dramatic terms of Christ's second coming, is one that has particular appeal in the wake of secular anxieties about humanity's future expressed in the rhetoric of climate change. It is also relevant ethically, since an epic-narrative reading will tend towards a sense of inevitability that can lead to political complacency, whereas a more dramatic reading stresses individual agency (Deane-Drummond 2010). This example draws out a further implication of the dynamics of an appropriate reading of science and theology, in as much as the way science is interpreted alongside a biblical text will influence ethical and political action or inaction. The secular discourse of climate change, with its almost explosive mixture of science and politics, becomes all the more intense when viewed through a religious lens according to apocalyptic imagery.

Finally, finding ways of giving a place to religious belief, and a sense of religious agency, but without suppressing science, is itself necessary in a political climate where tendencies for suspicion on both sides may run amok. But this in itself opens up other questions about how to think of ways of connecting what it means to be human with scientific interpretations of human origins and human behavior (Deane-Drummond 2014). Further, if, as I have argued, a dramatic reading of human history is coherent with evolutionary accounts, how and in what sense might a pneumatology be developed that is both faithful to Christian theological traditions but appropriates science wisely? One of the reasons for hostility felt by Christians towards evolutionary understandings of human origins is a reaction against the idea that human lives are simply a result of evolutionary events based on contingency. While this is, of course, an oversimplified view of the science, correcting that impression does not seem to alleviate the anxiety. The question becomes, how can the God-human encounter, or more specifically, my encounter with the Divine, be both personal and yet pay attention to God as Lord of all in a universe that seems to be the result of evolutionary processes? What kind of doctrine of the Holy Spirit would be meaningful in such a context? Is there any place at all for an evolutionary reading of human history, or is it simply irrelevant? Can we, with von Balthasar (von Balthasar 1990, 95), make the following claim that "It is the Spirit who perfects the self-illumination that takes place in theo-drama; he reveals its meaning retrospectively, from the end, and at the same time proclaims its universal scope forward, into an ever-new future"?

NOTES

I would like to express my sincere thanks to Professors Gerald McKenny and Phillip Sloan and Dr. Kathleen Eggleson for the formal invitation to speak at the conference on Darwin held at the University of Notre Dame in 2009. I would also like to thank members of the audience and Professor William Carroll for their response to my address. This essay is an adaptation of the paper delivered on that occasion and it draws on Deane-Drummond (2009). I would like to thank Professor Matthew Ashley for helpful conversations on this topic and for allowing me to read his essay for *Theological Studies* (Ashley 2010) while still in press.

1. See other works by Peacocke that develop the idea of Darwin as disguised friend, for example, Peacocke (2004).

2. This is not what Peacocke meant when he spoke of Darwin as a friend. See Dawkins (2006). It is, however, implied by Tom Greggs's approach to the issue in Greggs (2008).

3. Terry Eagleton lumps together Richard Dawkins and Christopher Hitchens, both advocates of what might be termed the new atheism, in a humorous title, "Ditchkins" (Eagleton 2009, 2).

4. See, for example, Graham Ward's book on Christology and culture, where in spite of the considerable merits of this volume, engagement between Christology and scientific discourse is missed out entirely, even though, arguably, the latter serves to shape Western culture ever since the Enlightenment (Ward 2005).

5. I elaborate on the meanings of these terms further in the section "Theodrama and Evolution" below.

6. An *anhypostatos physis* is a human nature that exists independently from an individual or person. In this scenario, Christ's personhood requires the assumption of human nature by the Word. From the moment of incarnation, there is *enhypostatos*, that is, human nature in a particular person. In some discussions, the human nature of Christ is seen as being taken up into the Word. See, in particular, Crisp (2007, 72–89).

7. Crisp devotes a whole chapter to considering this issue (Crisp 2007, 2–33).

8. Of course, the idea that Christ might become known as divine through his obedience and openness to God reflects a liberal tradition that goes as far back as Albert Ritschl. The point is that the evolutionary story of emergence is made paradigmatic, and the work of God, including the description of the meaning of theological terms, such as grace, as well as an understanding of Christology then becomes compatible with this, rather than the other way round.

9. This brief summary highlights aspects of his thought. For a much fuller critical appreciation see Deane-Drummond (2006).

10. It is worth noting that while the evolutionary emergence of religion is an area of more recent debate, Teilhard believed that the psychic elements in evolutionary processes were connected to divine action that beckoned from the future, expressed eventually as Omega.

11. His views are perhaps more in line with physicists and cosmologists, who speak of the Anthropic Principle, though in reality this means the principle of life. In addition, he remains a prophetic visionary in as much as he seemed to anticipate the World Wide Web through his understanding of a further evolutionary stage of human consciousness and communication. Such global cultural shifts can hardly be identified with the coming kingdom of God in the way that he anticipated.

12. The basic interpretation of theodrama as applied to human history that I am using here follows Quash's very helpful summary.

13. This is a tendency felt in some interpretations of the work of Thomas Berry, whose appeal to the story of the cosmos is especially influential among ecotheologians. See Berry (1988) and Swimme and Berry (1992). Matt Ashley offers a penetrating critique of the way in which a narrative reading of scripture and a narrative reading of the cosmos might serve to inform and influence each other. See Ashley (2010). Ernst Conradie is perhaps an exception, in that, while he shows his affiliation with Berry's work, he is also heavily influenced by the Reformed tradition. Given this, it is worth asking if his theology is more consistent with theodrama than with an epic narrative approach. See, for example, Conradie (2005, 27–28, 58–59). It should also perhaps be pointed out that many ecotheologians who are biblical scholars are seeking to develop new readings of the Bible that are informed by an ecological context understood in terms that go beyond a scientific understanding of ecology in as much as it informs a particular political stance towards the natural world. Elaine Wainwright is a feminist biblical scholar who is deliberately attempting to read the book of Matthew ecologically (Wainwright 2011).

14. There are, of course, disadvantages in such an approach, especially as much of the cultural history of early humans is dependent on speculation.

15. While the theory of punctuated evolution championed by Stephen Jay Gould is still debated, there seems to be agreement among evolutionary biologists that there are distinct phases of evolution, some of which are relatively static, and others leading to more dramatic changes. Controversy rages over whether there are genuinely macroevolutionary processes at work in addition to what might be termed microevolutionary events taking place at the level of organisms, shaped through natural selection and other contributory processes such as genetic drift. I discuss this further in Deane-Drummond (2009, 1–23).

16. Von Balthasar's dismissal of the work of liberation theologians was rather too hasty. The almost uncritical absorption of not only Thomas Berry's Universe Story, but Lovelock's Gaia hypothesis by liberation theologian Leonardo Boff is, nonetheless, problematic. See Boff (1997). For further comment see Deane-Drummond (2008).

17. Schloss takes this idea from Evelyn Hutchinson. See Schloss (2002, 58).

18. Schloss, for example, questions more saccharine accounts of evolution in a number of ways, not the least challenging the myth that infectious agents coevolved with hosts in order to "minimise pathogenicity"; this view, he suggests, is false on both "theoretical and empirical grounds. . . . Pathogenicity often increases over evolutionary time, depending on infectiousness and host density" (Schloss 2002, 73).

19. The three basic life history strategies are those that evade competition, those that win competition by efficiency or defense, and those that win

by sabotage, that is, taking more than they might "need" physiologically and in this sense enhancing their potential competitive impact. Some evolutionary biologists argue for an escalation rather than an attenuation of competition over time. See Vermeij (1987).

20. I have drawn out the specifically Christological aspects of this approach in *Christ and Evolution* (Deane-Drummond 2009).

21. Here Ashley draws on Aldo Leopold's *A Sand County Almanac* (Leopold 1989) as well as Paul Ricoeur.

REFERENCES

Ashley, J. M. (2010). "Reading the Universe Theologically: The Contribution of a Biblical Narrative Imagination." *Theological Studies* 71:870–902.

Barbour, I. (1990). *Religion in an Age of Science*. London: SCM Press.

Berry, T. (1988). *The Dream of the Earth*. San Francisco: Sierra Club Books.

Boff, L. (1997). *Cry of the Earth: Cry of the Poor*. Maryknoll, NY: Orbis Books.

Conradie, E. (2005). *An Ecological Christian Anthropology*. Aldershot: Ashgate.

Crisp, O. (2007). *Divinity and Humanity*. Cambridge: Cambridge University Press.

Darwin, Charles. ([1871] 2004). *The Descent of Man, and Selection in Relation to Sex*. Introduction by James Moore and Adrian Desmond. London: Penguin Books.

Dawkins, R. (1986). *The Extended Phenotype: The Gene as the Unit of Selection*. Oxford: Oxford University Press.

———. (1999). *The Extended Phenotype: The Long Reach of the Gene*. Oxford: Oxford University Press.

———. (2006). *The God Delusion*. London: Bantam Press.

Deane-Drummond, C. (2008). *Ecotheology*. London: Darton Longman and Todd.

———. (2009). *Christ and Evolution: Wonder and Wisdom*. Minneapolis, MN: Fortress.

———. (2010). "Beyond Humanity's End: An Exploration of a Dramatic versus Narrative Rhetoric and Its Ethical Implications." In *Future Ethics: Climate Change and Apocalyptic Imagination*, ed. S. Skrimshire, 242–59. London: Continuum.

———. (2014). *The Wisdom of the Liminal: Evolution and Other Animals in Human Becoming*. Grand Rapids, MI: Eerdmans.

Deane-Drummond, C., ed. (2006). *Teilhard de Chardin on People and Planet*. London: Equinox.

Eagleton, T. (2009). *Reason, Faith and Revolution: Reflections on the God Debate.* New Haven, CT: Yale University Press.

Greggs, T. (2008). "The Dawkins Delusion." Unpublished lecture delivered to the Centre for Religion and the Biosciences, University of Chester, 26 November 2008.

Haught, J. F. (1996). "Ecology and Eschatology." In *And God Says That It Was Good: Catholic Theology and the Environment*, ed. D. Christiansen and W. Grazen, 47–64. Washington DC: U.S. Catholic Conference.

———. (2010). *Making Sense of Evolution: Darwin, God and the Drama of Life.* Louisville, KY: Westminster John Knox Press.

Leopold, Aldo. (1989). *A Sand County Almanac and Sketches Here and There.* New York: Oxford University Press.

Numbers, Ronald. (2006). *The Creationists: From Scientific Creationism to Intelligent Design.* Cambridge, MA: Harvard University Press.

Peacocke, A. R. (1979). *Creation and the World of Science.* Oxford: Clarendon Press.

———. (2004). *Evolution: The Disguised Friend of Faith?* Philadelphia: Templeton Foundation Press.

———. (2005). *The Palace of Glory: God's World and Science.* Adelaide: ATF Press.

———. (2007). *All That Is: A Naturalistic Faith for the Twenty-First Century.* Edited by Philip Clayton. Minneapolis, MN: Fortress Press.

Pleins, J. David. (2013). *The Evolving God: Charles Darwin on the Naturalness of Religion.* New York: Bloomsbury Academic.

Quash, Ben. (2005). *Theology and the Drama of History.* Cambridge: Cambridge University Press.

Schloss, J. (2002). "From Evolution to Eschatology." In *Resurrection: Theological and Scientific Assessments*, ed. T. Peters, R. J. Russell, and M. Welker, 56–85. Grand Rapids, MI: Eerdmans.

Swimme, B., and T. Berry. (1992). *The Universe Story: From the Primordial Flaring Forth to the Ecozoic Era—A Celebration of the Unfolding of the Cosmos.* San Francisco: Harper.

Teilhard de Chardin, P. (1966). *Man's Place in Nature: The Human Zoological Group.* Translated by René Hague. New York: Harper and Row. Orig. pub. 1956.

———. (1969). *Let Me Explain.* Translated by R. Hague. London: Collins.

———. (1971). "Christology and Evolution." In *Christianity and Evolution*, translated by R. Hague, 81–92. London: Collins.

Vermeij, G. J. (1987). *Evolution and Escalation: An Ecological History of Life.* Princeton: Princeton University Press.

von Balthasar, Hans Urs. (1989). *Explorations in Theology*, vol. 1, *The Word Made Flesh.* San Francisco: Ignatius.

———. (1990). *Theodrama*, vol. 2, *Dramatis Personae: Man in God*. Translated by Graham Harrison. San Francisco: Ignatius Press.

Wainwright, E. (2011). "Beyond the Crossroads: Reading Matthew 13, 52 Ecologically into the Twenty-First Century." In *The Gospel of Matthew at the Crossroads of Early Christianity*, ed. D. Senior, 375–88. Leuven: Peeters.

Ward, G. (2005). *Christ and Culture*. Oxford: Wiley-Blackwell.

Wilson, David Sloan. (2011). *The Neighborhood Project: Using Evolution to Improve My City, One Block at a Time*. New York: Little, Brown and Company.

PART FOUR

Past and Future Prospects

FIFTEEN

IMAGINING A WORLD WITHOUT DARWIN

Peter J. Bowler

What would have happened if Charles Darwin had not lived to write *On the Origin of Species*? Perhaps his bad health caused the early death he feared, or maybe he fell overboard while on the voyage of the *Beagle*. Would the world have still experienced the Darwinian Revolution under another name, or would the history of science, and of Western culture, have unfolded in a significantly different direction? The technique of using counterfactual histories, where we imagine what would have happened if a key event had turned out differently, has recently begun to attract more attention, and it may be worth exploring in the history of science. As we celebrate the bicentenary of Darwin's birth and the 150th anniversary of the publication of the *Origin*, it may be an appropriate time to consider just how crucial his influence was. We live in a world where evolutionary theory has gained a secure position in science, but where Darwinism is roundly condemned by a variety of cultural opponents. Religious thinkers blame the theory of natural selection for destroying faith in a designing God, thus promoting materialism and atheism. They join hands (rather paradoxically) with critics from the political left in blaming Darwinism for inspiring malign ideologies of "social Darwinism." In the most extreme case illustrated by Richard Weikart's *From Darwin to Hitler*, the theory is seen as a significant factor in the rise of racism, eugenics, and Nazism (Weikart 2004).[1] The future of Darwinism may well depend on how well its supporters can uncouple the scientific theory from these extreme claims. Here, I suggest, is a potential use for counterfactual history. If we can construct a plausible counterfactual scenario in which the negative social forces identified with Darwinism would have arisen *even if the selection theory had not emerged in science*, then the credibility of these rather simpleminded accusations can be undermined.

This approach will imply complex issues centered on the relationship between science and ideology that I cannot explore in a short paper. Let me make it clear that I shall not be seeking to undermine all connections between the two realms. My aim is not to whitewash Darwinism but to show that there were other scientific theories in play that had very similar social applications, so we need to think much more carefully about how theories and their wider implications are related. The claim that Darwinism is somehow "responsible" for the subsequent excesses of the Nazi regime would only make sense if one could argue that there is a direct causal link, which would imply that without the selection theory there would have been no race science and no eugenics. It's hard to believe that even Weikart wants to put it so bluntly, given the many other cultural and social forces at work leading to the rise of Nazism in Germany. But the claim that Darwin's theory played a crucial role in the mix of factors leading to Nazism must imply that there would be significant (and presumably beneficial) differences if that one factor had not been present. To explore a counterfactual history in which Darwin did not publish the selection theory will allow us to identify exactly what effect his theory had on the development of social thought. If it turns out that much of what came to be known as "social Darwinism" can plausibly be seen emerging from a non-Darwinian science, then the modern critics need to rethink their position. If, on the other hand, we can identify some elements in the later ideologies that are more difficult to explain without the input from Darwinism, then this is where we have to focus our attention in future discussions of the theory's implications.

The bulk of this chapter will offer my ideas on what might have happened if Darwin had not published the *Origin of Species* in 1859, focusing first on religious debates and then on ideological issues. My presumption is that an evolutionary movement would have emerged in the course of the 1860s, exploiting non-Darwinian mechanisms that were proposed even in our own world as alternatives to Darwin's theory. But I shall suggest that the theory of natural selection would not have been an important component of the debates until much later, leaving the crucial question: how much difference would this make to the wider implications attributed to evolutionism?

Before offering these suggestions, however, I first need to say something about the technique of counterfactual history and its implications for the history of science. I also need to address what many would regard as the most obvious objection to the use of Darwin as a case study. If, as is widely believed, the theory of natural selection was somehow "in the air" in the mid-nineteenth century, then if Darwin had not published it, someone else, most obviously Alfred Russel Wallace, would have taken his place and the Darwinian Revolution would have unfolded under another name. Note that such a view would severely limit Darwin's originality and might be used to question why we are making so much of a fuss over the bicentenary. I shall challenge this interpretation by arguing that Darwin was an original thinker and that no one else was in a position to write anything that might have had an impact equivalent to that of the *Origin*. Wallace's theory was significantly different from Darwin's, and his isolated position—professionally, and at first geographically—make it hard to see how he could have made much of an impact on his own.

It is also worth noting that no one ever talked about "Wallace-ism," and the clumsiness of this term alerts us to the fact that Darwin's iconic status may have been derived in part from the ease with which his name translated into the noun "Darwinism" and the adjective "Darwinian." As Janet Browne points out, it also derives from the ease with which his features could be caricatured as ape-like (Browne 2005). The closest rival here is Herbert Spencer, who was less often caricatured but became an icon more potent than Darwin for followers whose main concern was social evolution. And, as we know, Spencer was a Lamarckian evolutionist long before he coined the term "survival of the fittest."

COUNTERFACTUALS AND SCIENCE

Counterfactual history is dismissed as pointless by many professional historians. Those who believe that history unfolds in response
to inexorable cultural or social forces will inevitably reject any suggestion that a single event could trigger consequences that would
lead to a world entirely different from the one we know. Counterfactuals have, however, gained some currency among military historians, who are aware of how often the outcome of a battle seems to be
determined by quite trivial events that unleash a cascade of ever more
significant consequences. There is also a substantial body of fictional
accounts of alternative universes in which, for instance, the Confed-
eracy won the Battle of Gettysburg or the Luftwaffe the Battle of
Britain. Battles are unique in that they have decisive consequences for
everything that happens thereafter, although imagining how those
subsequent events unfold is much more difficult, especially over the
long term. This is presumably why such reconstructions are often
presented as forms of science fiction.[2]

Can we make a case for the possibility of equally crucial events
changing the historical development of science and its implications?
I want to argue for a modified form of the counterfactual technique
that accepts that there are some general trends but that does leave
room for alternative histories where the sequence of events would be
different from the one we experienced. I do think that the general
idea of evolution was almost certainly going to become popular in
the late nineteenth century, but I will try to show that it might have
emerged even if the theory of natural selection had not been proposed. I shall then argue that most of the cultural and social implications associated with the theory would have emerged very much as
we know them, because here too there were wider trends at work
leaving little room for alternatives to emerge. But the consequence of
this is the conclusion that most of what has been called "social Darwinism" is a product of these wider trends, not of the specific influence of the selection theory.

Most counterfactual histories are little better than science fiction,
but I shall argue that the situation is different in this case, permitting

a somewhat more academically respectable model of the alternative universe to be created. Even in our own world, most late nineteenth-century evolutionism was non-Darwinian in character, so it is much easier to imagine how things might have looked had the theory of natural selection been absent altogether. We are simply taking out what most scientists and other thinkers regarded as the least convincing and most controversial part of Darwin's thinking.

It is important to emphasize that supposing the possibility of alternative ways in which science could have developed does not undermine the assumption that science is based on an objective study of the real world.[3] Human interests and priorities may determine which areas of research get opened up and which theories are explored, without implying that the work that *was* done in our own world was somehow led up a blind alley with no objective significance. Without Darwin, evolutionists simply wouldn't have been interested in, for instance, artificial selection and would have focused on other things (as indeed most of them did). This might have slowed down the investigation of heredity and variation, but other areas of research would have received more attention. Topics that have reemerged with the advent of evolutionary developmental biology might have flourished and would not have been swept under the carpet, as they were in our world, by the transition to a genetics focused solely on the transmission of characters. I believe that by the late twentieth century, scientific biology would be employing more or less the same battery of theories and methods that we have in our world—but the sequence of discovery and development would have been different, with immense consequences for our perception of the wider implications of the various components.

DARWIN'S ORIGINALITY

The claim that the theory of natural selection was somehow "in the air" in the mid-nineteenth century rests on the assumption that since all of the components of the theory had become widely available, it was only a matter of time before someone put them together and articulate the idea. It is argued that others besides Darwin did indeed articulate it, most obviously Patrick Matthew and Alfred Russel Wallace, and there are minor literary industries devoted to arguing that these people should be recognized as the true discoverers of the theory. But some of these claims are based on a huge oversimplification of what the full theory entails. Darwin discovered natural selection after he had already recognized that evolution must be represented as a branching tree in which populations diverge from a common ancestry when they are exposed to different environments, most obviously due to separation by geographical barriers. He realized that this model could explain the way in which we classify organisms into groups, with common ancestry being the basis of the underlying similarities that unite superficially divergent forms. In addition he looked to artificial selection, both as a means of helping to understand how the natural equivalent worked, and as a source of information about variability and heredity.[4]

The case for Matthew breaks down because although he conceived the idea of selection, he had no idea how to use it. Wallace came to his theory by a route that is in many ways very similar to Darwin's, having independently articulated the model of the tree of life in 1855 and then recognizing the implications of Malthus's principle of population. Darwin himself thought that Wallace had independently discovered the whole theory, and this assumption has underpinned most subsequent interpretations of Wallace's 1858 paper. But several historians have pointed out that there were significant differences between the ways in which the two discoverers articulated the idea.[5] Wallace does not seem to have conceived the process of selection in terms of a ruthless struggle between individual variants within a population. I would argue that he was unclear about the role of individual selection and if left to himself would have preferred to focus on a kind of group selection. He did not accept a relationship between natural and artificial selection, a key resource in Darwin's efforts to explain how individualistic natural selection worked, and was critical of the theory of sexual selection. It seems clear that a theory developed by Wallace alone would have been significantly different from the one the Darwin had developed. We must also note that in 1858 Wallace was a relatively unknown naturalist working in the Far East, and he did not return to Britain until 1862. Even if he had been able to get his paper published, its influence would have been minimal—remember that in our world the joint Darwin-Wallace papers attracted very little attention. It would have taken Wallace years to prepare anything that could have had an impact equivalent to the *Origin of Species.*

If one discounts Wallace, it is hard to think of any other naturalist who might have been in a position to develop the theory of natural selection instead of Darwin. There was certainly a growing sense that some form of natural transmutation would be preferable to the old idea of divine creation, so it is plausible to argue that the general idea of evolution was "in the air." James Secord's work on the reception of Robert Chambers's *Vestiges of the Natural History of Creation* is perhaps the best evidence we have that the basic idea was beginning to be widely discussed by the 1850s (Secord 2000). Herbert Spencer had written in support of Lamarck in 1851, and his *Principles of Psychology* of 1855 explained the evolution of the human mind in terms of the inheritance of acquired characteristics. He would have written his *Principles of Biology*, published in 1864, even without Darwin, because it too promoted a Lamarckian approach to evolution. Spencer may have coined the phrase "survival of the fittest," but he always thought the process was secondary to the Lamarckian effect. In Germany it seems probable that Ernst Haeckel's commitment to a naturalistic philosophy would have led him toward the morphological evidence for evolution, perhaps a little more slowly without an input from Darwin. We should also bear in mind that it was Haeckel's *Generelle Morphologie* of 1866 that persuaded T. H. Huxley to begin using evolution in his scientific work, not the *Origin of Species* (see Desmond 1982).

Haeckel's enthusiasm for Darwinism in our world opens up another intriguing prospect. It is normally assumed that Darwinism is a product of the Anglophone tradition in natural history, a transformation of Paley's natural theology in the light of Lyell's geology, Malthus's principle of population, and the British obsession with breeding pigeons, dogs, and other domesticated species. But Robert J. Richards and other scholars have noted an influence from German biology in Darwin's thinking, while in more recent work Richards has argued against the efforts of those historians (the present writer included) who have suggested that Haeckel's Darwinism was a very distorted version of the program articulated in the *Origin of Species* (Richards 2008).[6] Is it possible, then, that in a world without Darwin, Haeckel could have independently discovered the theory of natural selection? I have to say that I regard this as highly unlikely, so unlikely that the exercise merely reinforces my suspicions about Haeckel's true sympathies. He certainly understood the selection theory when he read Darwin's book, but he had little interest in biogeography and, like most German readers, little understanding of the significance of Darwin's discussions of artificial selection and the variability of domesticated species. Haeckel would have developed an evolutionism on his own, but it would have relied on Lamarckism to explain adaptive evolution, and have appealed even more strongly to morphological evidence.

I thus suggest that Haeckel and Spencer, in their very different ways, would have articulated a theory of progressive evolution based on Lamarckism, and that their work would have encouraged a gradual transition to evolutionism in the course of the 1860s. It might have been a slower process at first, without the input from Darwin's theory and evidence, but it would also have been less traumatic, at least in Britain. Those naturalists who did not believe that adaptation was the whole story would have developed theories of orthogenesis based on the idea that variation was controlled by forces programmed within the organism. The recapitulation theory would have flourished, allowing embryology to be used as a tool for understanding the evolution of life in addition to the fossil record. The reconstruction of the whole history of life on Earth, never one of Darwin's priorities, would have become the dominant evolutionary research program.

All of these things, of course, happened in the European and American context, and most of the research was done without appeal to the theory of natural selection (Bowler 1983, 1996). The one thing that would be missing in a world without Darwin would be the theory of individual natural selection. The possibility that the less successful species would eventually be driven to extinction would almost certainly have been recognized, since this was accepted even by many opponents of Darwinism in our world. But there would be no agonized discussions of the nature of individual variation, and no debates about blending heredity and selection, since such discussions were prompted by Darwin's work and many of them were intended to discredit it. I suspect that the idea of natural selection would not have been introduced until the turn of the century, when growing concerns about the social implications of heredity would have focussed attention onto this issue and prompted a belated recognition of the significance of artificial selection. Only then would it have been recognized that a natural equivalent might serve as an alternative to the hitherto unchallenged Lamarckian theory.

EVOLUTION AND RELIGION

The question then remains: in a world that was developing an evolutionary program very similar to the one we are familiar with, but without the theory of individual natural selection as an explanation of adaptation, how would the wider implications of the theory have been perceived? One major component of the religious debate would remain unchanged. It would still be difficult for conventional Christians to accept the evolution of humans from an animal ancestry— and there would still have been plenty of cartoons highlighting the incongruity of the link between humans and apes. Original sin would still have to be replaced with the idea that God's plan allowed for the upward progress of life, mind, and society.

Nevertheless, in our own post-Darwinian universe we know that many religious thinkers were able to reconcile themselves to the idea of a natural origin for humankind providing they were able to believe that the process of evolution itself was progressive and purposeful (J. Moore 2009; Bowler 2007). Lamarckism was widely accepted as just the kind of process that a wise and benevolent God would have instituted to do His will. Liberal Protestant clergymen became, in James Moore's words, "Herbert Spencer's henchmen"—and it was Spencer's Lamarckian version of self-improvement (discussed below) that was most attractive (J. Moore 1985, 1979). Charles Kingsley's *Water Babies*, widely perceived as a Darwinian text because of its many references to the scientific debates, also promotes a form of self-help Lamarckism adapted to the ideals of muscular Christianity. Deeply religious scientists such as the Quaker palaeontologist Edward Drinker Cope also saw Lamarckism as the ideal basis for a theology of evolution. Outside the churches, writers such as Samuel Butler and later George Bernard Shaw all expressed a passionate preference for Lamarckism as the way to retrain a sense of purpose in the world.

It was natural selection that served as the clearest target of the religious and moralistic thinkers' ire. The problem was that Darwin had confronted them from the beginning of the debate with evolutionism in its most materialistic form. Natural selection seemed to be the antithesis of design—it was trial and error based on a ruthless struggle for existence, undermining faith in both the wisdom and the benevolence of the Creator. But now imagine a world in which Bishop Samuel Wilberforce did not have to face a T. H. Huxley armed with the selection theory, but could instead move comfortably toward the model of theistic evolutionism being developed by Richard Owen. This would be a world in which Asa Gray did not have to agonize over how to reconcile selection with design, but could simply assume that the Creator would somehow load the dice to give more adaptive than nonadaptive variants. Imagine Butler never having to confront the "nightmare of waste and death" represented by natural selection. Even those naturalists who did not focus on adaptation, such as St. George Jackson Mivart, were appalled by the haphazard and open-ended nature of Darwin's theory and appealed instead to a more orderly creative pattern unfolded through laws of predetermined variation. Natural selection became a kind of bogeyman invariably called in to frighten those who worried about the religious and moral implications of evolutionism, and a great deal of work had to be done to convince everyone that it wasn't the only form of evolution available. In a world without Darwin, evolution would have appeared in its most palatable form, as a vision of orderly, purposeful development, and much of the trauma would have been avoided. Most late nineteenth-century religious thinkers would have found it far easier to make the transition to an evolutionary worldview.

This raises an interesting prospect, if we dare to extend our counterfactual world a bit closer to the present. The revival of religious enthusiasm in America, which we know as fundamentalism, would presumably have happened whatever the status of evolutionism, since it was a reflection of deep social and cultural tensions (Larson 1998). But it is conceivable that without the Darwinian bogeyman as so visible a target, the fundamentalists' attention would have been deflected elsewhere as they sought to identify the foundations of the modernism they distrusted. The automatic assumption that evolutionism equates with atheism, so prevalent in the creationist literature of the twentieth and early twenty-first centuries, would be harder to sustain if the world did not have its attention fixated on the Darwinian confrontations of the 1860s as symbols of the "war" between science and religion. Even if we assume that some form of selection theory would eventually have been developed around 1900, it would have been absorbed into a more secure evolutionism that had not been looking over its shoulder at the spectre of Darwinism for half a century. I submit that there is at least a possibility that in a world without Darwin, the claim that evolutionism must necessarily entail the destruction of religious belief would not be haunting our concerns about the future development of the theory.

SOCIAL DARWINISM

Turning to the pernicious social implications of Darwinism, the counterfactual approach will allow us to ask how many of these supposed derivatives would have arisen in a world without the selection theory. I will argue that we have good reasons, based on what actually happened, for believing that most of the social implications of evolutionism would have been the same even if the movement was based on non-Darwinian theories. Indeed, it can be argued that Darwin's theory actually mitigated the worst excesses of biological racism. The problem is that much uncritical literature has been written on the assumption that any reference to evolutionary progress or the struggle for existence must imply an input from Darwinism. But once we recognize that there are other theories of evolution, which played an important role even in our world, then we are forced to examine more clearly the references to evolutionary models in ideological discourses. The concept of the struggle for existence was introduced by Malthus long before Darwin used it as a component of his theory and was thus available even to those who did not believe in evolution. It is also clear that the evolutionist can exploit several different ways of arguing that struggle can be the driving force of progress. These are often confused with the Darwinian selection model, but that confusion could not happen in a world without Darwin, and the ideological implications of the other theories would be more clearly apparent.

One underlying problem is the many different levels at which the model of struggle can be employed. Is the competition between individuals in the same population, or between rival populations—varieties or species in biology, nations or races in social theory? Although the *Descent of Man* introduces a theory of group selection to explain the origins of the moral sense, Darwin's basic theory focused on individual competition, for which the most obvious social model is the free-enterprise ideology associated with the rise of capitalism. This is the classic form of social Darwinism most often linked to the influence of Herbert Spencer. But later in the century the attention of social writers switched to nationalism, militarism, and imperialism, for which the most obvious biological model would be competition between species seeking to exploit similar resources in the same location. This was certainly part of Darwin's theory, but it is not the main form of natural selection, and this level of competition can be recognized without believing that individual competition works within the rival groups. At the social level, ideologies of individual competition were usually seen as antithetical to those of nationalism and imperialism because what benefits the individual may not benefit the nation (think of an arms manufacturer). The fundamentalists of the early twentieth century claimed that Darwinism had inspired German militarism during World War I—but it is quite conceivable that the worship of struggle would have emerged as an ideology even without the Darwinian model. Where Darwinism does seem to have played a role was as a convenient symbol of the struggle and death in nature, and hence how this might be justified in society. Our problem is to decide whether the absence of the symbol provided by Darwin's scientific theory would have limited the impact of ideologies based on progress and struggle.

Let us begin with the classic version of social Darwinism, the supposed use of the basic theory of natural selection to justify free-enterprise individualism. This link was noted both by Marx and Engels and has long been a feature of left-wing critics of Darwinism who claim that the biological theory is no more than a projection of the ideology onto nature. It has long been recognized that this link was, at the time, associated most openly with the social philosophy of Herbert Spencer, which was often popularized through an apparent analogy with the Darwinian theory. The fact that Spencer coined the term "survival of the fittest" seemed to justify the assumption that he was a Darwinian in his biological thinking, and hence that he simply translated the process of natural selection into his social thinking. Since Spencer was presented as the most extreme advocate of individualism, it was natural to assume that he must have thought that social progress would result from the continual elimination of the unfit.

The recent account of Spencer's thought by Mark Francis has challenged his identification with an ideology of "progress through struggle," but recognizes that his followers, especially in America, seem not to have been aware of the extent to which he backed away from some of his early ideas (Francis 2007).[7] This means that Spencerianism did exist in the form we have long imagined—but I have long argued that when we unpack those aspects of Spencer's thought that gave rise to his reputation as a social Darwinist, we find that their biological components are more properly interpreted as a form of Lamarckism. Like many of his contemporaries, Spencer did occasionally worry about the prolific breeding of unfit members of society, but he certainly did not advocate the constant elimination of a significant proportion of the population through starvation. It's hard to think of any so-called social Darwinist in the individualist mode who did openly welcome the prospect of the wholesale death of significant numbers in every generation—which a simple analogy with natural selection would entail. In fact, the usual result anticipated by the advocates of unrestrained free enterprise was that the competition between individuals would encourage the least able (the "unfit") to greater efforts. The threat of starvation was the best teacher, and only a very few degenerates would fail to benefit from the lesson. And when we recall that Spencer's *Principles of Psychology* had already explained the evolution of human mental faculties in purely Lamarckian terms, we can see how he could develop an evolutionary application of his ideology without any recourse to natural selection. As organisms (and humans) struggled to improve themselves, they would pass their acquired characters on to their offspring, thus improving the race as well as themselves.

Given Spencer's interests in biology, psychology, and sociology, it is hard to believe that he would not have developed something very similar to the social evolutionism we associate with him, even in a world without Darwin. All those American capitalists identified by Richard Hofstadter as "social Darwinists" would still have been able to hail him as a hero—although they would not have had that pejorative label attached to them by their opponents. But what about the other ideology of "progress through struggle," which focussed on national conflict, militarism, and imperialism? These were all factors that Spencer strove to eliminate, yet the later decades of the century saw a growing enthusiasm for this model, culminating in the efforts to justify the policies of national rivalry that generated the Great War. The very fact that this development took place decades after the original Darwinian debate suggests that the introduction of the selection theory probably did little to inspire it, although catch-phrases such as the "survival of the fittest" could certainly be adapted to this new use. Darwin's theory did include an element of competition between varieties and species seeking to exploit similar resources, although this is not the primary mechanism of natural selection, and it is perfectly possible to believe that the least successful products of evolution will be eliminated without seeing natural selection as the process that actually produces them. Indeed, most of the non-Darwinian evolutionists in our own world accepted the negative role of natural selection as a weeding-out mechanism at this level.

There were several alternative sources from which this secondary level of selection could be derived. Malthus had originally introduced the term "struggle for existence" in the context of competing tribal groups. The most obvious scientific application of this insight is biogeography, where most biologists used terms such as "conquest" and "colonization" when describing episodes in which a species gained access to new territory and displaced the original inhabitants.[8] Even Alfred Russel Wallace, certainly no friend to imperialism, slipped into this kind of terminology on some occasions. In Germany, Ernst Haeckel would almost certainly have developed similar views on interspecies struggle, even if his evolutionism had been completely focussed on Lamarckism as the cause of the original differentiation of species. It is worth noting that the early twentieth-century paleoanthropologists who celebrated the elimination of "unsuccessful" human species, such as the Neanderthals, including W. J. Sollas and Arthur Keith, were not enthusiasts for the theory of individual natural selection (Bowler 1986).

This leads us to another element in the social implications attributed to Darwinism, focused on the sensitive topic of race. There is no doubt that Darwinism did become associated in many people's minds with the claim that some human races are closer to our ape ancestors than others. But as the latest work by Adrian Desmond and James Moore reminds us, Darwin was passionately opposed to the more extreme versions of race biology that were actually promulgated by polygenist creationists such as Louis Agassiz and non-Darwinian physical anthropologists such as James Hunt (Desmond and Moore 2009). It was the theories of parallel evolution promulgated by Darwin's opponents that allowed the races to be seen as entirely distinct, while Darwin and his closest supporters argued for the unity of the human family. They did concede that some races had advanced more rapidly than others, but this was an almost universal attitude at the time. It was closely associated with the hierarchy of cultural stages created by anthropologists and archaeologists such as E. B. Tylor and L. H. Morgan. And, as the work of George Stocking and many others tells us, that development took place almost entirely independently of the Darwinian revolution in biology (Stocking 1987). It fostered a linear, hierarchical view of culture and race, which the Darwinians accepted, but which was really at variance with the branching model of evolution at the heart of their biological theory. I think it can very easily be argued that in a world without Darwin, there would have been more rather than less misapplication of biology to the race question. To suggest that the racism of the Nazi party in Germany was primarily inspired by Darwinian natural selection is to misunderstand completely the role of evolutionism—and other areas of science—in the creation of such attitudes.

My final topic is eugenics, the policy of restricting the reproduction of those deemed "unfit," which became so popular at the end of the nineteenth century and remained popular until discredited by the excesses of the Nazis. There is no doubt that Darwin's theory contributed to this movement. Francis Galton was Darwin's cousin and was clearly inspired by the analogy with artificial selection that Darwin used to illustrate his theory. But there were many other factors that contributed, and it is worth noting that Galton himself favored saltations as the source of entirely new species. Galton was largely ignored at first, and the growing enthusiasm for policies based on the rigid inheritance of human characteristics toward the end of the century had wide social and cultural foundations. This enthusiasm could have drawn on several sources for scientific inspiration if Darwin's theory had not been available. John Waller has shown that there had long been a concern about hereditary illnesses in medical circles. He notes how easily this could be fused with the enthusiasm for artificial selection that everyone knew (even without Darwin) to be the foundation of animal breeding (Waller 2001; generally Kevles 1985). Medical writers also talked of the need to excise diseased parts of the body as an equivalent to preventing the spread of unfit characters in the population. If the surge of interest in human heredity led to the discovery of a particulate model of inheritance, it should be noted that the geneticists of our world were not at first enthusiasts for natural selection and drew the inspiration for their science's human applications directly from the breeders' activities.[9]

What Darwinism did provide was the assumption that the elimination of the unfit was a natural process that, if checked by civilization, might lead to the degeneration of the race and eventual extinction. In many people's minds this was linked more with the ideology of race competition than with individualist natural selection, but few would have worried about this distinction at the time. I suspect that if writers such as Weikart who seek to portray Darwinism as a direct cause of Nazism have any hope of sustaining their case against the counterfactual arguments outlined above, this is where they have their best chance. In a world without Darwin, there would certainly have been a eugenics movement, but without the idea that artificial selection had a natural equivalent, it would have been less easy to justify claims that the unfit should be eliminated (as opposed to merely discouraged from breeding). Even here, though, it should be remembered that a good pigeon breeder only improved his flock by killing a large number of birds, so artificial selection could provide an alternative source for the ruthlessness of the more extreme eugenic policies. But it has to be conceded that artificial selection was the one part of Darwin's theory that did not resonate well with German culture, so this is an area where more research could be done to assess the potential inputs from various sources.

CONCLUSION

In this all too brief survey of the broader implications of my imagined counterfactual universe, I have tried to show that many of the alleged implications of Darwinism can be associated equally well with other forms of evolutionism. The theory of natural selection, long proclaimed as a typical product of mid-nineteenth century thought, was in fact very hard for most people of the time to accept. Without Darwin, they may well not have thought of it at all, at least for several decades, yet they would have developed an evolutionary model based on the non-Darwinian alternatives that were actually preferred by most thinkers even in our own world. Natural selection would only have appeared in the early twentieth century—which is when it did, in fact, at last become a key feature of scientific biology. My argument is that such a belated appearance of the selection theory would have left the way clear in the late nineteenth century for a much less traumatic emergence of the evolutionary movement. The theory would not have been associated so immediately with irreligion and it would not play so iconic a role in the rhetoric of the "war" between science and religion. This would allow the selection theory to be absorbed more easily into the culture of the later twentieth century. But most of the ideological implications attributed to Darwinism, including those which fed into the confused situation in German culture that gave rise to Nazism, would have been in place anyway. For us, Darwinism has come to serve as a symbol of all that is most disturbing about evolutionism. There can be little doubt that this symbol had played a powerful role in shaping the way our culture responds to the theory. But we must beware of attributing too much power to the symbol, and I have tried to argue that counterfactual history is a promising way of trying to identify the other scientific and cultural agents whose presence has been masked by our focus on the image of Darwin as the centrepiece of the evolutionary movement.

NOTES

1. Weikart makes no secret of his commitment to intelligent design theory. For an earlier and equally controversial exposition of a similar thesis see Gasman (1971). For a response to Weikart, see for instance Richards (2008, 2013) and chapter 8 in this volume.

2. A noted opponent of counterfactualism was the historian E. H. Carr; see the new edition of his *What Is History?* (Carr 2001). Collections of counterfactual histories by working historians include Ferguson (1997) and Cowley (1999, 2001). These include numerous examples of the technique applied to military history; a fictional example of the technique applied to the outcome of the Battle of Gettysburg is W. Moore (1955). For a brief outline of my own views see Bowler (2008) and more fully Bowler (2013).

3. For a brief introduction to the philosophical issues raised by counterfactual histories of science, see Radick (2008), which introduces the "Focus Section" in which Bowler (2008) appeared.

4. For a slightly more detailed outline of the points raised here see Bowler (2009). A useful collection of the primary sources relevant to the "precursors" is McKinney (1971).

5. Malcolm Kottler provides a useful outline of the debates over Wallace's theory in Kottler (1985).

6. Sander Gliboff argues that Haeckel did accept the contingency of Darwin's branching model of evolution, but concedes that he tended to think of variation as a positive response to the environment; see Gliboff (2008).

7. The classic account is Hofstadter (1959). This has been challenged by Bannister (1979). On the idea of struggle at the national level see Crook (1994). For my own views on social Darwinism see Bowler (1993).

8. On the metaphors of imperialism in biogeography see Bowler (1996, chap. 9). For Malthus and the struggle for existence, see Bowler (1976).

9. On the role played by non-Darwinian evolutionary ideas in the foundation of Mendelism, see Bowler (1989).

REFERENCES

Bannister, R. C. (1979). *Social Darwinism: Science and Myth in Anglo-American Social Thought*. Philadelphia: Temple University Press.

Bowler, P. J. (1976). "Malthus, Darwin and the Concept of Struggle." *Journal of the History of Ideas* 37:631–50.

———. (1983). *The Eclipse of Darwinism: Anti-Darwinian Evolution Theories in the Decades around 1900*. Baltimore: Johns Hopkins University Press.

———. (1986). *Theories of Human Evolution: A Century of Debate, 1844–1944*. Baltimore: Johns Hopkins University Press.

———. (1989). *The Mendelian Revolution: The Emergence of Hereditarian Concepts in Modern Science and Society*. London: Athlone.

———. (1993). *Biology and Social Thought, 1850–1914*. Berkeley: Office for the History and Philosophy of Science and Technology, University of California.

———. (1996). *Life's Splendid Drama: Evolutionary Biology and the Reconstruction of Life's Ancestry, 1860–1960*. Chicago: University of Chicago Press.

———. (2007). *Monkey Trials and Gorilla Sermons: Evolution and Christianity from Darwin to Intelligent Design*. Cambridge, MA: Harvard University Press.

———. (2008). "What Darwin Disturbed: The Biology That Might Have Been." *Isis* 99:560–67.

———. (2009). "Darwin's Originality." *Science* 323:223–26.

———. (2013). *Darwin Deleted: Imagining a World Without Darwin*. Chicago: University of Chicago Press.

Browne, J. (2005). "Commemorating Darwin." *British Journal for the History of Science* 38:251–74.

Carr, E. H. (2001). *What Is History?* New ed. London: Palgrave Macmillan.

Cowley, R., ed. (1999). *What If? The World's Foremost Military Historians Imagine What Might Have Been*. New York: Scribners.

———. (2001). *More What If? Eminent Historians Imagine What Might Have Been*. New York: Scribners.

Crook, P. (1994). *Darwinism, War and History: The Debate over the Biology of War from the "Origin of Species" to the First World War*. Cambridge: Cambridge University Press.

Desmond, A. (1982). *Archetypes and Ancestors: Palaeontology in Victorian London, 1850–1875*. London: Blond and Briggs.

Desmond, A., and J. R. Moore. (2009). *Darwin's Sacred Cause: Race, Slavery and the Quest for Human Origins*. London: Allen Lane.

Ferguson, N., ed. (1997). *Virtual History: Alternatives and Counterfactuals*. London: Picador.

Francis, M. (2007). *Herbert Spencer and the Invention of Modern Life*. Stocksfield: Acumen.

Gasman, D. (1971). *The Scientific Origins of National Socialism: Social Darwinism in Ernst Haeckel and the German Monist League*. New York: American Elsevier.

Gliboff, S. (2008). *H. G. Bronn, Ernst Haeckel, and the Origins of German Darwinism: A Study in Translation and Transformation*. Cambridge, MA: MIT Press.

Hofstadter, R. (1959). *Social Darwinism in American Thought*. Rev. ed. New York: George Braziller.

Kevles, D. (1985). *In the Name of Eugenics: Genetics and the Uses of Human Heredity*. New York: Knopf.

Kottler, M. (1985). "Charles Darwin and Alfred Russel Wallace: Two Decades of Debate over Natural Selection." In *The Darwinian Heritage: A Centennial Retrospect*, ed. D. Kohn, 367–432. Princeton: Princeton University Press.

Larson, E. J. (1998). *Summer for the Gods: The Scopes Trial and America's Continuing Debate over Science and Religion*. New York: Basic Books.

McKinney, H. L., ed. (1971). *Lamarck to Darwin: Contributions to Evolutionary Biology, 1809–1859*. Lawrence, KS: Coronado Press.

Moore, J. R. (1979). *The Post-Darwinian Controversies: A Study of the Protestant Struggle to Come to Terms with Darwin in Great Britain and America*. Cambridge: Cambridge University Press.

———. (1985). "Herbert Spencer's Henchmen: The Evolution of Protestant Liberals in Late Nineteenth-Century America." In *Darwinism and Divinity: Essays on Evolution and Religious Belief*, ed. J. R. Durant, 76–100. Oxford: Basil Blackwell.

———. (2009). "Religion and Science." In *The Cambridge History of Science*, vol. 6, *The Modern Biological and Earth Sciences*, ed. P. J. Bowler and J. V. Pickstone, 541–62. Cambridge: Cambridge University Press.

Moore, W. (1955). *Bring the Jubilee*. London: Heinemann.

Radick, G. (2008). "Introduction: Why What If?" *Isis* 99:547–51.

Richards, R. J. (2008). *The Tragic Sense of Life: Ernst Haeckel and the Struggle over Evolutionary Thought*. Chicago: University of Chicago Press.

———. (2013.) *Was Hitler a Darwinian? Disputed Questions in the History of Evolutionary Theory.* Chicago: University of Chicago Press.

Secord, J. A. (2000). *Victorian Sensation: The Extraordinary Publication, Reception and Secret Authorship of "Vestiges of the Natural History of Creation."* Chicago: University of Chicago Press.

Stocking, G. W. (1987). *Victorian Anthropology.* New York: Free Press.

Waller, J. (2001). "Ideas of Heredity, Reproduction, and Eugenics in Britain, 1800–1875." *Studies in the History and Philosophy of the Biological and Biomedical Sciences* 32:457–89.

Weikart, R. (2004). *From Darwin to Hitler: Evolutionary Ethics, Eugenics, and Racism in Germany.* New York: Palgrave Macmillan.

SIXTEEN

WHAT FUTURE FOR DARWINISM?

Jean Gayon

What future for Darwinism? I will propose some criteria for exploring this question in the domains of both evolutionary biology and the human sciences. Do not expect me to tell you where we will stand thirty years from now. It will be enough to identify a few general tendencies. For the sake of brevity, I will not devote a preamble to explain why I include both evolutionary biology and the human sciences in my enquiry, and why I distinguish the two.

EVOLUTIONARY BIOLOGY: EXTENSION, REPLACEMENT, OR EXPANSION?

Let us begin with evolutionary biology, using freely a proposal that Stephen Jay Gould made in his 2002 book, *The Structure of Evolutionary Theory* (Gould 2002). In his introduction, Gould compares the theoretical framework of contemporary biology with the former classical, or purely Darwinian, framework. He maintains that the new theoretical framework is neither an "extension" of the older Darwinian theory, nor a "destruction" or "replacement" of it. In Gould's terms, "extension" means that the same principles have been applied to a wider spectrum of phenomena. "Replacement" means that other principles are now at the heart of evolutionary theory. Instead of these two terms, Gould prefers "expansion." "Expansion" means that the same principles remain central to the theory, but they have been "reformulated" in a way that gives a truly different aspect to the entire edifice.

The trilogy—extension/replacement/expansion—is rather unorthodox in terms of current philosophy of science. Philosophers of science are familiar with the first two conceptions of scientific change: "extension" connotes the idea of a wider descriptive and explanatory scope; "replacement" evokes Thomas Kuhn's "paradigm shift." But the notion of "expansion" does not seem to be a classical one in the philosophy of science. Although Gould was quite clear in explaining what "expansion of the Darwinian framework" meant in practice, he was not perfectly explicit about what "expansion" could mean in general for scientific theories. However, two criteria seem to emerge from his analysis:

1. Generalization of the principles of a theory
2. Addition of new principles

I will return later in this paper to the manner in which Gould applied these general criteria to evolutionary theory, but will not go further here into the analysis of Gould's proposal (see Gayon 2008).

I will instead freely use Gould's distinction between "extension," "replacement," and "expansion" in order to interpret major changes that are presently visible in evolutionary theory at a level of abstraction higher than that considered by Gould. The perspective defended echoes the late Ernan McMullin's thoughts about the necessity for philosophers to take history of science into account if they want to correctly appreciate the issue of "fertility" of theories (McMullin 1976).[1] The future outcome of these tendencies is presently unpredictable, but exploring an uncertain future is precisely what I want to do.

Darwin designates his theory as a theory of "descent with modification through natural selection" (Darwin [1859] 1964, 343, 359).[2] This formula suggests that Darwin's theory was made of two components: "descent with modification" and "natural selection." Although Darwin did not organize his *Origin of Species* in line with this distinction, he was perfectly aware of its importance. In a letter to the American botanist Asa Gray, who reacted quite favorably to the *Origin*, he wrote this revealing sentence in 1863: "Personally, of course, I care much about natural selection, but that seems to be utterly unimportant compared to the *Creation* **or** *Modification*" (Darwin to Gray, 11 May 1863, in Burkhardt et al. 2007, 403; italicized and bold characters as in text). It is worth examining whether or not the most fundamental doctrines at these two levels undergo deep changes today.

DESCENT WITH MODIFICATION

Let us first consider the "descent with modification" part of the theory. Darwin had a rather restrictive interpretation of this common idea—we might well say a model of it. This model is clearly expressed, not in a definition, but in the unique illustration given in the *Origin*, the famous branching diagram given in chapter four (fig. 16.1).

This diagram was so important for Darwin that he felt the need to comment on it at length three times in his book, using no less than eleven full pages in the "Natural Selection" chapter four, three pages in chapter ten on the geological succession of organic beings, and three pages in chapter thirteen devoted to classification and embryology (Darwin [1859] 1964, 116–26; 331–34; 420–22).[3] This diagram may be interpreted in two ways. Both interpretations are Darwinian in spirit, but Darwin explicitly developed only one in his comments on the famous diagrams.

One way of understanding Darwin's diagram is to see it as an idealized representation of what his followers called "the general fact of evolution." Many biologists and paleontologists commonly adopted this interpretation after Darwin. According to it, Darwin's diagram expresses a certain number of plausible generalizations about several big classes of facts that constitute evolution, as, for instance, modification of species, gradualness of this modification, extinction and splitting of species as the common fate of most species, indefinite divergence of those species that survive, and so forth.[4] This aspect of Darwin's theory is one of the most spectacular examples of a "paradigm shift" in the history of science.

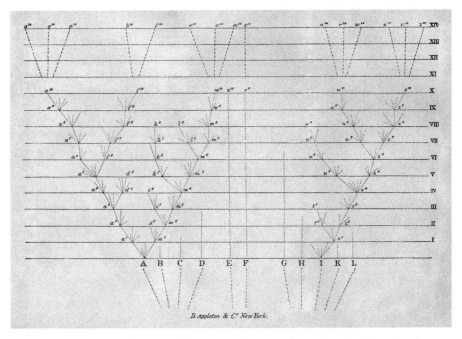

Figure 16.1. Diagram from *On the Origin of Species*, chapter 4, 6th edition (1871) showing the branching pattern of evolutionary development from common points of ancestry. Since the horizontal (diversity) and vertical (time) coordinates are purely relative, the diagram represents a picture of evolutionary relationship extending from that of varieties within species to that of major group relationships in the whole history of life. Courtesy University of Notre Dame Library.

Not all post-Darwinian evolutionists adhered to each of these postulates. For instance, the postulate of the gradualness of change has been much debated. If one does not accept it, then one will prefer a kind of branching tree with a succession of horizontal and vertical lines, rather than oblique lines. But, on the whole, Darwin's branching tree remained as an exemplary model of how evolutionary data should be summed up prior to any explanatory hypothesis. In modern terms, we might say that Darwin's tree was an attempt to grasp the general "pattern" of evolution, this "pattern" being a necessary preamble to whatever hypothesis upon the processes that produce such a pattern.

However, Darwin did not choose to comment upon his diagram in such terms. In fact, all his comments in the *Origin* bear upon processes. It is not haphazard that the diagram is introduced in the chapter devoted to natural selection. In reality, Darwin never speaks of his diagram as an idealized way of representing data, but as an idealized way of representing *what should happen in evolution if his explanatory hypotheses were correct.* Here are the key points, given in the order of the text itself.

Darwin first explains that "the modified descendants of any one species will succeed by so much the better as they become more diversified in structure, and are thus enabled to encroach on places occupied by other beings" (Darwin [1859] 1964, 116). Then Darwin conjectures that the most divergent variations should be preserved, because they allow the species to maximize the number of places occupied in nature: "as a general rule, the more diversified in structure the descendants from any one species can be rendered, the more places they will be enabled to seize on, and the more their modified progeny will be increased" (119).

Darwin also explains that the more improved branches will destroy the less improved ones and replace them: "this is represented in the diagram by some of the lower branches not reaching to the upper horizontal lines" (Darwin [1859] 1964, 119). In other words, Darwin supports the claim that competition and differential elimination occurs not only within local varieties, but also between "well-marked varieties" or "races."

Darwin extends the previous principles to the level of larger taxonomic groups: species, genera, families, etc.:

> In each genus, the species, which are already extremely different in character, will generally tend to produce the greatest number of modified descendants; for these will have the best chance of filling new and widely different places in the polity of nature . . . . As in each fully stocked country natural selection necessarily acts by the selected form having some advantage in the struggle for life over other forms, there will be a constant tendency in the improved descendants of any one species to supplant and exterminate in each stage of descent their predecessors and their original parent. . . . These two species (A) and (I), were also supposed

to be very common and widely diffused species, so that they must originally have had some advantage over most of the other species of the genus. (Darwin [1859] 1964, 121)

In time, this entails that the number of highly diversified groups is limited. One can easily see on the diagram that only a few among the initial groups reach the superior level: "very few of the original species will have transmitted offspring to the fourteen-thousandth generation. We may suppose that only one (F), of the two species which were least closely related to the other nine original species, has transmitted descendants to this late stage of descent" (Darwin [1859] 1964, 122).

Finally, Darwin considers the anomalous case of species that do not change and still persist (Darwin [1859] 1964, 124). Such cases raise a delicate question for taxonomists. For example, observe lineage "F": this species does not change, but finds itself in an intermediate position between the mean types of descendants from "A" and "I." This means that the adaptive story of species does not coincide with their genealogical history. Adaptive differentiation in the most successful groups tends to cloud over the underlying purely genealogical sequence (Darwin [1859] 1964, 421). This is why modification through natural selection is a methodological obstacle to the reconstitution of the genealogy of species ("community of descent"): "We have no written pedigrees; we have to make out community of descent by resemblances of any kind. Therefore we choose those characters which, as far as we can judge, are the least likely to have been modified in relation to the conditions of life to which each species has been recently exposed" (Darwin [1859] 1964, 425).

We see therefore that Darwin's comments on his diagram are laden with hypotheses about evolutionary processes: variation, natural selection, inter-group competition, and differential extinction. Moreover, Darwin repeatedly utilizes the term "process" all through this section of the *Origin*. It is then clear that the tree-diagram presented in the chapter on "natural selection" (not a coincidence) can definitely not be interpreted as being just a "pattern" in the sense utilized by modern evolutionary biologists.

I have carefully commented on Darwin's branching diagram because it is perhaps on this issue that we observe today the most spectacular departure from the traditional Darwinian orthodoxy. Until a few years ago, Darwin's diagram had not been seriously challenged except for details relative to its shape: should the branches be oblique or not (gradualness of evolution)? Is it true that higher taxa differentiate in the same erratic way as varieties and species? These have been recurrent criticisms of the tree from the beginning. They do not affect the topology of genealogical relationships. They modify only the shape of the diagram. The members of the cladist school have, however, developed another sort of criticism: they have reproached the Darwinian tradition for confusing processes and patterns.[5] In the cladists' perspective, a phylogeny should not aim at representing more than the hierarchical structure of embedded taxa, with no allusion to the processes and events that have led to the diversification of species. None of these criticisms, however, challenge the overall vision of evolution as a genealogy of species undergoing irreversible differentiation.

Several major classes of phenomena challenge this view today. The main phenomena are lateral gene transfer and symbiosis. Since the end of the 1990s, lateral gene transfer has been known to be a major phenomenon among prokaryotes (Eubacteria and Archaea). The interesting point is that it does not occur only between members of the same species, but also between members of lineages that may have diverged a very long time in the past (for instance a hundred million years or more). Prokaryotes are single-celled organisms without a membrane-bound nucleus. Since Carl Woese's pioneering work in the 1970s and 1980s, they have been divided into two "domains": bacteria and archaea. Eukaryotes constitute the third "domain": they include all single-cell and multicellular organisms with a membrane-bound nucleus. At the beginning, the application of molecular techniques to prokaryotes generated the hope of reconstructing the unique and universal tree of life (Woese 1987). But an increasing number of anomalies led to the discovery that gene transmission among prokaryotes is not only vertical, but also horizontal. The magnitude of this phenomenon has been fiercely debated for almost twenty years. It seems now clear that it is important enough to cast serious doubts on the very existence of a unique tree representing the natural relationships among all cellular organisms or, at least, the possibility of reconstituting this tree if it exists (for a review, see McInerney, Cotton, and Pisani 2008). Whether this turnover of genes is also significant among eukaryotes remains a debated issue.

But even if lateral gene transfer is evolutionarily significant only for prokaryotes, this is enough to make unsatisfactory the idea of a unique "tree of life" representing the totality of the genealogical relationships between living beings: systematists will have to manage some combination of tree-like and network-like modes of representation.[6] It is likely that both the "tree" and the "network" will have to coexist in a pluralistic view of evolutionary patterns, where *several* histories are told by several methodologies (Doolittle and Bapteste 2007). In fact, evolutionary biologists now seem to admit that Darwin's diagram, which focuses exclusively at the level of organisms and species, encapsulates only one fraction of the history of life. In the case of prokaryotes at least, lateral gene transfer has been important enough to obliterate and, perhaps, to overcome "vertical transmission" at a large historical scale.

Another process that jeopardizes Darwin's representation of genealogy is symbiosis. Symbiosis has probably been a major evolutionary process, especially at the level of cell evolution (Margulis and Sagan 2002, chap. 16). Symbiosis is much more rare than lateral gene transfer; but when it occurs, it has dramatic effects because it implies the coexistence of two full genomes. Symbiosis seems to have been a key process in the emergence of a number of major groups of unicellular eukaryotic organisms. Just like horizontal gene transfer, but at a higher level, symbiosis introduces complications that cannot be assimilated by an exclusively tree-like phylogenetic pattern. The effects of both lateral gene transfer and symbiosis on the "tree of life" are schematically represented in figure 16.2.

These recent challenges addressed to Darwin's tree of life can easily be interpreted in terms of the three modes of theoretical change mentioned earlier—"extension," "replacement," and "expansion." We cannot say that Darwin's tree has been *replaced* by something else (the "network"), because the tree remains an appropriate representation for a huge number of phylogenetic relationships between species. But it must be combined with a network approach, especially

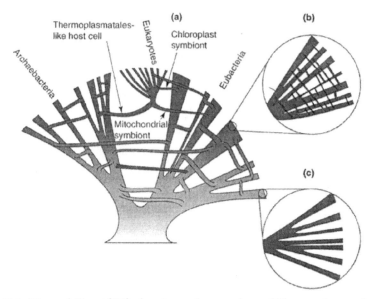

Figure 16.2. Network Tree of Life showing entire genealogy of life as a mixture of tree and network. Source: McInerney, Cotton and Pisani (2008, 278); used with permission. Original legend: "A network of life. (a) Eukaryotes are known to be chimeric, with chloroplast and mitochondrial genes having a different origin from nuclear genes, and originating from different groups of Eubacteria. There is more debate over whether lateral gene transfer between Eubacteria, Archaeobacteria and Eukaryoa and between major groups of prokaryotes is common or rare. (b) Lateral gene transfer is known to be so common between members of some groups of bacteria that they are effectively panmictic, whereas (c) it can be completely absent from other groups."

for prokaryotes, that is to say, for the major part of the history of life both in terms of time span (the first three billion years) and in terms of present biomass. Neither can we say that Darwin's notion of descent with modification has been "extended" in the sense of applying Darwin's tree-like model to new classes of phenomena. In reality, what has occurred is that new sources of variation that were not considered by Darwin (horizontal transfer of infra-organismic entities such as genes, membranes, organelles, etc.) dramatically affect the topology of phylogenetic relationships. Gould's notion of "expansion" seems here to be a plausible interpretation: Darwin's genealogical approach remains, but new sources of variation force us to consider other models of descent with modification.

With reference to Gould's two criteria for "expansion" of a theoretical framework—generalization of the principles of a theory and addition of new principles—this is exactly what we observe in the present case:

1. The principle of "descent with modification" is being generalized, in the sense that it applies not only to lineages of organisms, species, and higher taxa, but also *and independently* (that is to say, not only in a transitive manner) to infra-organismic levels.
2. We also observe an addition of new principles: new sources of variation (lateral gene transfer, symbiosis) result in a different way of representing phylogeny. Note in passing the interference between *pattern* and *process*. Just as Darwin's tree relied upon his idea of the processes of variation and natural selection, the contemporary vision of phylogenies relies upon new knowledge about variation.

EVOLUTION THROUGH NATURAL SELECTION

Let us now consider the other side of "Darwin's theory," the explanatory side. Recall what Gould meant when he proposed to see modern evolutionary theory as an "expansion" of the traditional Darwinian framework (Gould 2002, 53 and 14–15). He meant:

1. Generalization of Darwinian (or more precisely synthetic) principles: not only individual natural selection, but also species selection, species drift, clade selection, and clade drift. In other words: application of Darwinian schemes to entities that were not considered by Darwin.
2. Addition of new principles, such as morphological, developmental, and historical constraints. These are not external limits but positive factors that canalize evolutionary pathways.

I would like here to concentrate on the first aspect: the generalization of Darwinian principles, and more especially the generalization of the principle of natural selection. This seems to have been a broad tendency in evolution theory since the 1970s. Again, I am not trying here to offer an exhaustive account of what is going on in evolutionary theory, but rather an argument about the possible future of Darwinism.

In contrast to Spencer, who proposed to reformulate Darwin's principle of natural selection under the form of a tautology ("survival of the fittest"), Darwin formulated his own hypothesis of natural selection in a way that was precise enough to make it testable. In modern philosophical terms, we might say that he proposed a particular model of natural selection among other possible models. This model relied upon several conditions that were quite restrictive. To recall these rather restrictive conditions:

1. The Malthusian principle (meaning that natural selection applies to a population with limited numbers)
2. Application of the principle to individual organisms only, or in certain cases, to colonies of insects (in the *Origin*) or "tribes" in the case of humans (in the *Descent of Man*)[7]
3. Existence of variation among individuals for a given trait
4. Differences in the chances of survival and reproduction of individuals as a function of their differences
5. Heritability of variation

As I have tried to show elsewhere (Gayon 1998), none of these conditions were empirically trivial. These preconditions were in fact so restrictive that it took decades before they were adequately open both to mathematical treatment and empirical test.

In the past decades, evolutionary biology has gone in the opposite direction of generalizing the principle of natural selection. There have been two major attempts to develop this alternative. Richard Lewontin explored a first possibility in his famous article of 1970 on "The Units of Selection" (Lewontin 1970). He proposed to generalize Darwin's natural selection through a formulation that avoids referring to any particular level of organization. Natural selection is then defined as a process that requires the existence of populations of entities sharing three properties: variation, reproduction, and heritability: "The generality of the principles of natural selection means that any entities in nature that have variation, reproduction, and heritability may evolve" (Lewontin 1970, 1).

This general formulation entails that the first two conditions mentioned above are unnecessary: (1) Natural selection does not require that the numbers of a population be limited; it can change the composition of a population of entities even if there is no limit to its growth. (2) Natural selection can apply to any kind of entities that exhibit the three properties formulated by Lewontin, from molecules to cells, organisms, demes, and, eventually, species and taxa. Gould's emphasis upon differential rates of speciation and extinction within larger lineages, and, correlatively, on species selection and clade selection, are in line with Lewontin's proposal.

There is however another way of generalizing the notions of natural selection and fitness. It consists in saying that differential reproduction is not a necessary condition for a process of natural selection to occur. This proposal has arisen among ecologists. A number of them (e.g., Van Valen 1991; Blandin 2007), followed by philosophers (Bouchard 2007, 2008), have defended the claim that ecosystems (and "other ecological entities" such as certain symbiotic organisms assembled anew at each generation)[8] can indeed evolve in spite of their inability to "self-reproduce" as such. But some sort of natural selection applies to them: what is maximized, then, is the survival and the persistence of the ecosystem (or of another kind of ecological entity).

There has been a rather harsh debate on the question whether such an ecological concept of natural selection and fitness is relevant or not. Richard Lewontin, for instance, has expressed the most serious doubts: for him, if the conditions of reproduction and heritability are suppressed, one has only a process of differential elimination, not an indefinite and open process of transformation (Lewontin 2009). I admit my perplexity in the face of such a generalization on the natural selection principle (which, incidentally, is incompatible with the previous one, based upon reproduction and heritability). But, simultaneously, I think that future evolutionary theory will have to deal more and more with ecological and terrestrial dynamics. Differential elimination and differential emergence of types of ecosystems or other ecological entities is certainly an important evolutionary phenomenon.

Whether one should consider this as a generalization of Darwinian schemes is uncertain. It is an open question whether Darwinian schema will be able to colonize geobiology. I cannot resist here quoting a few words from James Lovelock's recent publication, *The Vanishing Face of Gaia: A Final Warning*:

> Biologists say a living thing is one that reproduces, and the errors of reproduction are corrected by natural selection. . . . Gaia fits the physicist's definition [of life][9] but fails the biologist's test because it does not reproduce. . . . But something that lives a quarter of the age of the universe surely does not need to reproduce, and perhaps Gaia's natural selection takes place internally as organisms and their environment evolve in a tightly coupled union. (Lovelock 2009, 127)

I know that most evolutionary biologists dislike this kind of declaration, but I tend to think that future evolutionary biology will have to cope with the delicate question of coupling its theories with the more physical theories of geobiology. For this purpose, one will probably need more than the traditional view of natural selection as an all-encompassing principle acting blindly and independently on innumerable species behaving as Leibniz's monads "without any doors or windows."

To conclude on the present fate of the explanatory part of Darwin's theory, we see therefore that we do not observe a replacement of the Darwinian schema, but an expansion consisting in generalizing the old principles and adding new ones.

DARWINISM AND THE HUMAN AND SOCIAL SCIENCES

In the second part of this chapter, I will speculate about the future of Darwinism in the human sciences. Let me first explain why evolutionary biology and the human sciences deserve separate treatment. The reason is that Darwinism addresses quite different challenges to these two academic fields. In evolutionary biology, the issue is: are we still "Darwinian" and in what sense? For more than half a century,

Darwinism, under the form of the Modern Synthesis, has functioned as a dominant paradigm in evolutionary biology. Since the 1970s, this paradigm has been criticized in many ways. Renowned evolutionary biologists have even proclaimed that this theory is "effectively dead, despite its persistence as text-book orthodoxy" (Gould 1980, 120).[10] Retrospectively, this proclamation was excessive. As Gould himself recognized later, his announcement of the death of Darwinism had probably been premature. On the whole, the old framework has been rather resilient, in spite of the hubbub around evolutionary-developmental biology (evo-devo) (see the introduction to this volume). Darwinism resembles a fortress that has endured many assaults but that has not been dismantled so far. If we now consider the human sciences, the situation is just the opposite. The stronghold here is not "Darwinism" but the human sciences themselves. In many disciplines, from psychology to sociology, linguistics, economics, philosophy, and religious studies, we observe a number of attempts to rebuild the theoretical framework of these disciplines upon an evolutionary and Darwinian basis. Correlatively, in each of the disciplines, a strong resistance to this process of "Darwinization" can be seen.

There is, then, a remarkable asymmetry. In evolutionary biology, Darwinism stands in a defensive position: it is challenged as a dominant and all-encompassing theoretical framework. In the human sciences, the situation is the reverse: Darwinism is on the offensive. This looks like a radicalization of the Darwinian framework, which is invading a number of fields of research that had previously ignored or refused to consider it as much as possible. This is why I have chosen to examine these two cases separately.

In so doing, I do not want to suggest that there exists any sharp dividing line between evolutionary biology and the human sciences. This would be nonsense, because some research programs in the human sciences can be seen as parts of evolutionary biology: perhaps the most spectacular example is evolutionary psychology. But this is not true of the majority of the research programs in the human sciences that claim to have a strong evolutionary dimension. For instance, evolutionary economics, evolutionary epistemology, and evolutionary ethics are very much "evolutionary" and "Darwinian" in spirit, but they can hardly be seen as parts of evolutionary biology.

In many cases, however, we cannot tell, since most of these evolutionarily oriented programs in the human sciences are interpreted in various ways. Cultural evolution illustrates this ambiguity: if one takes it in the sense of an epidemiology of ideas, the relation with evolutionary theory is mostly analogical (Cavalli-Sforza and Feldman 1981; Sperber 1996a, 1996b); but if one takes cultural evolution in the sense of gene-culture co-evolution, then this program, even if controversial, is part of evolutionary theory (Gayon 2005).

There is another reason for considering separately the fate of Darwinism in evolutionary biology and in the human sciences. Evolutionary biology is today a well-constituted discipline, both intellectually and institutionally, whereas the human sciences are a conglomerate of disciplines that are extremely diverse in terms of subjects, methods, theories and institutional bases. We are no longer in the intellectual conjuncture of the eighteenth and nineteenth centuries, when it was admitted that there was room for a science named "anthropology," the aim of which was to embrace all possible aspects of the human phenomenon. At the end of the nineteenth century, this idea of a unified Science of Man broke down, and the use of the word "anthropology" was abandoned in favor of the expression "human sciences,"[11] which themselves became later the familiar "human and social sciences," not to speak of the "human, social, and economic" sciences.

An objective sign of this fragmentation of the human sciences is the near absence of international (or even national) periodicals specifically devoted to the human sciences in their totality. Nothing corresponds in this field to major journals in mathematics, physics, chemistry, biology, or geology. The move from "anthropology" (singular) to "human sciences" (plural) was not a trivial change: it meant that the sciences studying the human have different methods and theories, and constitute genuinely different disciplines, with different curricula, different scientific norms, and different institutions. The plural in the expression "human sciences" is a decisive part of the *ethos* of these sciences, as they exist today. It means reciprocal tolerance and non-overlapping magisteria. I think that this aspect of the human sciences is extremely important if one wants properly to assess the signification of the spread of Darwinism in these sciences today.

What then is at stake in the present proliferation of evolutionary and Darwinian approaches in a number of human sciences today—evolutionary psychology, historical linguistics, physical and cultural anthropology, evolutionary economics, evolutionary epistemology, evolutionary ethics, cultural evolution, and even evolutionary religious studies, to mention just the most spectacular examples? I think that there are two different issues to be considered, depending on the disciplinary perspective that we choose.

From the viewpoint of evolution, we can contemplate a spectacular *extension* of the Darwinian framework. It is no question here of a theoretical *expansion* of Darwinism (in the sense discussed above). The issue is not to reconstruct the foundations of Darwinian science, but rather to apply standard Darwinian concepts to as many dimensions of the human condition as possible, with as many bridge statements as needed. Sometimes, this extension of scope is founded upon analogy (e.g., as employed in evolutionary economics, in historical linguistics, or in cultural epidemiology); sometimes it is literal (e.g., as in evolutionary psychology). In all cases, however, the idea is that evolutionary methods and concepts provide fertile heuristics for analyzing the human sciences.

But we can also consider the extension of Darwinism from the viewpoint of the human sciences and of their pluralistic *ethos*. Then Darwinism comes as a threat: the threat of the dissolution of their pluralistic, tolerant, and antireductionist ethos. This threat is indeed very real, but it constitutes also an exciting perspective: the perspective of some sort of reunification of the human sciences, upon the basis of at least one possible common language—evolution. In spite of all the excesses associated with the penetration of Darwinism into the human sciences, evolution does offer today a means of overcoming the dramatic disunity of these sciences, and their no less dramatic separation from the natural sciences.

I do not mean here that evolution is today the sole possible common language for the human sciences (widely speaking), or even that it can apply to all of them. Convergences can be observed on other grounds, such as the cognitive sciences, for instance. But there are clear signs that a number of human sciences are presently undergoing partial convergence upon the basis of evolution. Think for instance of the multiple interactions between evolutionary economics, cultural evolution, and evolutionary psychology. Therefore, I believe that evolution contributes today to the rehabilitation of the very idea of "anthropology" in the older sense of this term.

FOUR PREDICTIONS

In conclusion, I venture to make four predictions. Firstly, I think that the most basic Darwinian principles (common descent, modification of species, variation, and natural selection) will remain as a definitive part of evolutionary theory. These principles may be reformulated, and complemented with other principles, but I see no reason not to consider them as part of what Pierre Duhem called the "positive" aspect of scientific theories, in contrast with their varying "metaphysical" aspect (Duhem 1914).

Secondly, because the persistence of Darwinism seems to rely heavily upon important generalizations of Darwinian (or Neo-Synthetic) principles (common descent, variation, natural selection, and random drift), we may expect a proliferation of models that will flesh out the abstract reformulations of these Darwinian principles.

Thirdly, I predict that the spread of Darwinism over the human sciences will stimulate important convergences and hybridizations. The day may not be very far off when the massive application of evolutionary methods to a number of problems in the human sciences will no longer be the object of an ideological battle over the issue of naturalism, but rather a way of redefining the theoretical domain of the human sciences.

Finally, I cannot restrain myself from comparing the fate of today's "universal Darwinism" with what was called in the seventeenth and eighteenth centuries, the "Mechanical Philosophy." In both cases, we observe a similar tendency to include virtually everything in an all-encompassing philosophy of nature. I predict that universal Darwinism as a philosophy of nature (biological nature and human nature) will die one day or another, just as the mechanical philosophy did approximately two centuries ago.

NOTES

1. I thank Phillip Sloan for this useful reference and the editors of this volume for their careful reading, linguistic corrections, and critical suggestions on this paper.

2. This formulation remained unmodified in the first five editions. In the sixth edition, Darwin wrote: "descent with modification through variation and natural selection" (Darwin 1872, 313, 404).

3. I mention here only the passages where Darwin comments on the detail of the diagram. There are other, less precise, references to "the diagram so often referred to" (Darwin [1859] 1964, 431).

4. For a more precise formulation see Gayon (2008, 2009).

5. For an overview of this school of taxonomic theory, see Lennox (2010).

6. For a recent and comprehensive review, see Ragan, McInerney, and Lake (2009).

7. For a thorough review and analysis of this controversial issue see Sober (2011).

8. I refer it to Niles Eldredge's distinction between the two kinds of "individuals" that evolutionary theory forces us to recognize: genealogical entities and ecological entities. Genealogical entities are those that are able to self-reproduce; they form lineages and allow a flux of information (e.g., genes, cells, organisms, demes, species, monophyletic taxa). Ecological entities are characterized as systems that remain approximately stable against the fluxes of matter and energy that cross them (e.g., proteins, organs, organisms, biotic communities, ecosystems). For Eldredge, organisms are both genealogical and ecological entities, because they do self-reproduce in spite of the fact they do not replicate, at least in sexual organisms. Eldredge's distinction is based on David Hull's notions of "replicator" and "interactor," the two kinds of entities that are required for natural selection to occur (Eldredge 1985).

9. Lovelock refers to Schrödinger's claim (1944) that a long–sustained dynamic reduction of internal entropy distinguishes life from its inorganic environment.

10. "I well remember how the synthetic theory beguiled me with its unifying power when I was a graduate student in the mid-1960s. Since then I have been watching it slowly unravel as a universal description of evolution. The molecular assault came first, followed quickly by renewed attention to unorthodox theories of speciation and by challenges at the level of macroevolution itself. I have been reluctant to admit it—since beguiling is forever—but if Mayr's characterization is accurate, then that theory, as a general proposition, is effectively dead, despite its persistence as text-book orthodoxy" (Gould 1980, 120).

11. In the sense of a unified and comprehensive science of man, including not only physical anthropology and ethnology, but also geography, linguistics, psychology, and everything that we now gather under the name of "human and social sciences."

REFERENCES

Blandin, P. (2007). "L'écosystème existe-t-il? Le tout et la partie en écologie." In Martin 2007, 21–46.

Bouchard, F. (2007). "Évolution d'écosystèmes sans réduction à la *fitness* des parties." In Martin 2007, 57–64.

———. (2008). "Causal Processes, Fitness and the Differential Persistence of Lineages." *Philosophy of Science* 7:560–70.

Bowler, P. J. (1984). *Evolution: The History of an Idea.* Berkeley: University of California Press.

Burkhardt, F., et al., eds. (2007). *The Correspondence of Charles Darwin,* vol. 11. Cambridge: Cambridge University Press.

Cavalli-Sforza, L., and M. W. Feldman. (1981). *Cultural Transmission and Evolution: A Quantitative Approach.* Monographs in Population Biology 16. Princeton: Princeton University Press.

Darwin, C. ([1859] 1964). *On the Origin of Species and Preservation of the Favoured Races in the Struggle for Life.* Reprint of 1st ed. Cambridge, MA: Harvard University Press.

———. (1872). *The Origin of Species by Means of Natural Selection, or the Preservation of Favoured Races in the Struggle for Life.* 6th ed. London: John Murray. Available at http://darwin-online.org.uk/.

Doolittle, W. F., and E. Bapteste. (2007). "Pattern Pluralism and the Tree of Life Hypothesis." *Proceedings of the National Academy of Sciences* 104:2043–49.

Duhem, P. (1914). *La théorie physique: Son objet—Sa structure.* Paris: Vrin. English trans., *The Aim and Structure of Physical Theory,* Princeton: Princeton University Press, 1954.

Eldredge, N. (1985). *Unfinished Synthesis: Biological Hierarchies and Modern Evolutionary Thought.* New York: Oxford University Press.

Gayon, J. (1998). *Darwin's Struggle for Survival: Heredity and the Hypothesis of Natural Selection.* Translated by M. Cobb. Cambridge: Cambridge University Press.

———. (2005). "Cultural Evolution: A General Appraisal." *Ludus Vitalis* 13:139–50.

———. (2008). "'Is a New and General Theory of Evolution Emerging?' A Philosophical Appraisal of Stephen Jay Gould's Evaluation of Contemporary Evolutionary Theory." In *Evolutionism: Present Approaches*, ed. W. J. Gonzalez, 77–105. La Coruña: Netbiblo.

———. (2009). "Actualité du darwinisme." *Bulletin de la Société française de Philosophie* 103:1–49.

Gould, S. J. (1980). "Is a New and General Theory of Evolution Emerging?" *Paleobiology* 6:119–30.

———. (2002). *The Structure of Evolutionary Theory*. Cambridge, MA: Belknap Press.

Lennox, J. (2010). "Darwinism." Available through the *Stanford On-Line Encyclopedia of Philosophy* at http://plato.stanford.edu/entries/darwinism/.

Lewontin, R. (1970). "The Units of Selection." *Annual Review of Ecology and Systematics* 1:1–18.

———. (2009). "Postface." In *Les mondes darwiniens: L'évolution de l'évolution*, ed. T. Heams et al., 1077–80. Paris: Editions Syllepse.

Lovelock, J. (2009). *The Vanishing Face of Gaia: A Final Warning*. Victoria, AU: Penguin Books.

Margulis, L., and D. Sagan. (2002). *Acquiring Genomes: The Theory of the Origin of Species*. New York: Basic Books.

Martin, T., ed. (2007). *Le tout et les parties dans les systèmes naturels*. Paris: Vuibert.

McInerney, J. O., J. A. Cotton, and D. Pisani. (2008). "The Procaryotic Tree of Life: Past, Present . . . and Future?" *Trends in Ecology and Evolution* 23:276–81.

McMullin, E. (1976). "The Fertility of Theory and the Unit for Appraisal in Science." *Boston Studies* 39:395–432.

Ragan, M. A., J. O. McInerney, and J. A. Lake. (2009). "The Network of Life: Genome Beginning and Evolution." *Philosophical Transactions of the Royal Society of London B* 364:2169–79.

Schrödinger, E. (1944). *What is Life? The Physical Aspect of the Living Cell*. Cambridge: Cambridge University Press.

Sober, E. (2011). *Did Darwin Write the Origin Backwards? Philosophical Essays on Darwin's Theory*. Amherst, NY: Prometheus Books.

Sperber, D. (1996a). *La Contagion des idées*. Paris: Odile Jacob.

———. (1996b). *Explaining Culture: A Naturalistic Approach*. Cambridge, MA: Blackwell.

Van Valen, L. (1991). "Biotal Evolution: A Manifesto." *Evolutionary Theory* 10:1–13.

Woese, C. (1987). "Bacterial Evolution." *Microbiology and Molecular Biology Reviews* 51:221–71.

CONTRIBUTORS

Gennaro Auletta is researcher at the University of Cassino, aggregate professor at the Pontifical Gregorian University, former scientific director of the STOQ Project, scientific director of the specialization "Science and Philosophy," fellow of St Edmund's College of the University of Cambridge, and associate of the Von Hügel Institute of the University of Cambridge. He is the director of five research projects and author and editor of sixteen books and more than seventy papers. His areas of interest include metaphysics, philosophy of nature, logic, foundations and interpretation of quantum mechanics, quantum information, system biology, cognitive biology, top-down causation in biology and neurosciences, and the mathematical definition of complexity.

Peter J. Bowler is professor emeritus of the history of science at Queen's University, Belfast. He is a fellow of the British Academy, a member of the Royal Irish Academy, and a past president of the British Society for the History of Science. He is the author of several books on the history of evolutionism, including *Darwin Deleted* (2013), *Evolution: The History of an Idea* (3rd ed., 2003), and *The Non-Darwinian Revolution* (1988). With Iwan Morus he is the author of the textbook *Making Modern Science* (2005). He has also published on science and religion, such as *Reconciling Science and Religion* (2001), and on popular science in the early twentieth century.

William E. Carroll is Research Fellow at Blackfriars Hall, Oxford, and member of the Faculty of Theology and Religion of the University of Oxford. His research concerns the reception of Aristotelian science in medieval Islam, Judaism, and Christianity, and the development of the doctrine of creation. He has also written extensively on the ways in which medieval discussions of the relationship among the natural sciences, philosophy, and theology can be useful in questions arising from developments in contemporary biology and cosmology. He is the co-author, with Steven Baldner, of *Aquinas on Creation* (1997); and the author of *La Creación y las Ciencias Naturales: Actualidad de Santo Tomás de Aquino* (2003); *Galileo: Science and Faith* (2009); and *Creation and Science* (2011).

Ivan Colagè is currently conducting his interdisciplinary research on the relations between brain and culture at the Pontifical University Antonianum, in Rome, where he also teaches courses on logic and philosophical anthropology. Recently, he co-edited, together with Gennaro Auletta and Marc Jeannerod, *Brains Top Down: Is Top-Down Causation Challenging Neurosciences?* (2013).

Paolo D'Ambrosio has written a doctoral dissertation under the guidance of Prof. Gennaro Auletta on evolutionary theorizing and its philosophical implications. He contributed to the book *Integrated Cognitive Strategies in a Changing World*, edited by G. Auletta, in collaboration with I. Colagè, P. D'Ambrosio, and L. Torcal (2011). Paolo is currently studying the evolution of higher cognitive abilities and the uniqueness of the human being in the context of a research project hosted by the Pontifical University Antonianum and supported by the John Templeton Foundation.

Celia Deane-Drummond is professor of theology at the University of Notre Dame. Her research interests are in the engagement of theology and natural science, including specifically ecology, evolution, animal behavior, and anthropology. Her research has consistently sought to explore theological and ethical aspects of that relationship. Her most recent books include *Future Perfect*, edited with Peter Scott (2006), *Ecotheology* (2008), *Christ and Evolution* (2009), *Creaturely Theology*, edited with David Clough (2009), *Religion and Ecology in the Public Sphere*, edited with Heinrich Bedford-Strohm (2011), *Animals as Religious Subjects*, edited with Rebecca Artinian Kaiser and David Clough (2013), and *The Wisdom of the Liminal: Evolution and Other Animals in Human Becoming* (2014), and *Reimaging the Diving Image* (2014).

David J. Depew is professor emeritus, University of Iowa. He is the co-author, with Bruce H. Weber, of *Darwinism Evolving: Systems Dynamics and the Genealogy of Natural Selection* (1995) and, with Marjorie Grene, of *The Philosophy of Biology: An Episodic History* (2004). He is currently working with John Jackson of the University of Colorado on an extended study of the interaction of genetic Darwinism and anthropology in twentieth-century America.

Kathleen Eggleson is a biological scientist and practical ethicist at the University of Notre Dame, Center for Nano Science and Technology (ND*nano*), and past associate director of the John J. Reilly Center for Science, Technology, and Values. Her research and teaching combine scientific and ethical analyses of the interface between emerging technologies and society. She has received awards from the National Science Foundation to support investigation of anticipatory governance of complex engineered nanomaterials in collaboration with David Guston at Arizona State University, life cycle–based ethics education in collaboration with Matthew Eckelman and colleagues at Northeastern University, and ethical and societal issues related to novel computer architectures on a multinational project led by Wolfgang Porod of the University of Notre Dame. Her published articles and presentations have broad impacts, as her insights have been quoted in publications such as the *New York Times* and *Science*.

Jean Gayon is professor of philosophy at Université Paris 1 Panthéon Sorbonne, where he is director of the IHPST (Institute of History and Philosophy of Science and Techniques), and a philosopher and historian of biology. He is the author of *Darwinism's Struggle for Survival: Heredity and the Hypothesis of Natural Selection* (French ed. 1992, English ed. 1998), and co-author of several books, including *L'Éternel retour de l'eugénisme* (with Daniel Jacobi, 2006), *L'épistémologie française, 1830–1870* (with Michel Bitbol, 2006 PRS), and *French Studies in the Philosophy of Science* (with A. Brenner, 2009). He is currently writing a book on the history of genetics in France.

Scott F. Gilbert is a Distinguished Finland Professor at the University of Helsinki and a senior research fellow at Swarthmore College. He is the author of *Developmental Biology* (10th ed., 2014) and *Ecological Developmental Biology: Integrating Epigenetics, Medicine, and Evolution* (2009), and is co-author of *Bioethics and the New Embryology* (2005). He is currently working on projects relating evolution, development, and ecology.

Paul E. Griffiths is University Professorial Research Fellow at the University of Sydney, Australia, and visiting professor at Egenis, the Centre for the Study of the Life Sciences, University of Exeter, UK. He is the author of *What Emotions Really Are: The Problem of Psychological Categories* (1997), and co-author with Kim Sterelny of *Sex and Death: An Introduction to the Philosophy of Biology* (1999), and with Karola Stotz of *Genetics and Philosophy: An Introduction* (2013). His current research focuses on the concept of information in biology and the application of evolutionary ideas to medicine.

Gerald McKenny is Walter Professor of Theology at the University of Notre Dame and past director of the John J. Reilly Center for Science, Technology, and Values. He is the author of *To Relieve the Human Condition: Bioethics, Technology, and the Body* (1997), and *The Analogy of Grace: Karl Barth's Moral Theology* (2010). He is co-editor of four books on bioethics, the ethics of biotechnology, and Continental moral philosophy, and he has published numerous articles and book chapters on theological ethics, bioethics, the ethics of biotechnology, and the philosophy of medicine. He is currently writing a book on claims regarding the normative status of human nature in theological and philosophical debates over the ethics of human biotechnology.

Alessandro Minelli is emeritus professor of zoology at the University of Padua, Italy, where he taught until his retirement in 2011. First active in biological systematics and phylogenetics, he moved later to evolutionary developmental biology. He is the author of *Biological Systematics* (1993), *The Development of Animal Form* (2003), *Forms of Becoming* (2009), and *Perspectives in Animal Phylogeny and Evolution* (2009).

Stuart A. Newman is professor of cell biology and anatomy at New York Medical College, where he directs a research program in developmental biology and evolutionary theory. He received a Ph.D. in chemistry from the University of Chicago. Newman has contributed to several scientific fields, including cell differentiation, theory of biochemical networks and cell pattern formation, protein folding and assembly, and evolutionary developmental biology. He is co-editor (with Gerd B. Müller) of *Origination of Organismal Form: Beyond the Gene in Developmental and Evolutionary Biology* (2003) and co-author (with Gabor Forgacs) of *Biological Physics of the Developing Embryo* (2005).

John O'Callaghan is an associate professor of philosophy at the University of Notre Dame, where he directs the Jacques Maritain Center. He is past president of the American Catholic Philosophical Association and a permanent member of the Pontifical Academy of St. Thomas Aquinas. His areas of research are medieval philosophy, Thomistic metaphysics and philosophical theology, ethics, and science and religion.

Robert J. Richards is the Morris Fishbein Distinguished Service Professor in the Departments of History, Philosophy, and Psychology, and in the Committee on Conceptual and Historical Studies of Science at the University of Chicago, where he directs the Fishbein Center for the History of Science and Medicine. His most recent books are *The Romantic Conception of Life: Science and Philosophy in the Age of Goethe* (2002); *The Tragic Sense of Life: Ernst Haeckel and the Struggle over Evolutionary Thought* (2008); and *Was Hitler a Darwinian? Disputed Questions in the History of Evolutionary Theory* (2013)—all published by the University of Chicago Press.

Phillip R. Sloan is emeritus professor, Program of Liberal Studies and Program in History and Philosophy of Science at the University of Notre Dame. His scholarly work has focused on the history and philosophy of life science in the Enlightenment period to the present. He is the primary editor of *Controlling Our Destinies: Historical, Philo-*

sophical, Ethical, and Theological Perspectives on the Human Genome Project (2000) and (with Brandon Fogel) of *Creating a Physical Biology: The Three-Man Paper and Early Molecular Biology* (2011). He is a fellow of Section L of the AAAS, and past director of the John J. Reilly Center for Science, Technology, and Values, and of the Notre Dame Program in History and Philosophy of Science.

John S. Wilkins is an honorary fellow of the School of Historical and Philosophical Sciences at the University of Melbourne and an associate member of the Department of Philosophy, University of Sydney. His books include *Species: A History of the Idea* (2009) and *The Nature of Classification* (2013).

Bernard Wood is University Professor of Human Origins at George Washington University and adjunct senior scientist at the National Museum of Natural History of the Smithsonian Institution. He is presently the director of the Center for the Advanced Study of Hominid Paleobiology and the editor of the Wiley-Blackwell *Encyclopedia of Human Evolution* (2011). His research focuses on hominin systematics and on ways to improve the reliability of hypotheses about the relationships among fossil hominins. He is particularly interested in how to reconstruct the structure and function of extinct hominins that have no obvious living analogue. He is also determined to improve access to information about the hominin fossil record.

Archbishop Józef Życiński earned his first doctorate in theology in 1976 at the Pontifical Faculty Theology in Kraków and a second doctorate in philosophy at the Academy of Catholic Theology in Warsaw. He established and held the Chair of Logic and Methodology at Kraków's Papal Academy of Theology from 1980 until his death in 2011, and served as the vice-dean and dean of the academy's faculty of philosophy from 1982 to 1985 and 1988 to 1990, respectively. In 1990 he became the Bishop of Tarnów, and in 1997 the Archbishop of Lublin and *ex officio* the Grand Chancellor of the Catholic University of Lublin, where he established and held the Chair of the Relationship Between Science and Faith. He held memberships in the Russian Academy of

Natural Sciences, the Committee for Philosophical Studies of the Polish Academy of Sciences (PAN), the European Academy of Science and Art, and PAN's Committee on Evolutionary and Theoretical Biology. He was the author of fifty books on philosophy of science, relativistic cosmology, and the history of the relations between natural science and Christian faith. He authored more than 350 published articles in journals in Poland and abroad. He founded the journal series *Philosophy in Science*, published by the Vatican Observatory and the University of Arizona. His more than fifty book publications included: in Polish, *The Cultural Odyssey of Man* (2005) and *Christian Inspirations in the Origins of Modern Science* (2000); in English, *The Structure of Metascientific Revolution* (1988), *Three Cultures: Science, the Humanities, and Religious Values* (1990), and *God and Evolution* (2006). Archbishop Życiński died unexpectedly on February 10, 2011. He is buried in the Cathedral of St. John the Baptist and St. John the Evangelist in Lublin.